SOFTWARE TOOLS AND TECHNIQUES FOR ELECTRONIC ENGINEERS

THE McGRAW-HILL
INTERNATIONAL SERIES IN SOFTWARE ENGINEERING

Consulting Editor

Professor D. Ince
The Open University

Titles in this Series

Portable Modula-2 Programming – Woodman, Griffiths, Souter and Davies
SSADM: A Practical Approach – Ashworth and Goodland
Software Engineering: Analysis and Design – Easteal and Davies
Introduction to Compiling Techniques: A First Course Using ANSI C, LEX and
YACC – Bennett
An Introduction to Program Design – Sargent
Object-Oriented Databases: Applications in Software Engineering – Brown
Object-Oriented Software Engineering with C^{++} – Ince
Expert Database Systems: A Gentle Introduction – Beynon-Davies
Practical Formal Methods with VDM – Andrews and Ince
SSADM Version 4: A User's Guide – Eva
A Structured Approach to Systems Development – Heap, Stanway and Windsor
Rapid Information Systems Development – Bell and Wood-Harper
Software Engineering Environments: Automated Support for Software
Engineering – Brown, Earl and McDermid
Systems Construction and Analysis: A Mathematical and Logical Framework –
Fenton and Hill
SSADM V4 Project Manager's Handbook – Hammer
Knowledge Engineering for Information Systems – Beynon-Davies
Introduction to Software Project Management and Quality Assurance – Ince,
Sharp and Woodman
Software System Development: A Gentle Introduction – Britton and Doake
Introduction to VDM – Woodman and Heal
An Introduction to SSADM Version 4 – Ashworth and Slater
Discrete Event Simulation in C – Watkins
Objects and Databases – Kroha
Object-Oriented Specification and Design with C++ – Henderson
Software Engineering Metrics Volume I: Measures and Validations – Shepperd
Software Tools and Techniques for Electronic Engineers – Jobes
Reverse Engineering and Software Maintenance: A Practical Approach –
Lano and Haughton
Coding in Turbo Pascal – Sargent
A Primer on Formal Specification – Turner and McCluskey
Design of Concurrent Systems – Mett, Crowe and Strain-Clarke
An Introduction to Software Engineering Using Z – Ratcliff

SOFTWARE TOOLS AND TECHNIQUES FOR ELECTRONIC ENGINEERS

Keith Jobes
University of Greenwich

McGRAW-HILL BOOK COMPANY

London · New York · St Louis · San Francisco · Auckland · Bogotá
Caracas · Lisbon · Madrid · Mexico · Milan · Montreal
New Delhi · Panama · Paris · San Juan · São Paulo · Singapore
Sydney · Tokyo · Toronto

Published by
McGRAW-HILL Book Company Europe
Shoppenhangers Road, Maidenhead, Berkshire SL6 2QL, England
Telephone 0628 23432
Fax 0628 770224

British Library Cataloguing in Publication Data

Jobes, Keith
 Software Tools and Techniques for
 Electronic Engineers.—(McGraw-Hill
 International Series in Software
 Engineering)
 I. Title II. Series
 005.3

ISBN 0–07–707720–2

Library of Congress Cataloging-in-Publication Data

Jobes, Keith
 Software tools and techniques for electronic engineers / Keith
Jobes.
    p.   cm. — (The McGraw-Hill international series in software
engineering)
 Includes bibliographical references and index.
 ISBN 0–07–707720–2 (pbk.)
 1. Electronics—Data processing.   2. Software engineering.
I. Title.   II. Series.
TK7835.J57   1993
005′.0246213—dc20   93–32582
                                                              CIP

12345 CL 9654

Typeset by MFK Typesetting Ltd

and printed and bound in Great Britain by Clays Ltd St Ives plc

CONTENTS

WHAT REQUIREMENT DOES THE BOOK ADDRESS?

This book aims to collect together most of the topics that come under the umbrella heading of *software*, and present them in a way electronic engineers would find useful. Much of the material can be found elsewhere, but mainly in texts aimed at computer science students and others, whose closer involvement in software requires detailed treatments not really appropriate to the needs of electronic engineers; they typically might need to be involved with the hardware design as well as the software design aspects of a computer system, or perhaps just as users of computer systems. Specifically, the text covers the areas of languages, operating systems, and software development tools and techniques. If the book can be said to have any particular focus, it is in the area of software engineering, and can to some extent be looked on as an introduction to this key activity. The treatment, although dominated by software-related issues, does from time to time cross the hardware–software divide in view of the needs of electronic engineers in both these areas.

WHAT PRIOR KNOWLEDGE DO YOU NEED?

The material covered would follow on naturally from a first course on programming and microprocessors, both these areas normally being covered in the early parts of most engineering courses. Some familiarity is assumed in the language Pascal or alternatively Modula-2. It should be stressed that no attempt is made to teach the principles of programming as such, there are no shortage of good books that aim to do just this, although chapters on C and FORTRAN are included to complement any previous background in programming that readers might have, and additionally to help support the critical comparison between programming languages which features in one part of the book. Some familiarity in the instruction set of the 68000 microprocessor would be useful, because this microprocessor is used as the vehicle for illustrating concepts in various parts of the book. However, an appendix outlining the 68000 is given for those readers with no prior background in this microprocessor.

HOW CAN THE BOOK BE USED?

It is not essential to read all the material in strict sequence. In particular the chapter on FORTRAN could be omitted entirely with little loss of continuity between the remaining chapters. The bulk of the material is largely independent of the discussion of the operating system UNIX given in Chapter 11, although it has to be said that UNIX is really too important to ignore! Chapters 6 and 7 assume some understanding of the material in Chapters 2, 3 and 5. The chapters on operating systems and real-time systems assume some understanding of algorithms as discussed in Chapter 5. Finally, Chapter 14 might usefully complement a practical course on microprocessors, dealing as it does with development and testing of software for microprocessors, although a prior reading of parts of Chapter 13 might prove helpful. Chapters 3 and 4 on C and FORTRAN support the discussion in Chapter 6, where languages are compared. Chapter 5 deals with aspects such as data structures, which also support arguments in Chapter 6. Some of this material in Chapter 6 is also relevant to the sections dealing with operating systems in Part III.

The text divides into four parts as follows:

Part I Introduction

Chapter 1 Introduction to the use of computers This is a quite basic general review of the role of the computer in processing information, which some readers might find useful if they have limited background in computing.

Chapter 2 The role of software in electronic engineering A quite general review of the different ways electronic engineers need to interact with software. The concept of real-time systems, a pervasive theme in many other parts of the book, is introduced, and also the role of software in relation to hardware is explained. The different types of hardware systems are described, which is of importance to electronic engineers in view of the fact that the types of systems they design often comprise hardware and software elements.

Part II Languages and algorithms Essentially, this section explores the role of computer languages and algorithms in computing, but strictly from the standpoint of an electronic engineer. In some respects the material can be regarded as forming the groundwork for the discussion of software engineering in Part IV. As elsewhere in the text, the aspects of techniques and languages of particular importance to real-time systems are given special emphasis.

Chapter 3 The language C This is a short, but nevertheless reasonably complete, guide to the language C, which is widely used in real-time systems programming. Although the chapter can be viewed as an introduction to the language, there should be enough information in it to allow you to write programs in C; however, a specialized text needs to be consulted if you need to do any serious programming in the language. Another function of this chapter is to allow material elsewhere in the book that makes reference to this language to be made more intelligible. The chapter on the operating system UNIX is a case in point.

Chapter 4 FORTRAN FORTRAN has distinctive features that make it a popular language in 'number crunching' types of applications. Some of the programming that electronic engineers do falls into this category, and so is one reason for including material on this language. Another reason is the vast amount of code that has been written in FORTRAN and that users will need to employ in applications.

Chapter 5 Data abstraction, data structures and algorithms Strictly, not a chapter on languages as such, but rather an attempt to give the basis for some problem-solving methods that can be exploited when using the relatively high-level techniques of software engineering described in Part IV. The role of data structures is central to this objective. Examples of algorithms likely to be encountered in practice, such as sort algorithms, are described. No attempt is made to give a comprehensive treatment of all the different types of sort algorithms or other types of algorithm.

Chapter 6 Languages—choices and characteristics The comparisons between languages in this chapter are restricted to those languages of interest to electronic engineers. These include assembler, which still has a role to play in the programming of real-time systems. The other main languages discussed are Modula-2 and Ada, although no attempt is made to give anything like a complete description of these languages, but rather to concentrate on some of their distinctive features. The concepts of data typing and information hiding are fundamental to the kind of comparisons being made, and get a moderately detailed treatment.

Part III Systems software The core of this section centres around operating systems and compilers. Similar emphasis is given to the needs of electronic engineers and of real-time systems as elsewhere in the book.

Chapter 7 Translators, linkers and loaders The two key areas covered here are compilers and assemblers. The section on assemblers is not intended as a detailed guide to actually writing code at this level, although the syntactic aspects of assembly language are covered in moderate depth, as is a comparison between the different types of assembler. The discussion should be seen as complementing the section in Chapter 6 where assembler and high-level languages are compared. One of the main aims of the material on compilers is to show the role of the compiler in controlling the quality of code that it generates. It is important—in real-time work especially—that program codes execute efficiently, and enough background is given into the internal structure of a compiler program to give an appreciation of the problems involved. The important area of compiler optimization is of especial relevance here.

Chapter 8 Introduction to operating systems This chapter aims to give an introduction to operating systems, in terms of what they are and how users interact with them. Material on the way the operating system MS-DOS is used in practice reinforces the material.

Chapter 9 Multitasking operating systems This chapter is a logical continuation of the previous chapter. Here, internal details of a particular class of operating system are discussed, a point of view carefully avoided in Chapter 8. The material is in some ways an introduction to the discussion of real-time systems given in Chapter 12, where specialized forms of operating systems play a particularly important role.

Chapter 10 Store systems—management and use This chapter is complementary to the material on operating systems, and is in some ways an extension of the previous chapter. A discussion is included of a specific memory management chip. This is because of the likely interest in memory management hardware, and in the ways the relevant software needs to interact with it, of those concerned with the hardware element in the design of a system.

Chapter 11 The operating system UNIX No book that includes material on operating systems should fail to omit reference to UNIX. This chapter uses this important operating system as a practical illustration of operating concepts in action. Enough information is given to form a practical introductory guide to using UNIX, although the chapter in no sense makes any pretentions to being a comprehensive treatment.

Part IV Systems design and development Part IV builds on the concepts developed in the first three parts, and essentially is an introduction to software engineering slanted mainly—but not exclusively—towards real-time systems. A general introduction to the design of real-time system software and algorithms in Chapter 12 is followed by material dealing specifically with software engineering in Chapter 13. The final chapter deals with development and testing techniques and presents some of the more practical aspects involved in the development of the hardware and software.

Chapter 12 Real-time systems—characteristics and algorithms This chapter looks at some of the principles behind the design of real-time software, and in so doing builds on topics discussed in Chapters 5 and 9, which form the groundwork for many of the relevant concepts involved.

Chapter 13 Introduction to software engineering In some ways this chapter is the culmination of the material in many of the preceding chapters, and gives—albeit at a strictly introductory level—the various stages in the development of a software product. This includes areas such as *requirements capture*, where the nature of a problem is determined, and the various ways software is tested. After reading this chapter, although you will not have a *detailed* idea of the methods used in software engineering, you should be aware of the fact that development of software is far from being merely a question of writing code. The chapter is intended to form an introduction to more advanced texts on software engineering.

Chapter 14 Development tools and techniques Testing software is an area covered in the discussion on software engineering in the previous chapter. This chapter presents some of the more practical aspects of the development and testing of software, concentrating specifically on real-time embedded systems. Development system hardware is described, including techniques for the use of instruments such as in-circuit emulators in debugging programs and hardware.

A section giving an example of debug techniques applied to a 68000 assembler program complements and reinforces material given earlier in the book on assembler language concepts, and for that matter might be useful in a complete course on assembler programming. It should be possible to read this particular section without a detailed understanding of the material in the software engineering chapter.

ACKNOWLEDGEMENTS

The author would like to thank colleagues at the University of Greenwich—Ken Davis, Geoff Allwood and Alan Soper—for reviewing various parts of the draft manuscript, and for the many helpful suggestions made. I would also like to thank numerous other colleagues whose ideas over the years have given me many useful insights.

TYPOGRAPHICAL AND OTHER CONVENTIONS

The bulk of the text is written using Times roman fount, the exceptions being Courier fount for code, references to quantities such as identifiers, etc., made inside the text, and also for operating system commands. Where possible, conventions adopted with particular languages have been followed. For example in the case of FORTRAN code, upper case is used, as is standard in FORTRAN 77. (Note however that the use of upper case is not mandatory in other versions of FORTRAN, notably FORTRAN 90.) Other conventions are: the first reference to a technical term or a definition of it is in bold, italics are used for emphasis within the text and single characters are enclosed in (single) quotation marks as in '*'.

Part I

Introduction

1

INTRODUCTION TO THE USE OF COMPUTERS

1.1 INTRODUCTION

The aim of this introductory chapter is to act as a bridge between the more detailed material in later chapters, and to help readers with a limited background in computing in general and software concepts in particular. Depending on your background, you may find it possible to skim through the material in this chapter, or perhaps even to skip it entirely. Specifically, we take a general view of how computers are used as tools in solving problems and in the processing of information. The nature of the hardware and software resources needed in a computer system to support these objectives is examined.

1.2 THE COMPUTER—ITS IMPORTANCE AND APPLICATIONS

One can hardly envisage a world without the computer, and it is only a slight exaggeration to say that in an industrialized society we are almost as dependent on computers as on the electricity supply. Computers are used for an enormous range of applications in everyday life; producing electricity bills, reserving seats with an airline and video games are just a few sample examples out of an extremely large and varied list of applications one could quote. It is difficult to imagine an office without the personal computer for processing documents, or for the preparation of reports using packages like spreadsheets. In industry no company could successfully survive without the use of computers for systems such as machine tools or robots of various kinds. Looking around the typical home also, one starts to appreciate how all-pervasive is the influence of the computer as components in video recorders, washing machines and the like.

The applications for computers have been growing at an ever-increasing pace, and this has been made possible by the extremely rapid improvement in the performance of computer hardware, and just as important the availability of computers at lower and lower cost in relation to performance. Computers have benefited especially from the advances in semiconductor technology and its application to the microprocessor in particular. Microprocessors are becoming ever more complex and capable, a trend which shows no sign of slackening for the foreseeable future. This has generated successive generations of ever more complex applications. It seems almost a law of computing that no matter how much more capable computer hardware becomes available, there never seems to be any shortage of new applications that can exploit the improvements in performance.

1.3 PROBLEMS, ALGORITHMS, LANGUAGES AND HARDWARE

1.3.1 Computers, information processing and programs

Irrespective of the type of processor, cost, or application area of a computer, there are two essential characteristics that all computers have in common (see Fig. 1.1):

1. A computer is a machine that processes information. This it does by manipulating data, which may be in the form of characters, digits, floating point numbers, lists of people's names and addresses, etc. An example would be to sort a list of names into alphabetical order, although a list of the various ways information could be processed by computer is almost infinite.
2. The computer is controlled by a step-by-step sequence set of instructions which collectively constitutes a **program**, the program being stored in the computer's memory. One of the most fundamental characteristics of computers is that they are general purpose machines which can be adapted to solve a wide number of problems, the function is set by the program.

A helpful comparison can be drawn with the way a pocket calculator is used to do arithmetic. One enters numbers, hits function keys like '+', and eventually the '=' key to

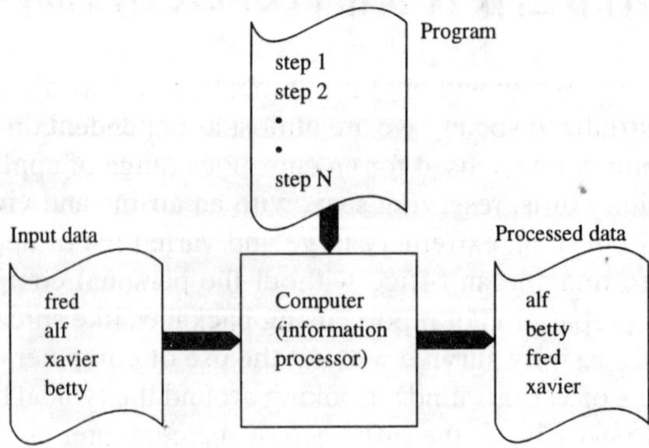

Figure 1.1 The computer as information processor.

get the answer. The sequence followed is comparable to the instructions a computer would have executed—albeit 'automatically'—to achieve a comparable result. Changing the sequence of actions allows the *same* hardware to carry out a different function. Computers are naturally much more adaptable in terms of the variety of information that they process; speed of execution, and the degree of complexity possible in programs are the other main ways computers differ in capability from simple calculators.

1.3.2 Elements of a computer

A computer consists of the following (see Fig. 1.2):

- A processor (often called a **CPU** or Central Processing Unit), which is the subsystem having the responsibility for actually controlling the processing of data. This contains: a **control unit** which uses the program codes to direct its operation; an arithmetic unit (commonly called an **ALU** or Arithmetic Logic Unit), which is comparable in principle to the arithmetic unit in a pocket calculator, although naturally much faster in operation and more versatile; and small-capacity, high-speed storage in the form of registers.
- Memory, to hold program codes and also any data used by programs. A computer's memory is—in nearly all cases—split into a primary, fast store based around semiconductor memory chips, and a slower, secondary store system which generally holds a much larger amount of information than the computer's primary store system.
- Peripheral devices such as VDUs and keyboards through which the computer interacts with the outside world.

The operation of a computer can be briefly summarized as follows. The control unit gets the next instruction code from memory. The instruction is then used by the control unit to set up a sequence of signals to implement the instruction; this involves control of the arithmetic unit, peripheral devices and memory. This sequence is often called the **fetch execute cycle**, and it is carried out repeatedly for all instructions in the program.

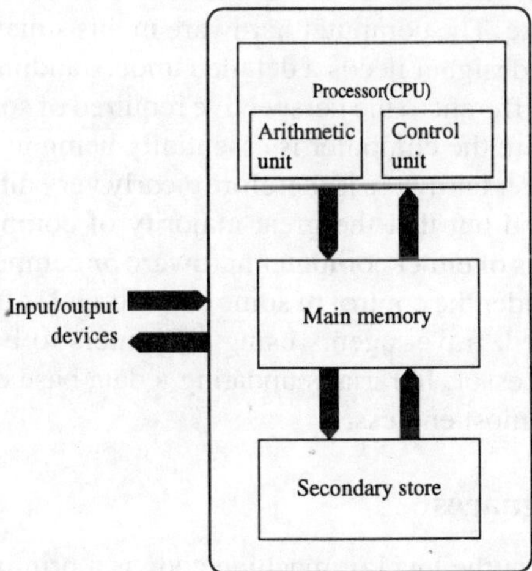

Figure 1.2 Conceptual view of a computer.

1.3.3 The computer as a tool in problem solving

So, given the nature of a computer, and the fact that we can control its actions by running some program which is useful to us, it is appropriate to look at how to use the computer as a problem-solving tool. Anyone using a computer to solve a problem needs to formulate some plan, or set of rules, in terms of which a solution can be obtained. A *plan*, in this sense, is usually termed an **algorithm**. *Recipe* would be an acceptable alternative term, and in the context of recipes as found in cookbooks gives a vivid—and highly relevant—illustration of the meaning of the algorithm concept. It is the function of people like programmers to translate an algorithm into a program that runs the computer when loaded into its memory. As far as a computer is concerned, a program, essentially, is a sequence of binary codes—generally referred to as **machine code**. These codes carry out functions such as control of the machine's arithmetic unit, sending information such as characters to a VDU, arranging for information to be stored in memory, etc.; sequencing such control actions—i.e. executing instructions in some given order—is inherent in the execution of a program.

Several questions need answering at this stage: firstly, how is someone's statement of a problem turned into a workable algorithm? Secondly, how can the algorithm then be translated into a program? In many ways the remaining chapters in the book can be viewed as providing answers to these two questions. In particular, the questions go to the heart of Chapter 13 on software engineering, where the main issue is how to turn a customer's wishes into a workable piece of software.

1.3.4 Hardware—how much do we need to know about it?

A further question most people just starting to use computers would probably want to ask is: how much need one know about the computer's hardware? The answer to this varies, depending on the nature of the application. Many electronic engineers these days are engaged on the development of software to run on embedded systems[1]—a computerized electronic ignition system for a car is a typical example—and design would involve a mixture of hardware and software elements: design of circuit board layout, choice of semiconductor components, etc. The computer hardware in this situation forms an integral part of the application, so the designer needs a detailed understanding of appropriate hardware principles. This is quite different to the perspective required of someone producing (say) an accounting package, where the computer is essentially being used as a development tool. Involvement with computer hardware is therefore clearly very different and more limited in scope. It might be pointed out that the great majority of computer users do not involve themselves in the internals of either computer hardware *or* computer software, but rather as users of computers run under the control of some commercial software package. Numerous examples could be quoted: travel agents using computers to book seats with an airline, typists using a word processor, librarians updating a data base containing information on books, etc.—the list is almost endless.

1.3.5 The role of languages

Preparation of programs at the level of machine code is a primitive activity programmers

---

[1] Embedded systems are discussed in the next chapter.

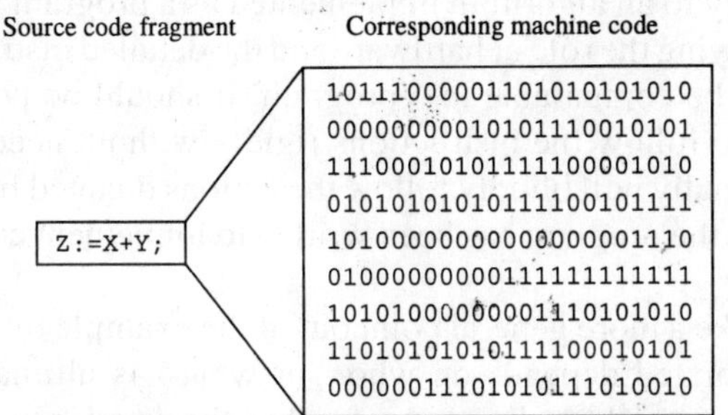

Figure 1.3 Source code and machine code.

would find extremely difficult for all but the simplest programs. One of the reasons for this is that a detailed understanding of a computer's architecture is required. Almost always, programmers use a language more closely related to the way in which a problem is stated, and then have this translated into a program in machine code, i.e. a form that is capable of being entered into a computer's memory and actually used to run the computer. The difference between the form of language used by programmers (often called **source code**), and the corresponding machine code is shown in Fig. 1.3. It should be clear from this why few people write programs at machine code level unless there is some compelling reason. (The source code here simply adds two numbers held in different parts of memory, and stores the sum in a third part of memory.)

To carry out the preparation of a program, and its translation into machine code requires the use of specialized programs, respectively called **editors** and **compilers**. A computer would commonly be provided with a number of compiler programs for each of the languages with which it has to deal.

1.4 ALGORITHMS IN MORE DETAIL

Algorithms are used to describe problems in all branches of science and technology, and one could also quote numerous examples of algorithms used in everyday life, for example: the sequence of steps that constitutes the 'execution' of a recipe or a knitting pattern. Sometimes a straight sequence of steps is involved in an algorithm, while in other cases something more complex is called for. Consider the following deceptively simple steps needed in preparing a part of a meal:

1. Take a slab of cheese.
2. *Then* grate cheese into fine pieces.

Steps 1 and 2 follow a simple sequence, but inherent in the second step is a form of repetition if spelt out in a little more detail. In computer jargon we would call this step a **loop**. Step 2 could be elaborated on by saying something like 'Take cheese grater *then* repeatedly grate until all the cheese has been processed'. Here, we have provided more detail, and also clarified somewhat the nature of the repetition involved. This is characteristic of the way an algorithm is developed to get it in a form where an implementation becomes possible.

Drawing an analogy to an algorithm implemented as a program running on a computer, the cook would be playing the role of hardware and the detailed instructions written down in the cookbook would be comparable to a program. It should be possible—in principle at least—for the cook to follow the instructions rigidly without needing to add any further ideas. A computer equally will blindly follow the actions dictated by its program, irrespective of whether or not the program has been written to implement correctly the algorithm it was based on.

It is possible to make a more general point out of the example by noting that just how far an algorithm is elaborated depends on who—or what—is ultimately implementing the algorithm. Few humans would really need more than the detail in the original statement, but if, for the sake of argument, we were thinking in terms of some kind of computer-controlled robot to carry out the same function, the amount of detail needed would be considerably more. Just think of the apparently trivial step to implement 'get cheese grater'. This could involve our robot locating the grater, using some vision subsystem, moving to where the robot arm could grasp it, and then manipulating the grater so as to make it possible to carry out the next step of grating the cheese. An algorithm for this would be complex, and the amount of programming involved to implement that algorithm considerable.

1.5 CHOICES IN LANGUAGES

Because there is such a wide variety of problems that a computer can be used to solve, a correspondingly large number of high-level languages have been developed,[2] some of the better known are:

- **FORTRAN**[3]—one of the earliest high-level languages (1955), and still in widespread use, mainly in scientific and engineering work where its features, which are convenient for dealing with computation intensive 'number crunching' type problems, makes its use attractive.
- **COBOL**—another early language used mainly in data processing work; unlike the areas for which FORTRAN would be preferred, COBOL is designed to deal effectively with applications that use large volumes of data, but for which the nature of the computation involved might be quite simple. The language makes use of statements that are close in form to everyday language, reflecting the likely background of programmers working in the business area, while FORTRAN is much more 'mathematically' orientated, reflecting the need to deal with mathematical expressions.
- **BASIC**—introduced originally as a simple language, suitable for teaching programming; it is still in wide use in this role, but is also quite widely used in industry. BASIC can—in its original form at least—be said to contain some rather primitive features in the interests of keeping the language simple to learn, but this means that programs of any length become more difficult to write.

---

[2] Languages are covered in Chapters 3, 4 and 5. This is just a short introductory review of some of the ones that are more commonly used. Comparisons between languages specifically of interest to the electronic engineer are discussed and compared in Chapter 5.

[3] More detail on FORTRAN is given in Chapter 4.

- **PASCAL**—early languages like FORTRAN contained somewhat primitive features—to some extent reflecting the simple nature of the hardware platforms available in the early days of computing—which were increasingly seen as an obstacle to writing programs of any size effectively. Languages of which Pascal is a quite early (1970) example, aim to help programmers write their code in a way more closely identified with the terms in which an algorithm would be expressed. So, for example, if the needs of a particular problem required one of seven different courses of action depending on the day of the week, in PASCAL we might devise a set of variables that could have values named as 'sunday', 'monday', etc. It would be much clumsier to carry out this type of programming in languages such as FORTRAN, where, in the example, simple integers would be used to represent the 'days of the week', in other words using a format more removed from the problem domain than we would like.
- **LISP**—this is one of the oldest computer languages, and is still in widespread use today. LISP is an acronym for LISt Processing, and this gives a clue as to its nature, namely as a language aimed at the manipulation of lists. This is of particular benefit in areas such as AI (artificial intelligence).

1.6 HARDWARE—A BRIEF OVERVIEW

IBM compatible PCs are familiar to most computer users, and feature in later chapters. This being so, it might prove helpful at this stage to use this type of computer as a vehicle for describing some of the more practical aspects of just what constitutes the hardware element in a practical form of computer system. Figure 1.4 shows a recent version of the PC; its components are as follows:

1. What might be regarded as the heart of the computer is a circuit board called a **motherboard**. This contains the following: the microprocessor chip, main memory and also chips to allow the processor to interface with peripheral devices such as the keyboard. The main memory is in the form of semiconductor chips (we noted earlier that this is much faster to access than the secondary storage devices; however, main memory is normally of much smaller in-storage capacity). The bulk of the main memory is volatile—something users might remember before switching off their machines! The motherboard in this computer is the large, vertically mounted board to the rear of the photograph. The memory chips are mounted in modules on the bottom right-hand side of the motherboard.
2. Devices with which users interact when they need to input or receive information to or from the machine. A VDU, a keyboard and a mouse are shown here, but numerous other input/output devices exist.
3. An **expansion board** which can be used to enhance the capability of the basic machine. A wide variety of functions can be carried out by such boards. For example there are boards that allow communication to other computers via phone links (called modems), and boards that allow the computer to be interfaced to sensors and other devices and thus enable a computer to be used as an element in a data acquisition system. Here the expansion boards are the vertically mounted boards that plug into the motherboard on the left of the photograph; they are here used to control (a) the video output to the VDU and (b) the disks.

Figure 1.4 Typical small computer system.

4. **Secondary storage** devices in the form of **hard** and **floppy disk drives**—to be described in a little more detail shortly—which can hold relatively large quantities of data and can do so even when the machine is switched off—and thus termed **nonvolatile**. A disk drive is a form of magnetic medium no different in principle to any tape recorder. The actual disk has a magnetic coating on which the digital information is stored after being recorded. The surface is divided into tracks, with each track subdivided into sectors; these can be regarded as the disk's elementary unit of storage. The read/write head is moved radially to the track that has the required information, in response to a request from the computer, and is then read (or written to) as the required sector passes under the recording head. This is shown in simplified form in Fig. 1.5. (Typically a sector would store 512 characters, and a disk might have 80 tracks with 18 sectors per track in the case of a floppy disk storing a total of 1.44 Mbytes of data—however, such figures differ quite a lot from one type of disk to another.) The machine

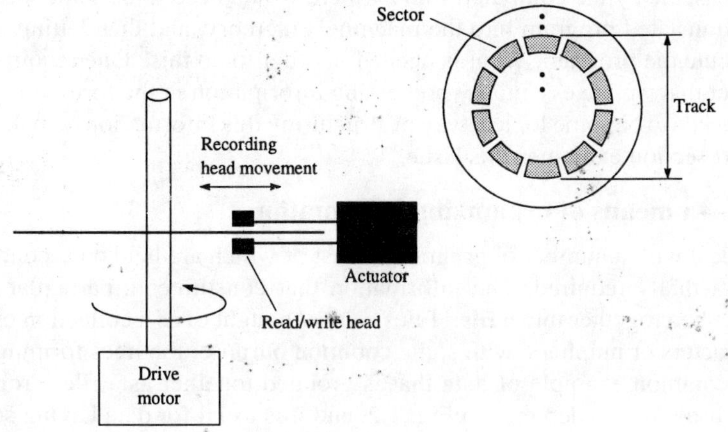

Figure 1.5 Disk drive.

has two disks, a single 3½ inch floppy disk and a hard disk; both are mounted in the drive
bays at the top right-hand side in the photograph. Note the ribbon cable connecting the
disk drives to an expansion board which is a disk controller.

While a program is being executed, it may reside in the computer's main memory.
However, secondary storage devices will be required to supplement main memory for
the following reasons:

(a) To give a way of holding the program, even when the machine is switched off.

(b) To provide a relatively large amount of storage space to hold what is usually the
considerable number of programs that a computer will commonly use during its
lifetime.

(c) In the case of floppy disks and also some other types, the nature of the media allows
for the situation where data needs to be removed from the computer, perhaps to be
used in another computer. Naturally such provision is essential if commercial packages are to be used on a computer. (With most types of hard disk drives, the disks
cannot be removed; their function is rather to allow a relatively large, high-speed
store for holding the bulk of the information in a computer, and they are really
complementary to floppy disks which are now seldom used in this role.)

5. Last but not least, the power supply. This is the unit at the top left-hand side of the
machine in the photograph.

1.7 HOW DO USERS INTERACT WITH HARDWARE?

This is not a book principally about hardware—although there are some notable exceptions—so we are justified in focusing attention more on how hardware is *used* than on how
it *works*. For example, if one adopts the position of someone developing a program, such an
individual would first need to create the program somehow and then translate it into the

kind of code usable by the computer. Finally there would need to be some mechanism for loading the translated program into the machine's memory, and then letting the machine actually execute the program. What is needed in order to do this? One requirement arises out of the fact that we are essentially processing information in the form of programs and data. There needs to be some logical way of structuring this information to make it manageable; the next section examines this issue.

1.7.1 Files—a means of organizing information

Computers deal with a number of programs, most of which are held on secondary storage media until actually required. The information that constitutes a particular program is normally grouped together into a **file**.[4] Files can be thought of as a collection of data in the form of characters or numbers, with some common purpose; the lines forming a program would be a common example of data that is grouped together as a file. Programs often generate (or use) *data* as depicted in Fig. 1.2, and it is usual for data having some logical common function to be grouped together in the form of files. Files relate to programs and data that users can generate, but the file concept also applies to packages such as editors and compilers—in fact to *any* software that is being stored on the computer. Files generated by users—including programs such as editors and compilers—can be viewed as **application programs**. They *use* the resources of the machine to carry out some function.

The tangible way a file manifests itself on a disk is as information recorded on a collection of sectors. There needs to be some provision for identifying which sectors constitute a given file, and also the order in which sectors are linked together. In some situations, consecutive sectors may be used to hold successive records in files, but, in many other cases, consecutive records will be scattered around the disk on a number of different tracks. In any event, there needs to be, in general, some way of identifying the track and sector number for all records in a file, and this information needs to allow records to be linked together in the proper order.

So far we have talked about the needs of an individual in developing programs as a way of illustrating how users interact with computers, and as a means also of establishing just what kind of resources the computer has to provide. The file idea is one part of this, i.e. as a way of organizing data. A variety of operations on files are needed: copying them (from one set of sectors on a disk to another), deleting them and, where appropriate, allowing operations on files such as editing and compilation. The computer needs some provision to make these operations possible.

1.7.2 The role of the operating system

Before directly addressing the point at the end of the previous section, we look at the problem of how an application program interacts with the hardware (see Fig. 1.6). It would be extremely difficult—and unusual—for an application program to contain code that was responsible—for example—for *directly* moving the arm of a disk drive to some specified track. Invariably, computers have a collection of programs to carry out functions like these to control the hardware. In total, such programs are called a machine's **operating system**. An application program would normally call on the services of a program within the operating system to carry out functions such as control of disks and other hardware devices. An

---

[4] Secondary store is often termed a **file store** for this reason.

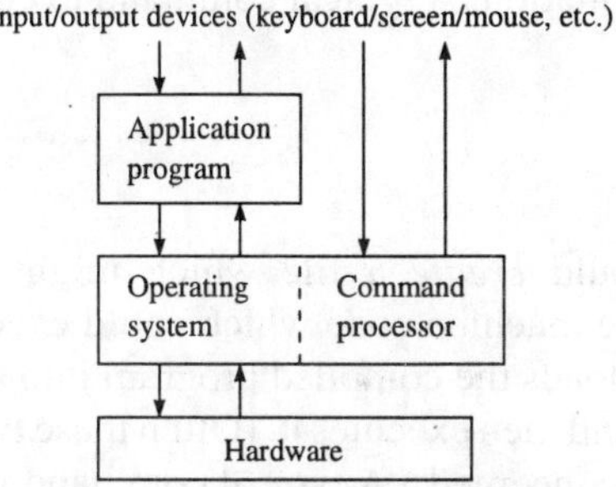

Figure 1.6 Relation between application programs, the operating system and the hardware.

operating system can be thought of as the software interface between an application program and the hardware, hiding much of the hardware's complexity from users.

Users can interact with a computer—via its operating system—either by an application program they are using or else, more directly, by entering a command using the keyboard. In the latter case, a **command program**, which forms part of the operating system, would be used. Essentially, this deals with characters entered by users on the keyboard, interpreting the string of characters collectively as a command. Most operating systems generate a **prompt** on the VDU (e.g. C:\), which is in effect an invitation for users to enter some command.

Having seen what the operating system's purpose is, we can now come back to the point raised at the end of Sec. 1.7.1. A program is to be created (by an editor), compiled and finally run on the computer; the operating system is needed to manage the hardware which allows all these operations on our program to be carried out. Just what commands are needed specifically? Each command will invoke a program, either what has been called an application program, or else a program forming part of the computer's operating system. Commands will be in the form of some unique character string, which in quite a few cases will need to have the name of a file included as part of the command line. Filenames are usually character strings, the format of which needs to satisfy whatever rules are imposed by the operating system. The example below shows how to invoke one of the editors in the operating system MS-DOS[5] to edit a file called `myfile.pas`.

```
C:\ edit myfile.pas
```

The C:\ does not form part of the command; recall that it is in fact the prompt generated by the operating system. Note that the '.pas'—an **extension**—distinguishes this particular file as Pascal source code. Following this command, the VDU would display some—or all—of the lines in the file depending on its size. Other information to assist users in the editing process would be displayed also. Note that if the file did not exist prior to invoking the editor, we would in effect be making the operating system—in cooperation with the editor program—*create* it, by the act of entering text and saving it after ending an editing session. The editing process would otherwise be modifying a file that has already been created.

---

[5] This is used with IBM PCs and is described in some detail in Chapter 8.

The second phase is compilation. A typical command to compile a program could look something like:

```
C:\ pas myfile.pas
```

The compiler program would *create* a file which might be called in this example `myfile.exe`, this being the machine code which could execute, after being loaded into memory. A final command loads the compiled program into the computer's memory perhaps from secondary store, and then executes it. (Often these two actions are combined into one as far as a command is concerned.) A typical command would be:

```
C:\ myfile
```

FURTHER READING

Dowsing, R. and Woodhams, F. (1985) *Computer Architecture—A First Course*, Van Nostrand, London.
Wilkinson, B. and Horrocks, D. (1987) *Computer Peripherals*, 2nd ed., Edward Arnold, London.
Willis, N. and Kerridge, J. (1983) *Introduction to Computer Architecture*, Pitman, London.

2

THE ROLE OF SOFTWARE IN ELECTRONIC ENGINEERING

2.1 INTRODUCTION

This chapter gives some indications of the various ways in which electronic engineers need either to develop or to apply software. The areas in electronic engineering that involve computing are diverse, although for many perhaps the largest single field of interest is in real-time systems (discussed in Sec. 2.2). **ECAD** (= Electronic Computer-Aided Design) is another important area with software techniques being increasingly relevant in chip design. Perhaps it should be added that in many cases 'software' is something that someone else has developed in the form of ready-to-use packages, examples such as the **Spice** circuit analysis package being one of several well-known examples one could quote. This book will not involve itself in this to any great extent; any design using a package such as Spice does not involve the application of software engineering principles as far as users are concerned, and so is not really relevant to a text on software techniques.

To set the material dealt with into some kind of historical perspective, one could point to microprocessor-based systems as perhaps the most visible, and commercially the most important, area of concern to electronic engineers. Interest in such systems dates back to the beginning of the 1970s when microprocessors were first introduced, at that time as a cheaper alternative to systems that had been based around MSI logic. Microprocessors were, by present standards, primitive, mostly 4-bit chips, and the programs they ran were very small by current standards, of the order of a few kbytes. Typical applications were systems such as the electronics inside a daisy-wheel printer, and petrol-pump controllers, where simple microprocessors were quite adequate. Coding was invariably done in assembler, often by individuals having an electronic engineering, rather than a computer science, background. It might be fair to say that, in a sense, the software was perceived as a kind of appendage to the hardware, and programming techniques tended to reflect this, being comparable to the kind of approach used in first and second generation computers of the 1950s. When programs are as small as they were in those days, this situation might be quite tenable, but developments in VLSI technology in general and in microprocessors in particular have meant that the complexity of applications, in both hardware and software terms, has grown out of all recognition; and now programs comprising 100 000 plus lines of code are not uncommon. The detailed implications of this increase in complexity are discussed in the chapters on software engineering, but the need is for electronic engineers to adopt the same systematic rigorous approach to the development of software that has long been used by people in other areas of computing such as data processing. They need to take the design of the software element in a system they are producing as seriously as that of the hardware element. To this end, electronic engineers need background in the following areas:

- Algorithm design
- A suitable language (or more probably several!) to turn an algorithm into something that can run on a computer
- Software tools such as editors and compilers to help actually produce the program
- An understanding of how to *apply* programming methods and software tools of various kinds in the study and application of the techniques of software engineering

2.2 REAL-TIME SYSTEMS—CHARACTERISTICS AND TYPES

2.2.1 What is a real-time system?

Since much of the interest of electronic engineers in computing systems centres around real-time systems, a definition might be in order at this point. Rather than giving some *formal* definition of what we mean by a real-time system, though, an illustrative example might more readily help focus on essentials. Consider a robot arm controlled by a computer. The computer might typically consist of several rack-mounted circuit boards, one of which would contain the controlling microprocessor chip. What are the essential requirements as far as the computer is concerned in a system like this?

- The arm needs to move between two points in some specified time, and if in fact it takes a longer (or shorter) time for any reason, the specification is not being met.
- There will usually be signals generated by sensors. (Examples are: tactile sensors used to indicate contact between themselves and other objects, force sensors—these *measure* force—used to help in the handling of objects of different sizes and optical sensors used as part of machine vision systems for viewing and interpreting its workspace.) In general, data from sensors can arrive at the computer in a nondeterministic order, possibly simultaneously. Some inputs will require more urgent attention than others and so demand a more speedy response.
- The arm is likely to be controlled from an instrument panel by an operator who—one needs to assume—has no computing knowledge beyond that needed to work the robot's controls.

- The robot will typically be used in a factory where there is likely to be a relatively hostile environment in terms of temperature variations, vibration, fumes and dirt, etc.

The first requirement points to the *essential* need to have the computer respond as expected to real-time deadlines. Contrast this with the way programs are handled by computer centres in a typical multiuser system, serving users connected to the computer by terminals. Users on such a system, running a program they have just prepared, say, might be happy getting output from their programs within the space of a few seconds or a few minutes perhaps, depending on the nature of the program. From time to time—if the computer is heavily loaded—the response time might be rather longer but, in any event, although users might grumble, they will be more or less satisfied with the service provided and, in any event, time delays of themselves will not result in any *incorrect* data being generated. With our robot arm on the other hand, response times not in accordance with that specified could cause serious problems, even physical damage to plant in some cases.

The second requirement in the list above is to do with meeting real-time deadlines but also points to the need to respond properly to outside stimuli no matter what order they arrive in, and to set up some priority system, if needed, for servicing deadlines.

The last two points are not concerned with *real time* in a strict sense, but with characteristics necessary in a class of real-time system termed **embedded real-time systems**. The term refers to types of computers that are viewed as components in a larger system of some kind, in the same way that the controller in the robot arm system is a component. This contrasts with computers such as PCs, mainframes, etc., which are general purpose machines running many different programs during their working lives. In the case of a mainframe, the machine operates in a very carefully controlled environment—perhaps air-conditioned, requires attention by operators from time to time and has data processing staff to deal with other aspects of the computer's operation. An embedded system on the other hand:

- Needs to be rugged enough to handle the environmental conditions it is going to be used in; and, taking this aspect further, reliability issues are likely to be a much more dominant part of the specification of many embedded real-time systems than with other types of computer. Few people would worry if their PC generates bad data occasionally when it is running a word processor package, but failures in the kind of computer systems found in an aircraft for example could lead to catastrophic consequences.
- Always runs the same program, which is only going to be changed if the system ever gets upgraded—this will be infrequently, perhaps once every year, every five years, or never!
- Requires a minimum of intervention from operators. For example, embedded systems should, in most cases at least, not require users to insert a floppy disk to boot the system; indeed, in the interests of robustness, embedded systems are frequently diskless.

2.3 TYPES OF MICROPROCESSOR-BASED SYSTEM

Since many areas in this book deal with embedded systems of various kinds, it might be helpful in understanding the software problems involved to take a very brief look at the different kinds of hardware that such systems comprise. Indeed, unlike many other areas in computing, both hardware and software feature in the design process; moreover, problems related to their interactive nature are frequently of great concern to a designer.

A logical starting point might be to compare the various types of microprocessor chips used, and we need to consider the suitability of a given type of microprocessor chip in terms of cost, reliability and performance. Microprocessor chips vary in terms of computational power from 4-bit chips used in simple control-type applications where computational speed may be of little importance to advanced 32- and 64-bit chips which can rival superminicomputers in terms of computational power; 8-bit chips like the 8085, Z80, 6502, etc., fall midway between these two extremes, being relatively cheap, simple to interface, but powerful enough to be usable in applications such as low-end home computers. Chips like the three cited above have been around for a long time but are still in common use.

2.3.1 Single-chip microprocessors

Single-chip microprocessors (Cahill, 1986; Lister, 1987) integrate the functions of CPU, memory and I/O on a single chip. They are of considerable commercial importance, and significantly outstrip general-purpose microprocessors in terms of dollar volume. Most semiconductor manufacturers produce such devices, which extend from simple 4-bit microcontrollers to comparatively powerful 16- and 32-bit devices. The applications are diverse, ranging over calculators, hand-held games, washing machine controllers, robotics and automotive electronics. The advantage of the single-chip microprocessor accrues from its low cost, this being of particular importance in products that are manufactured in extremely high volume such as calculators, where numbers running into millions are involved. However, since single-chip microprocessors are frequently dedicated to one application, they are not economic in small production volumes, e.g. less than 20 000 units.

The simplest type of single-chip microprocessors are 4-bit devices. (The '4' refers to the numbers of bits in the data path and ALU inside the chip, buses may in fact not always be brought out to pins.) Chips like these are used in calculators and small logic controllers, and the cost is typically less than $1. 4-bit microprocessors can handle binary-coded decimal (BCD) arithmetic operations one digit at a time, so that data comprising several BCD digits has to be processed in digit serial form. By implication, such devices are relatively slow, although this is no disadvantage in applications like calculators. In terms of production volume, 4-bit microprocessors outstrip all other types of microprocessor by a considerable margin. (Texas claim to have shipped 60 million of their TMS1000 devices alone.)

8-bit single-chip microprocessors are used for more demanding applications, and can carry out 8-bit arithmetic without needing to resort to digit serial arithmetic. The precision of an 8-bit word (i.e. 1 part in 256) is adequate for many applications, including those involving the use of 8-bit A/D and D/A converters. The 8-bit word also allows a more convenient size for op codes. Examples are the Intel 8048 and 8051, the Motorola 8604 and 6805, and the Zilog Z8 families.

16- and 32-bit devices offer a minicomputer-type architecture on a small scale, and they give a big step up in computational power compared to 8-bit microprocessors. A typical application is in the robotics field as robot arm controllers. Commonly, robot arms have six independent joints, and control of the arm may involve feedback of the position and possibly velocity of each joint. The controller has to work out the actual trajectory of the arm and compare it to the desired value. Appropriate drive signals would then have to be calculated and sent to each joint. The amount of computation involved is often sufficiently intensive to require a separate 16-bit microprocessor to act as the controller of each joint. An additional 16-bit controller may also be used to coordinate activities of the other microprocessors.

2.3.2 Example of the Texas TMS1000 single-chip micro

This is a very commonly used 4-bit microprocessor. Its data paths and ALU are 4 bits wide, although the memory is 8 bits wide. The I/O has been designed with display and keyboard multiplexing in mind (most applications need a display and/or keyboard, an example being the vacuum fluorescent display used in petrol-pump controllers—and are invariably based around 4-bit microprocessors). As with most single-chip microprocessors, the instruction set is primitive and orientated towards logic-type operations. The TMS1000 is slow, by virtue of its 4-bit data paths, and also because of the relatively low clock frequency of 330 kHz. All instructions execute in 18 microseconds.

Figure 2.1 shows, in much simplified form, a typical controller type application for the TMS2372 chip, this being a member of the TMS1000 family. The TMS2372 can accept two analog inputs directly, which it can convert to digital form. In this case, signals derived from two sensors are supplied to the analog inputs. There is a vacuum fluorescent display which is supplied *directly* from I/O lines on the chip. Further I/O lines, used as outputs, supply signals to fans, heaters and other points in the system. The user inputs are a multiplexed array of keys, the key multiplexing being driven from the display digit–drive I/O lines. This approach of multiplexing both input and output with the same software and hardware is common, to economize on I/O lines.

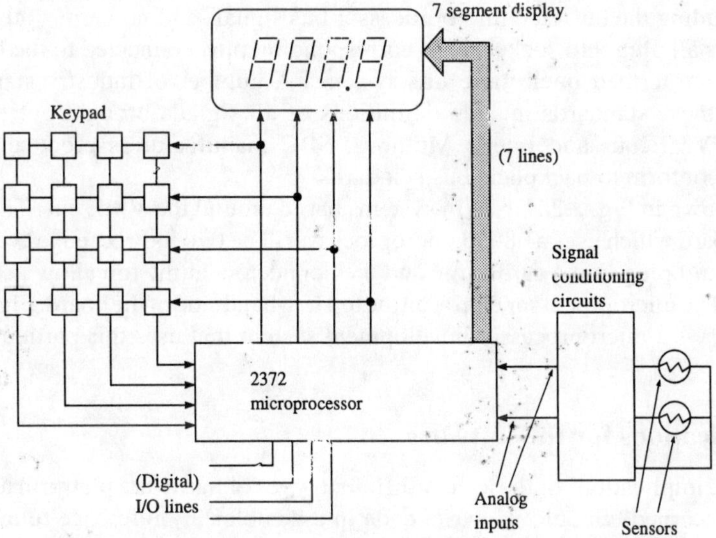

Figure 2.1 2372 chip used in simple controller system.

2.3.3 Single-board computers

Single-board computers (SBCs) are circuit boards which include most, if not all, of the elements needed to make a complete working computer. This means that the principle functions implemented would be: a CPU in the form of a microprocessor—often a standard chip such as a Z80 or 68000—nonvolatile memory (EPROM or PROM) for storing code, volatile memory (RAM) for holding data, I/O in the form of parallel and serial interface chips which provide access to the board's facilities via parallel and serial ports. Some SBCs might include analog/digital and digital/analog converters, although commonly functions like these might be hived off on to a separate board coupled to the SBC in some way. Systems based around SBCs are very commonly used in industry, particularly in applications such as instrumentation and control. Such systems provide the designer with ready-to-go hardware—and frequently also firmware—allowing a fast, cost-effective approach to the development of a microprocessor-based system.

In some respects, SBCs can be viewed as complementary to single-chip computer systems, and used where the limited computational power or high development cost of the latter would rule out their use. In fact, single-chip systems are only really viable in high-volume applications, and many types have very limited computational power. SBCs are often available in the form of 'off-the-shelf' circuit boards which can be purchased from companies specializing in their manufacture. A designer then only has the problem of interfacing the board to his or her system, and can concentrate his or her efforts on the higher-level aspect of a systems design. One type of SBC is shown in Fig. 2.2.

2.3.4 Backplane bus systems

Many types of microprocessor-based systems are built up from a number of circuit boards such as SBCs, graphics boards, disk controller boards, etc., which plug into a common backplane. All boards in such systems have connectors which are brought to a common set of signals including the buffered microprocessor bus signals and numerous other signals. The connectors all plug into sockets with corresponding pins connected to the backplane. Such systems are termed **backplane bus** systems. A number of industry standards are available and these standards include definitions of all signals brought out to the bus; examples are VME bus and Intel's Multibus. SBC manufacturers produce numerous boards which conform to backplane bus standards.

The SBC shown in Fig. 2.2 is used in systems based around the VME bus. The SBC is a Force VME board which uses a 68030 microprocessor. The two indirect edge connectors at the bottom would plug into a cardframe and the connectors at the top allow access to the board's ports. The microprocessor chip is on the top left-hand side of the board. Figure 14.3 in Chapter 14 shows a microprocessor development system that uses this particular type of SBC.

2.3.5 Ramifications for the software

So what are the implications of these very different types of hardware platform as far as the software is concerned? Should we write code in assembler in preference to a high-level language—or vice versa? To what extent is extensive study of the principles of software engineering needed before embarking on a design? How much does someone developing

Figure 2.2 Force VME single-board computer card.

the code need to know about the hardware? Some aspects of these questions are addressed in later chapters, but a brief overview at this stage might be helpful.

Take the most primitive type of system first—small single-chip microprocessor-based systems. Code in such systems needs to fit into the ROM element which forms part of the microprocessor chip, and which is going to be of rather limited capacity, perhaps as small as a few kbytes. This means that we have the problem of 'shoehorning' as much code as possible into whatever limited storage area is available. For this reason, programs are usually coded in assembler, because of the greater effectiveness with which a programmer can produce compact code, as compared to code generated by a compiler (this topic in fact gets further airing in Chapter 7). We quoted earlier the example of the TMS1000, there the development systems used to produce TMS1000 applications do not even support high-level languages. As available memory space for code gets greater, the need to produce compact code becomes less, and it then becomes possible to develop programs using high-level languages, with all the advantages which accrue from their use.

Robustness was mentioned as a vital feature in embedded systems, and it should be stressed that this applies as much to the software in a system as to the hardware. We will say more on this in the later discussions on software engineering, but, essentially, the message is that the critical nature of many applications—such as the aerospace example quoted earlier—does not allow failure be tolerated under any circumstances. Developers need to spend a lot of time and effort in the design of both the hardware and software elements in a system to ensure that this is so. This can be horrendously difficult if the software is very complex, however; indeed it may not always be possible to guarantee absolutely that code is completely bug free when programs get beyond a certain size.

Concurrency is an issue running right through the design and operation of just about any real-time system. As we have already noted, in real-time systems, the computer needs to deal with a number of inputs which in many cases can arrive in any sequence and at times that often cannot be predicted in advance. The computer then has the seemingly impossible task of running in parallel a number of programs that are structured to deal with data arriving at its input ports in this way—and for that matter *producing* other data at its output ports also. In some cases, computers might comprise more than one CPU, in which case several programs can operate in true concurrency. But, if there is only one CPU, a mechanism is needed for switching the CPU from one program to another, allowing each to run eventually. This is one of the most important issues discussed in Chapters 9 and 12. Concurrency is one of the more obvious reasons why real-time concurrent programming differs from simple sequential programming, where—on many types of personal computer at least—there really *is* only a single program running. (It should be added that with multiuser computer systems, as with many computer centres, the CPU is being shared between several users, each of whom has the illusion that he or she has exclusive access to the machine—this is discussed in Chapter 9.)

2.4 VLSI DESIGN AND THE IMPACT OF THE SOFTWARE APPROACH

VLSI design has long been inextricably bound to the computer used as a design tool (Hill and Coelho, 1987; Perry, 1991). Chip and circuit design using simulation and other packages running on workstations of various kinds is the norm for all but the simplest systems. A fact of life that hardware designers have to cope with is the rapid and continuing increase in complexity of electronic circuits, and the consequent need for continuing improvements in design techniques to match this complexity. If one examines the methods used by hardware designers over the years, the striking feature is the increasing resemblance in the techniques used to those traditionally employed by software developers. This is a reflection of the fact that complexity has long been a fact of life in large software projects, and that now, unlike the situation, say, 10 or 15 years ago, the complexity of hardware is now of the same order. Techniques that might have been suitable for the development of an MSI logic circuit comprising a few thousand transistors are quite unworkable in developing chips of the size of microprocessors, some of which can have several million transistors.

One of the most significant ways lessons from the software world have been used for hardware design is the use of a hierarchical approach (Hill and Coelho, 1987). It has long been common to divide software into a hierarchy of modules, in the interests of controlling the level of complexity that designers have to handle. In hardware too, it is often possible to reduce a design in this way. A simple example is a parallel adder, which can be decomposed into a number of full adders, which in turn can be decomposed into half adders. A design of the more primitive elements such as the half adder can then be used as a building block as part of a larger design. In other words design is occurring at a number of levels, with the internal details of the elements at a particular level being hidden from the levels above it. Data needs to be communicated on a need-to-know basis in much the same way that data is communicated between procedures in a program.

Figure 2.3 gives a simple illustration of the way a design in the adder example can be decomposed into a number of levels. The circuit can be viewed at three different levels, depending on the point of development reached in the design. Further layers could be added if, for example, we wanted to incorporate the parallel adder into an arithmetic unit, which in turn could comprise part of a microprocessor.

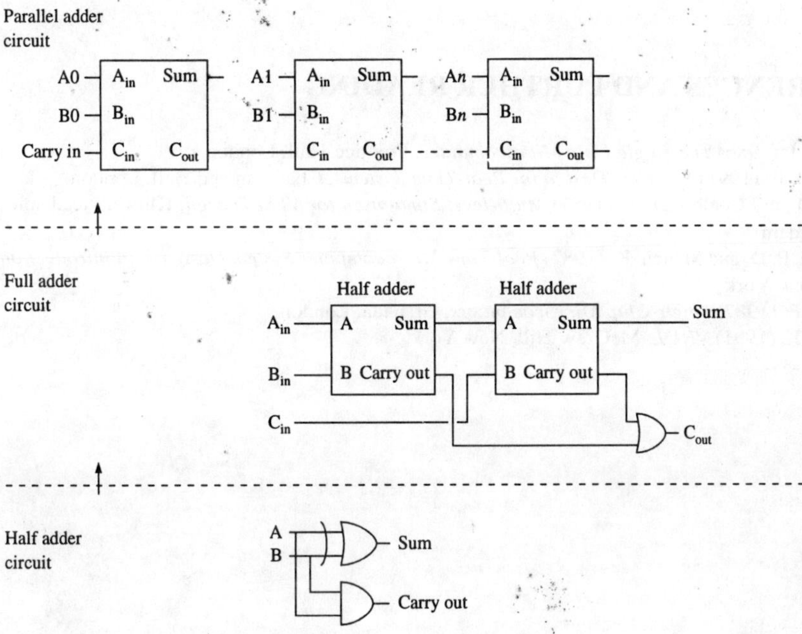

Figure 2.3 Hierarchical design of an adder circuit.

This particular example is somewhat simple, and it might be argued that using the same approach to design with more complex circuits is inappropriate. Nevertheless, with this cautionary remark in mind, what more can we say about possible approaches to design? In particular, how would we be modelling hardware in a software-based approach, and how would this approach differ from more traditional methods? The representation of a circuit in diagrammatic form, as shown in Fig. 2.3, might be fine for designing circuits such as parallel adders, but we have been saying that this is inappropriate in more complex cases. With a software-based approach, one would be thinking of a textual description. Within this form of representation there would be a network of elements such as half adders, each being described as a procedure, with some mechanism for interconnecting the elements together as a network to allow them to exchange data.

Conventional (sequential) high-level languages such as Pascal, FORTRAN, C, etc., are not well suited to the modelling of electronic circuits. Circuits such as our parallel adder exhibit a high degree of parallelism, something that does not feature in sequential languages in their standard form. Although it is not impossible to use sequential languages for this type of work, it is of considerable assistance to the designer if the conceptual model offered by a language supports the characteristics of the circuits being modelled. Thus, support for the kind of concurrency found in circuits is very important, as are features that model the interconnection of circuit elements. A number of **hardware description languages** have been developed that incorporate these features; examples are VHDL (Perry, 1991), HHDL (Hill and Coelho, 1987) and ELLA, with the first of these three looking to become the hardware description language of choice for many hardware developers in the future.

REFERENCES AND FURTHER READING

Cahill, S. J. (1986) *The Single Chip Microcomputer*, Prentice-Hall, London.
Cooling, J. E. (1991) *Software Design for Real-Time Systems*, Chapman and Hall, London.
Hill, D. D. and Coelho, D. R. (1987) *Multi-level Simulation for VLSI Design*, Kluwer Academic Publishers, Amsterdam.
Lawrence, P. D. and Mauch, K. (1987) *Real-Time Microcomputer System Design—An Introduction*, McGraw-Hill, New York.
Lister, P. F. (1987) *Single Chip Microcomputers*, Granada, London.
Perry, D. L. (1991) *VHDL*, McGraw-Hill, New York.

Part II

Languages and algorithms

3

THE LANGUAGE C

3.1 INTRODUCTION

C is a high-level language widely used in programming of the kind of real-time systems of interest to electronic engineers; it is a general purpose language used in many other types of applications also. One of the most attractive features of C, particularly with regard to its use in real-time applications, is that it combines the advantages of a high-level language with the ability to implement the kind of low-level operations possible with assembler; in other words where complex programs that also require close control over the hardware are involved. For example, 'bit twiddling' instructions such as shift and rotate are provided in C, operations that would be clumsy to code in languages like BASIC or Pascal, and would therefore possibly require fragments of assembler code to be embedded in the program. C is a block structured language like Pascal, lending itself to convenient program development with structured programming techniques. In the real-time programming area C has now to a large extent replaced assembler as a more productive way of writing code of any size. This chapter aims to give sufficient detail of the language to enable other chapters which refer to C to be understood, and also to support the discussion in Chapter 6 in which comparisons between several languages are made. The treatment is somewhat informal, and is in no sense a comprehensive guide to C, for which other texts should be consulted (Kernighan and Richie, 1988). Emphasis is given to the more distinctive features of the language.

C was introduced in the early 1970s to program the operating system UNIX, the code that, until that time, had been implemented entirely in assembler.[1] C has been called a

---

[1] The background to this is discussed in more detail in the chapter on UNIX, but for the present it might be noted that there has always been a close relation between C and UNIX.

portable assembly language, in that it is comparatively simple to install (**port**) a C program written for use on one machine to another, in spite of any differences in architecture. Portability is achieved by having a relatively small central core which is the language proper, and a library of utility routines called the **standard library**, carrying out functions such as input/output (I/O).

3.2 INTRODUCTORY PROGRAM

Books that introduce the reader to a computer language often give a short demonstration program at an early stage, partly as a confidence builder, and this book follows that tradition. The following shows a fairly minimal C program.

```
/* this is a simple program to display the message
   'Hello World' on the screen                        */
 main()
 {
     printf( "Hello World" ) ;
 }
```

Note that text enclosed by the delimiters '/*' and '*/' is taken as a comment, and ignored when a program is compiled. Pascal uses '{' and '}' for this purpose.

As with many block structured languages, C comprises modules which here are called **functions** (roughly equivalent to procedures and functions in Pascal but with some important differences). One function is called `main`, and any C program starts its execution from the function `main`. Functions can return a value to a calling program as in a Pascal function, however it is not mandatory to make use of the value so returned. The 'empty' brackets could in other programs or functions enclose parameters, and examples where this is so appear later. In more complex programs than this, other functions would be called from `main`. Functions are described more fully later. Blocks which comprise functions are enclosed by '{' and '}' and serve a comparable purpose to `begin` and `end` in Pascal. The program above has just a single statement; its meaning is reasonably self-evident, but note the ';' which acts to *terminate* statements in C. This is subtly different to the use of ';' in Pascal where it there serves to *separate* statements, so that the last statement in a block in a Pascal program need not be ended with a ';'.

If you have access to a computer with a C compiler, a strong suggestion is that at this point you try compiling and running the program above, this will prepare you for running the programs described later in the chapter and also in the examples at the end of the chapter.

3.3 VARIABLES

Variables are basic data objects in any program. In the case of C, as in Pascal, we are dealing with a strongly typed language.[2] Amongst other things this means that all variables

---

[2]Some would argue with this statement and early versions of C might be described as weakly typed (Kernighan and Richie, 1988), but type checking was considerably tightened up in the 1988 ANSI standard which gives a comprehensive definition of the language.

must be *declared*, i.e. the allowed data type such as integer, floating point, etc., must be stated. If this is not done, the program will fail to compile, the offending variable then being identified by the compiler. Declarations must appear at a point in the program ahead of the place the associated variables are actually invoked. Examples of typical declarations are given in 3.4.

Variable names can be made up of upper or lower case letters, digits, and also the underscore '_' character, often used in the interests of improving legibility, as in: `a_valid_variable`. It is common, but not mandatory to make variable names start with a letter. There is a distinction between upper and lower case letters. Finally, only the first 31 characters in a variable name are significant.

3.4 DATA TYPES

C supports the following data types:

- **int**—these are integers, the length of which is implementation dependent. There is control over the length of an `int` : `long int` (often) referring to 32 bits, `short int` to 16 bits, while `int` on its own can be 16 or 32 bits.
- **char**—8-bit integers. Although the name `char` may seem suggestive of characters, `char` can be used for *any* 8-bit integer quantity.
- **float**—single-precision floating point.
- **double**—double-precision floating point. The number of bits in this type are, as for objects of type `int`, implementation dependent. There is also a `long double` type which is an extended-precision floating point.

In the case of `int` and `char`, it is possible to choose between *signed* and *unsigned* values. As the name infers, unsigned means numbers that can assume values from zero upwards only. For example an `unsigned short int`, assuming a 16-bit two's complement, could have values between 0 and 65535, while its signed counterpart would have values that could range from 32767 to −32768. Clearly, if an `int` is to be used as an unsigned quantity in a program it makes sense to declare it as unsigned, otherwise its range will be restricted to about half that otherwise possible.

In the interests of speed it might be desirable for a programmer to force a compiler to hold particular variables in registers rather than memory, the latter involving longer access; C allows a special form of declaration that does this. Some compilers will try to make use of registers in preference to store-to-hold variables, but the provision of a special kind of declaration give the programmer more control over which variables are in fact to be held in registers.

Finally, variables can be initialized to a specified value in declarations, quoting the constant in one of the forms described above. This avoids the need for a separate assignment statement to do the same thing.

Rather than giving a lot of formal detail on the syntax involved in declarations, the following examples should adequately illustrate most of the previous points:

```
/*
  32-bit integer given an initial value BUFSIZ, this is a
  constant which would need to have been itself given a value
*/
  long int clock_cnt = BUFSIZ ;

/*
  letter is an 8-bit quantity, given an initial value which
  is the ASCII code for '%'
*/
  char letter = '%' ;

/* 16-bit integer held in a register */
  register short int fast_val

/*
  unsigned byte given an initial value of 0x10 ie 16 decimal!
*/
  unsigned char i_max = 0x10 ;
```

Casts—a mechanism for mixing different types

C was earlier described as a strongly typed language, but in one respect this is not really so, in that there is considerable latitude given in mixing types in expressions. Many C programmers value the freedom of action this allows them. To use one type as another in an expression it needs to be **cast**; for instance if an `int` variable needs to be used with floating point variables or constants in an expression, it may—depending on the operation required—need to be cast to `float`. Thus some `int` variable `tofloat` is cast by quoting it as `(float) tofloat`, and thereafter is a floating point number as far as the remainder of the program is concerned. As an illustration of where this kind of conversion is useful, consider the division of one `int` variable `numr` by another `int` variable `devr`. If `numr` had a value 1, and `devr` a value 2, the result would be 0, since, in integer division, the fractional part of an expression is discarded. To get a true result, both variables need to be cast to float as in:

```
  reslt = (float)numr/(float)divr ;
```

A rule in C is that any operation with two differing types will first **promote** the smaller type to the same type as the larger. Thus if a `float` is divided by an `int`, the `int` will be promoted to a `float` *before* the division occurs, giving the correct `float` value as the result.

3.5 CONSTANTS

Constants essentially are quantities in a program that are fixed at compile time. Constants in C can be expressed in the following ways: decimal, octal, hexadecimal, (single) characters, or strings. Constants can be assigned symbolically in the form of names that represent them, or else used literally in expressions and elsewhere; in principle this is much

like the provision in Pascal and many other languages. The examples below should clarify
the notational aspects of constants:

- Decimal numbers, e.g. -1 5678
- Octal numbers, e.g.   033   077
  (note the leading '0', which is mandatory)
- Hexadecimal numbers, e.g. $0 \times 0a$ $0 \times ffff$
- Character, e.g.   'h'
- String, e.g.   'a sample string'

A string constant is actually an array of characters terminated in a null byte (i.e. 0×00).
String and character constants interpret the backslash '\' character to specify some special
characters in conjunction with a key character following the '\' as detailed below:

- \n carriage return character
- \f ASCII form-feed
- \t ASCII tab
- \b ASCII backspace
- \\ the backslash character itself
- \0 the null byte

If a constant is to be associated with a symbolic name, there must be an appropriate declaration, as in:

```
#define BUFSIZ 1024
```

BUFSIZ can subsequently be quoted in expressions, declarations and elsewhere, and the
value it represents, i.e. 1024, will then be used. It is a common convention that symbols
representing constants are written in upper case.

Enumerations

Enumerations are an extension of the idea of associating constant values with names in the
interests of legibility and convenient program maintenance. By way of illustration, consider the situation where a variable is to model an alarm quantity which can have three
values: 'high', 'normal' or 'low'. These values could be defined literally using the names
given, and variables of such a type declared. An advantage of enumerations is that illegal
values are easy to identify. The declaration in this example could be an enumeration called
(say) alarm_type as:

```
enum alarm_type {
    low    ,
    normal ,
    high
} ;
```

A variable with these enumerated values could then be defined as:

```
enum alarm_type oil_pressure_alarm
```

Typical code that used this enumerated type might be:

```
oil_pressure_alarm = high ;
```

or:

```
if(oil_pressure_alarm == low) {
/* code to service low oil pressure condition */
 }
```

By default, the actual numerical values associated with the enumerations start at 0 for the first enumeration, 1 for the second, etc. In some cases, however, different numerical values might be required. In the `alarm_type` example this could be done as follows:

```
enum alarm_type {
    low    = 10 ,
    normal = 25 ,
    high   = 50
 } ;
```

3.6 EXPRESSIONS AND OPERATORS

Most high-level languages have operators like +, −, /, *, etc., to be used in assignment statements or elsewhere; but, additionally, C has bitwise logic operators, that are used in low-level operations such as **shift** and **rotate**. A list of these operators is:

- Left shift   <<
- Right shift   >>
- Bitwise AND   &
- Bitwise OR |
- Bitwise XOR   ^
- One's complement ~ ('~' is called **tilde**)—this is a bitwise NOT

The bitwise operators above should be compared to their counterparts in assembler. It may be noted that there are also a set of Boolean operators AND (&&), OR (| |), NOT ('!') provided, which have a similar meaning to comparable operators in Pascal and many other high-level languages. These must not be confused with the bitwise operators above—all too easy to do in some circumstances, in view of the somewhat similar symbols used. In this connection note that the == operator is the equality operator in C (i.e. it serves the same role as '=' in Pascal, Modula-2, Ada and some other languages), and not to be confused with '=' which is C's assignment operator (i.e. the same as ' : =' in Pascal).

The provision of bitwise operators of various kinds makes it particularly convenient to do the type of 'bit twiddling' operations commonly associated with assembler level programming. The examples below should adequately illustrate the use of the bitwise logic operators:

- If we have two variables X and Y with values 01001111 and 00110001 respectively, the assignment Z = X & Y gives 00000001—(equivalent to a *single* comparable assembler AND instruction). For example, assuming X and Y have already been loaded into data registers D1 and D2, then the 68000 assembler instruction directly equivalent to the expression X & Y is AND.B D0,D1.[3] Separate code would be needed to store finally the value of the expression in the memory location reserved for Z.
- ~1 = 11111110, ~2 = 11111101, etc., e.g. if num = 10101111 then num & ~1 is 10101110
- Consider a 68000 code sequence, where we read a port with address PORTC, and loop until the byte is an ASCII carriage return (= 0x0D):

```
LOOP: MOVE.B PORTC,D0
      AND.B   #$0D,D0
      BNE     LOOP          ;Wait for carriage return
```

The effect of the AND operation can be achieved in C by code such as portc & '/n', and this could be embedded inside a loop equivalent to the assembler code above.

Commonly used operations at assembler level are increment and decrement; special operators for these are provided in C, written respectively as ++ and −−. Additionally, it is possible to increment or decrement variables either before, or after, the assignments with which they are associated. For example the assignment y=x++; increments x *after* assigning its original value to y (= post-increment), while y=++x; increments x and *then* assigns the resulting value to y (= pre-increment). Comparable operations are allowed for the decrement operator. In a similar vein there are += and −= operators which can respectively increment or decrement variables by any specified amount. For example x +=y increments x by an amount y. Many other versions of these operators are available such as *=, /+, ^=, etc.

3.7 STATEMENTS

In languages like C, functions comprise blocks, being sequences of (for example) simple assignment statements, conditional statements of which the simplest type is if, to be described below, or looping statements such as while. if and while have broadly similar meaning to comparable statements in Pascal. A compound statement is a sequence of simple statements. The examples below show some typical statements:

1. x = r+7 ; /* simple assignment statement */

2. while(portc & 1);
   (see earlier 68000 example: Reads portc, waiting for carriage return)

---

[3] See Appendix 1 for more on the 68000 microprocessor.

```
3. if( x == 0xff )
        y = 2 ;
   else
        y = 1 ;⁴
```

The examples above are all simple statements, the one below features compound statements:

```
4. if( hit != 1 ) {
        f = 9 ;
        g = 2 ;
   } else {
        f = 1;
        g = 3 ;
   }
```

This carries out the assignments f=9 & g=2, if hit is *not* equal to 1, otherwise the alternative two assignments in the else clause are made. Note the use of '{' and '}' to enclose a block, in this case a sequence of two simple assignment statements. In other cases, what here were assignment statements, could be conditionals or loops, which themselves may contain compound statements.

```
5. cnt = 10;
     while( cnt >= 0){
         cnt--;
         sum = sum + cnt;
     }
```

This executes the loop until cnt is negative. Notice the use of the post decrement operator to decrement cnt. The assignment cnt=10 assigns the initial value of cnt used as a loop counter, the terminating condition, i.e. cnt >= 0, appears in the next clause, while the final clause determines the way the loop counter is to change between successive iterations of the loop. More complex forms are possible for the three clauses of a for loop. However, this is rather clumsy; an alternative, somewhat neater way of coding this is:

```
   for( cnt=10; cnt >= 0; cnt-- )
       sum = sum + cnt ;
```

This is a **counted loop**. Most languages have something similar but in C the counted loop is comparatively more powerful. The code above is equivalent to the previous example.

---

⁴Note the different use of the '==' and '=' operators. Had the Boolean expression following the if been x=0xff (perhaps in error), the effect would have been to assign the value 0xff to x, and *then* produce a TRUE or FALSE value depending on the result. Values greater than 0 are TRUE in C, so the statement following the if would always be the one to execute! Mistakes involving incorrect use of '==' are a common source of bugs in C, and can be especially interesting when loops are involved.

6. In keeping with its tradition of producing terse code, C provides a **conditional expression**, as a more succinct alternative to the `if` statement. If we wanted to make one of two possible assignments depending on the result of some Boolean test, a conditional statement of the form shown below could be written:

```
test ? assignment1 : assignment2
```

For example in `a > b ? x = 8 : x = 98;` the `x = 8` assignment would be carried out if `a > b`, otherwise `x = 98`. More complex forms of the conditional expression than this are possible, and a further example is given later (Sec. 3.13).

The `break` **and** `goto` **statements—terminating a loop early**

It is possible to terminate a `while` or `for` loop before the terminating condition is met. This is done with the `break` statement, which will terminate the innermost loop in the case where nested loops are involved. Note however that `break` does not give a jump to the outside of several nested loops, only to the next outermost loop. To achieve jumps out of several layers in a set of nested loops, the `goto` statement is available, as in the following example—`finito` is a label:

```
goto finito;
...

...
finito: <statement>1
```

A label must be defined in the same function as the associated `goto`, and for a jump from one function to another, the routines `setjmp` and `longjmp` are used. This kind of code is useful when a program is to be terminated due to some error condition, in which case the normal structure of the program can be bypassed. Error recovery is an important topic in real-time programming, and is considered in Chapter 6.

3.8 ANOTHER SHORT C PROGRAM

The program below features a while loop, and also a `printf` statement to display output from the program. The program simply displays integers from ten down to one on separate lines, followed by the message 'Blast off'.

```
main()
{
    int count = 10 ;

    printf("Counting down:\n") ;
    while ( count > 0 ) {
        Printf("%d\n",  count) ;
        count = count - 1 ;
    }
    printf("Blast off\n") ;
}
```

In the `printf` statement the `\n` in the first `printf` does a carriage return, while the `%d` in the first `printf` is associated with integers that are to be displayed, in this case `count`. Note the way `count` is initialized in the declaration.

The `while` loop could alternatively be coded as:

```
while( count > 0 )
    printf( "%d\n",  count-- ) ;
```

or:

```
for( count = 10;  count > 0;  count-- )
    printf( "%d\n", count ) ;
```

These two methods of coding the loop should be contrasted with the original version, which is not too dissimilar to comparable code that could be written in Pascal and some other languages. The economy of means which it is possible to achieve in C is well illustrated by this simple little example.

3.9 STRUCTURED TYPES

3.9.1 Arrays

Simple variables as discussed above are inadequate to deal effectively with many programming problems, for example writing a program that multiplied two matrices would be difficult with matrices of any size, just using the simple data types described above. An **array** in C is a **structured object** devised to deal with problems like this and is a collection of variables referenced by a common name, with individual elements referenced by an index notation. Each member of an array is of the same type, be this `int`, `char`, etc. Like (unstructured) variables, arrays need to be declared, and the example below declares `demo` to be an array of `char` with at most 11 elements (*not* 10 notice).

```
char demo[10] ;
```

The program block(s) for which this declaration is recognized would refer to the array elements as `demo[0]`, `demo[1]`, etc. The example below illustrates a simple use of the array, the value 55 is assigned to all elements of `demo`.

```
for( i=0; i<=10; i++ )
    demo[i] = 55 ;
```

The examples above used one-dimensional arrays. Multidimensional arrays are possible in C also. For instance a two-dimensional array called `matrix2d` could be declared as:

```
char matrix2d[10][10]
```

A common use of arrays is in the representation of strings. Essentially strings are arrays of type `char`; they contain printable ASCII characters and certain others such as <Carriage Return>, and must be terminated by a null '\0' character. Therefore some string name with a value 'roy' would have four elements:

```
name[0] ='r'
name[1] ='o'
name[2] ='y'
```

and finally

```
name[3]  ='/0'
```

Strings are declared as arrays of char as one would expect, and can also be given an initial value in the declaration, as in:

```
char name[4] ={ 'r', 'o', 'y', '/0' } ;  (note the array size)
```

This is rather clumsy; an alternative, more succinct form of declaration is possible, as in:

```
char name[] = 'roy' ;
```

Notice that the first Hello World program used a string quoted in this way. A number of routines are provided in C for manipulations involving strings; including concatenation, comparison, copying, and also I/O (input and output) of data in string form. This rich provision reflects the importance of strings in many C programs.

An example of a program which exercises some of the string handling routines is given below. The program simply reads a line of text into an array. The text is assumed to consist of at least two words separated by space characters. The first two words are copied into two strings and displayed. Additionally the first word is compared to the string "CC". The following string manipulation functions are used:

- `gets()` returns a string entered at the keyboard, which is terminated by a carriage return.
- `strcpy( copy_string, in_string )` makes a copy of `in_string`, this being the name of an array of `char`, in the form of the array of `char` called `copy_string`.
- `strcmp( a_string, b_string )` compares the strings `a_string` and `b_string` one character at a time starting from the left. The function returns a value of 0 if all pairs of characters agree, otherwise the function returns the difference between the values of the first nonmatching characters.

```
#define  LINE_BUF    80
#define  WORD_BUF    20
#define  SPACE       0x20
char CC[]        = "CC" ;
```

```
main()
{
    char  line[LINE_BUF], i = 0, k, j ;
    char  word[WORD_BUF], first[WORD_BUF], second[WORD_BUF] ;

    /*  -- put a line of text into 'line' -- */
    gets( line ) ;

    /*  - copy first two words in line to first and second - */
    for( k=0 ;k<2; k++ ){
       j = 0 ;

        /* copy characters from line up to a SPACE, into word */
        while( line[i] != SPACE )
           word[j++] = line[i++] ;
        word[j] = 0x00 ; /* terminate string with a NULL */

        if( k == 0 )
           strcpy( first, word ) ;
        else
           strcpy( second, word ) ;
        i++ ;    /* -- step past space character -- */
    }

    if( strcmp( first, CC ) == 0 )
       printf( "\nfirst word : %s\n", first ) ;
    else
       printf( "not CC" ) ;

    printf( "\nsecond word : %s\n", second ) ;
}
```

3.9.2 Structures

Array elements all need to be of the same type, but this restricts their use in cases where there is a need to model real-life objects, comprising several elements which are of different types. For instance, if a program is to model an object like a book, its attributes such as publication date, number of pages, title, etc., are of disparate types, and should be capable of being modelled as such in a program. A **Structure** (or **struct**) is the provision in C for cases like this; it is the counterpart of a Pascal record. For example, a structure called demo, comprising a variable of type int and another of type char would be declared as:

```
struct {
    int number ;
    char letter ;
} demo;
```

Following this declaration, individual elements of the struct could be accessed as in the following examples, the dot notation being the mechanism by which a member of a structure is picked out:

```
demo.number = 56 ;
demo.letter = 'a' ;
```

Arrays of structures are possible: for example using the previous idea of a book it is possible to think of a book collection being modelled as an array of books, i.e. objects that can be modelled as being of identical types, with each book in the collection being modelled as a suitable structure. To create an array of struct for demo above, a suitable declaration might be:

```
struct {
    int    number ;
    char   letter;
} demo[10] ;
```

This could be typically accessed by statements such as demo[0].number = 56 ;.

3.10 POINTERS

A pointer is a form of variable that can be viewed as representing an address; to be more precise it is the address of another variable, that which is being *pointed to*, in the form of the data stored at the address in question. The term pointer has usage outside its specific application in C as a generic term for an address, and pointers also feature in Pascal and Modula-2, serving a roughly comparable role. Pointers are used in Chapter 5, where the subject of data structures is discussed. Pointers in C have some quite distinctive features, and are heavily used by most C programmers. For example using pointers it is possible, in some systems at least, to access physical store locations, something not possible with any other type of variable, since compilers normally allocate storage locations for variables without making the addresses visible to programs. Assembler programmers commonly need to access store by address however, input/output from a port for example and the provision of pointers in C allows this type of access within a high-level language. Without such provision it would be necessary to embed fragments of assembler code to carry out the required operation.

A special form of declaration is needed for pointers. Recall that two quantities are involved: a pointer and the object (data) being pointed to. For example for some pointer called y, the contents of y need to be accessed to read or store a value at the address involved, as shown in Fig. 3.1. The value being pointed to is written as *y, '*' is a **pointer de-referencing operator** in this context. In the example in the diagram below, y is the pointer that has a value 0x80000 assigned to it. The value stored at the corresponding address, i.e. 0x1234, is then subsequently assigned to x. Note that y is a pointer to int, the same type as the variable to which the value pointed to is being assigned.

A common requirement in C is to have access to the address of a variable, as opposed to a variable that happens to be declared as a pointer. To access the address of some variable x

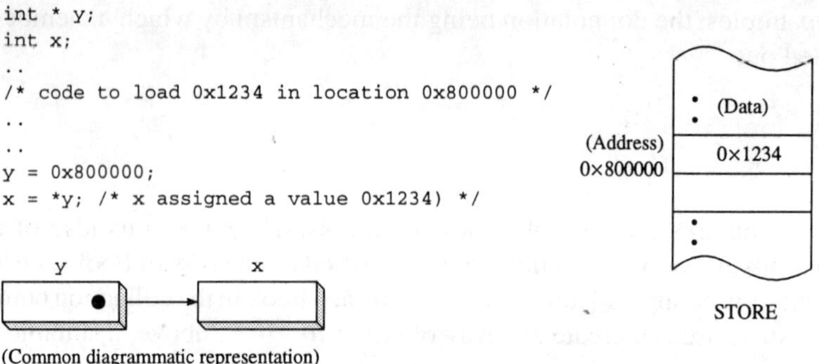

Figure 3.1 Pointers in C.

we could write code such as y = &x; that is the address of x is here being assigned to y. The '&' can be thought of as literally *the address of*. In a somewhat related way, the address of the first element of an array can be accessed and used by simply quoting its name, as in fact was done in the example in Sec. 3.9.1. Thus the assignment x = demo, where demo is the array used in earlier examples would assign to the variable x the address of the first element of the array demo. This is useful since we can, by incrementing x, point to subsequent elements of the array. In other words it is possible to do arithmetic with array indices involving pointers: this is often more efficient than an equivalent operation on the array, possibly an important consideration in the context of a real-time program where execution speed is critical.

The example below demonstrates the use of pointer arithmetic. The program uses a version of the string copy operation which featured in an earlier program. The function my_strcpy simply copies the input string character by character, incrementing cpy and orig, which are effectively pointers, each time so that they point to the next character. The value assigned becomes the object of the Boolean test in the while loop, and when the end of the string is reached, as signified by the null character, the value assigned is zero which is equivalent to FALSE, and the loop then terminates. This approach can be contrasted to that in the program in Sec. 3.9.1 where a copy of a string was made, using a loop which stepped through the array, incrementing the *array index* each time.

```c
#define   MAX_STR_LEN    80

/* -- function prototype -- */
void my_strcpy( char *cpy, char *orig ) ;

main()
{
    char clone[MAX_STR_LEN] ;
    char  string[] = "string to be cloned" ;

    my_strcpy( clone, string ) ;
    printf( "%s", clone ) ;
}
```

```
void my_strcpy( char *cpy, char *orig )
{
    while( *cpy++ = *orig++ )
        ;
}
```

3.11 FUNCTIONS

Functions were mentioned briefly earlier, being the building blocks from which a C program is made up. Programs in C comprise **declarations** and a list of functions, which have the same general format as the function main discussed previously. The simple example below, although not a particularly useful program, illustrates some aspects of program structure:

```
void demo_funct() ;

main()
{
    printf( "program starts in main\n" ) ;
    demo_funct();
    printf( " returned from demo_funct\n" ) ;
}

void demo_funct()
{
    printf( "now inside the demo_funct routine\n" ) ;
}
```

This example uses no variables. Had it done so, some decision would have been required about the **scope**[5] required for the variables. Variables that are only required inside a particular function should be declared inside that function, in which case they would not have any meaning to other functions; they are called **local variables**. Alternatively it is common for several functions to need access to the same variable. This can be done in several ways. One approach is to place the declaration outside the functions, commonly near the start of the program ahead of the main function. Such variables are called **global** and have a common meaning for the complete file containing the functions. An alternative method involves communicating data directly from function to function by passing values from one to another as **parameters**. An example of a function that does just this is printf, the values passed there being data such as the string 'Hello World' in the first program. To pass data to a function, the variable concerned must appear between the parentheses in the heading. The earlier example passed no data in this way, which is the reason the parentheses were 'empty'.

Note the declaration of demo_funct just ahead of main, this states that the function whose body appears later does not return a value and is thus of type void, also it uses no

---

[5] Scope refers to the range of code in functions or files for which variables are recognized.

arguments. This form of declaration is called a **function prototype**, and it has to agree with the definition and use of the function. This type of declaration facilitates error checking when the function definition and its prototype are not consistent. In the case of functions defined in the standard library, function prototypes often appear in separate header files.

Extending the earlier example somewhat, say we want to make demo_funct display the value of some integer variable passed to it by a calling program. Code such as that below could be used:

```
void demo_funct2( int x ) ;
main()
{
    int a=9 ;

    printf( "program starts in main\n" ) ;
    demo2_funct( a ) ;
    printf( " returned from demo_funct\n" ) ;
}

void demo_funct2( int x )
{
    printf( "demo_funct2 was called from main - %d\n", x ) ;
}
```

Note the declaration of x inside demo_funct2. Note also, that had we, for the sake of argument, chosen in demo_funct2 to call the variable a instead of x, this would have no connection with the a in main, both variables being local to the functions in which they appear. To show a simple use of global variables, consider a third version of the program:

```
int glob=9 ;
void demo_funct3() ;
main()
{
    printf( "program starts in main\n" ) ;
    demo_funct3();
    printf( " returned from demo_funct\n" ) ;
}

void demo_funct3()
{
    printf( "demo_funct3 shows the global - %d\n", glob ) ;
}
```

glob has a common meaning to both main and demo_funct3, so there is no need to transmit this variable as a function parameter.

Commonly, functions carry out some computation, and return a result to a calling function. One way this can be done is by including a statement with the keyword return followed by the value to be returned. There will be a value to the function as far as the calling function is concerned, and in fact the function can appear as part of an expression. For example if a function called `calc_root` calculated the root of an equation, returning its result, then we could write code like `printf("root=%d",calc_root());`. (The empty brackets after `calc_root` mean that no parameters are being passed to the function in this particular case.) It is implicit in all this that a function must be associated with a certain data type, which would also need to be declared in the associated function prototype, as in:

```
float calc_root();
```

Returning a value like this works for a single quantity, but in other circumstances more than one quantity could be involved. In such cases, it might be thought that one could pass back the value of a parameter *to* a calling function in the same way as values can be passed *from* a calling function. However things are not quite so straightforward; the mechanism by which C passes parameters is termed **call by value**, that is, the function being called makes copies of variables passed to it and it then proceeds to use the copies. The implication is that the original value of parameters will not be changed. If in fact we want to allow a function to change the value of a parameter, a calling function must provide the *address* of the variable(s) concerned. The program below demonstrates this principle, it calculates the two roots of a quadratic, passing the addresses of the variables with the roots back to the calling function. The coefficients a, b, and c are passed in the form of parameters to the function:

```
void root( float a,float b,float c,float *r1,float *r2) ;

main()
{
    float r1, r2, a = 1.2, b = 4.7, c = 7.9 ;

    printf( "roots of quadratic\n" ) ;
    root( a, b, c, &r1, &r2 ) ;
    printf( "%f %f\n", r1, r2 ) ;
}

/* -- roots of a quadratic - real roots assumed! -- */
void root( float a, float b, float c, float *r1, float *r2 )
{
float temp1, temp2 ;

    temp1 = sqrt(b*b-(4.0*a*c))/(2.0*a);
    temp2 = -b/(2.0*a)
    *root1 = temp1-temp2;
    *root2 = temp1+temp2;
}
```

Note the way `&r1` and `&r2`, i.e. the *addresses* of r1 and r2, appear in the call to root in `main`. The program in its present form does not allow for complex roots; as a simple exercise you could try to modify it to deal with this case.

Functions may also require **local variables**, declared inside the function body, and by default, the values of local variables would be lost from one function call to the next: they are referred to as **auto** variables. In some cases, however, it could lead to simpler programming if the values between successive function calls *could* be 'remembered', and there is provision for this in C (unlike Pascal); such variables need to be declared as **static** as in:

```
static   char perm_loc_var ;
```

3.12 MORE ON SCOPE AND DECLARATIONS

As well as controlling the scope of variables by declaring them as local or global, control is also allowed over the scope of variables used in a program comprising parts split between several files. Large programs are often divided in this way, reflecting the various logical parts into which the program is structured, in the interests of effective program development. Careful control of the scope of variables is then most important. It is possible to restrict scope to the file in which a variable is declared, or alternatively to allow the scope to extend globally and thus allow functions in certain other selected files access to it; this idea extends to functions as well as variables. The examples below illustrate most of these points.

The following declaration makes `another_place` refer to a common variable in all files where this declaration appears. Note that the keyword `extern` does not actually *create* the variable, it simply informs the compiler that the variable exists in some other file and that the variable can be used in the present file:

```
extern int another_place ;
```

This next example makes the scope of the variable `only_here` local to the file in which it appears. In other words, even if another file makes reference to a variable given the same name, it will in fact refer to a completely separate variable:

```
static   int only_here ;
```

The example below, which is a program split into three files, puts all these ideas together:

```
/* ---- (A) file master.c ---- */
static   int   only_here;
int      glob1;
/* -- function prototype for fun1() - defined in slave1.c -- */
extern   char fun1( <parameter list> ) ;
main()
{
    int    only_main;
 ...
}
```

```
/* ---- (B) file slave1.c ---- */
extern   int  glob1;
/* -- function prototype definitions -- */
char fun1( <parameter list> );
void fun2( <parameter list> );


/* -- function bodies -- */
char fun1( <parameter list> )
{
   ...
   ...
}
void fun2( <parameter list> )
{
   ...
   ...
}
   ...
```

```
/* ---- (C) file slave2.c ---- */
static  int  glob1 ;
/* -- function prototype definitions -- */
static void fon1( <parameter list> ) ;
static void fon2( <parameter list> ) ;
/* -- function bodies -- */
void fon1( <parameter list> )
{
   ...
   ...
}
void fon2( <parameter list> )
{
   ...
}
   ...
   ...
```

In the `master.c` file, `glob1` is a global variable defined in this file, to which access is allowed by other files, as well as by all functions within `master.c` itself. The scope of `only_here` on the other hand only applies within `master.c`, extending to all of the functions within this file. The variable `only_main`, declared inside `main` in `master.c`, is a local variable whose scope is confined within the function `main`, and does not extend to any other functions that might form part of `master.c`.

The file `slave1.c` has an `extern` declaration involving `glob1`, and this refers to the variable of this name both declared and defined in `master.c`. The file `slave2.c` also has a declaration involving a variable `glob1`. However, this is different to the variable of the same name defined in `master.c` and used by `slave1.c`. Note that `glob1` here refers to a variable defined inside `slave2.c` and only used inside this file.

The function `fun1()` declared in file `slave1.c` is required in the file `master.c` and the form of function prototype used allows this function to be shared between the two files. The functions `fon1()` and `fon2()` defined in file `slave2.c`, however, are to be kept private to that file, so the use of `static` in the two function prototype definitions ensures that no other file can access these two functions.

3.13 ALLOCATION OF STORAGE SPACE FOR CODE AND DATA

Part of the function of a compiler is to allocate storage space for programs.[6] This comprises: (1) the code, (2) the stack—used to store data such as function return addresses, function parameters, and also local variables, (3) space for global variables, and (4) space to hold dynamic variables associated with pointers. In the latter case, storage space must be allocated when a program runs, by a call to a library function which makes some requested amount of store available from a pool of unused store. This free pool of store, referred to as the **heap**, is located between that needed by the store allocated by the compiler for code and global variables and the stack. This is illustrated in Fig. 3.2.

The operator `g_val` will be stored in the area reserved for global variables. Depending on the compiler, `l_val` could have space reserved for it in the stack, or else be stored in a register. In any event, the stack is used when the call to `fun` is made to store the address to return to.

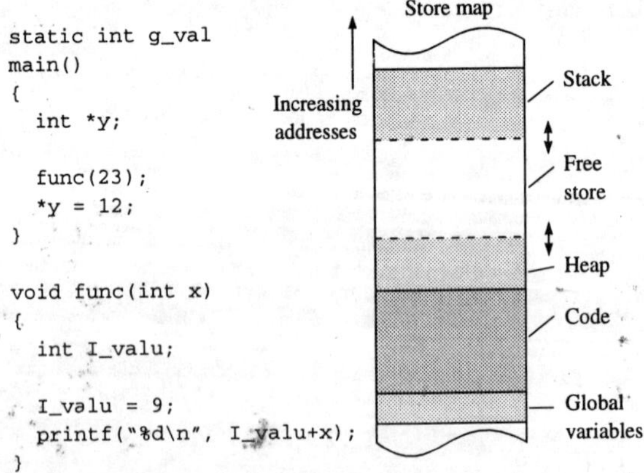

Figure 3.2 Store allocation in C.

---

[6] Questions relating to store allocation in a more general sense appear in Chapter 7.

Clearly it is essential that the stack and the heap are never allowed to overlap; were this to happen, the results would inevitably be fatal as far as a program is concerned. The library function `malloc`, used to request store, will in fact return a value of NULL if there is none left in the pool of free store. The code fragment below gives a way of requesting 1000 bytes of store from the free pool, and responding appropriately if there is none available.

```
    ..
    char  *mem ;
    ..

    ..

       if(( mem = malloc( 1000 ) ) == NULL )
          /* -- handle case of no free store -- */
       else
          /* -- proceed normally -- */
```

3.14 TYPEDEF—NAMES FOR USER-DEFINED TYPES

User-defined types are an important feature of Pascal and some other languages as a way of facilitating the process of modelling real-life quantities in a program. Constructs such as `struct` and `enum` also do this. The `typedef` construct carries this idea further, and allows meaningful names to be used for data types such as `struct`, arrays, arrays of `struct`, etc. The general form of the `typedef` statement is:

```
typedef typ  type_name ;
```

`typ` is a predefined type of some kind while `type_name` is the name that can be used in later code to refer to that type. For instance if we wanted a class of variable associated with data to be read from an 8-bit port, the use of `unsigned char` might suggest itself. To give the type a more meaningful name we could alternatively use a definition such as:

```
typedef  unsigned char  port_data_t
```

variables could then be defined, as in:

```
port_data_t   oil_alarm_port  ,
              water_alarm_port,
              speed_alarm_port ;
```

All this may superficially seem little more than a cosmetic embellishment to the language, but in practice is of considerable assistance in producing portable code, since alterations may involve no more than changing a `typedef` statement.

The somewhat more complex example below shows how the `struct` demo in Sec. 3.9 could be defined with `typedef`:

```
typedef struct{
   int    number ;
   char   letter;
} demo_t[10];

/* -- variable defined in terms of demo_t -- */
demo_t  demo_variable;
```

Elements in `demo_variable` could then be accessed, as in:

```
demo_variable[0].number = 56 ;
```

3.15 THE C PREPROCESSOR

Prior to compilation, a C source program is treated by a so called **preprocessor**, referred to as `cpp` in UNIX environments. The various functions carried out by the preprocessor are as follows:

1. **Constant evaluation**—one of the roles of the preprocessor is the conversion of names defining constants in `define` statements into the numerical values actually used by the program. It may be that, depending on factors such as the particular machine on which the program is eventually to run, the numerical values will need to change. However if constants are represented inside the source program by names, these will be independent of the characteristics of the computer, or of other factors. Essentially we are insulating the source program from the effects of changes that might be made during the working lifetime of that program. The preprocessor can be applied in other ways to carry this process further.

2. **File inclusion**—the contents of a file, perhaps containing definitions, including `define` statements in many cases, can be included in a C program by statements of the form `#include "filename"`. The entire contents of `filename` would then be inserted into the source program at the point at which the `include` statement occurs. The file is assumed to be in the same directory as the original C source program (an alternative form: `#include <filename>` would assume `filename` is located in a standard directory for such files). Such files are termed **header files**, and a number are provided as part of the C programming environment with most C compilers (Kernighan and Richie, 1988). As well as `define` statements, header files commonly contain definitions of data structures such as `struct`s or `array`s, possibly defined as user-defined types with `typedef`. As with `define`s, putting such things in header files, insulates the source code file from the effects of changes which may eventually be ultimately required when the program needs to be maintained. Users often write their own header files, in order to tailor programs to individual requirements; this can be useful if, as is usual in large software projects, source code is divided into separate files, for more flexible program development and maintenance. If the files share a common set of definitions it makes sense to put them in a common header file which can then be *included* in those files which require them, rather than having duplicated definitions.

3. **Macros**, which are sometimes used to replace function definitions, are treated by the preprocessor. This might be done for reasons of efficiency since the code for the macro

is embedded into the source code where it is quoted, rather than being called as for a function; the additional overhead in calling is avoided. This might be useful if the macro is to be heavily used in a tight loop in a program. Macros take the following form:

```
#define   name   replacement
```

`name`, when quoted later by a program causes `replacement` to be invoked, much as a function call. For example, a macro to return the larger of two arguments could read as follows:

```
#define   max( x,y )   (x > y ? x : y)
```

Having defined the macro `max`, the lines of code below would assign a value of 8 to x:

```
y = 8 ;
x = max(y,3)
```

4. **Conditional compilation**—in many cases, parts of a program are specific to the particular computer on which it runs, typically machine code fragments which might be embedded in the C source code. Another example is a debugging code which would not be required when the program had been fully tested. Conditional compilation can then be useful, it takes one of the following forms:

```
(a) #if   condition1
       /* code if condition1 applies */
    #elif condition2
       /* code if condition2 applies */
    #endif
```

```
(b) #if condition
       code if condition applies
    #endif
```

For example, if we had a program that would be required to run on either 68000 or 8085 microprocessors with suitable modifications, a code such as that below might be appropriate:

```
#if 68000CHIP
#define ADDRESS_BITS = 24
#elif 8086CHIP
#define ADDRESS_BITS = 16
```

The constants `68000CHIP` or `8085CHIP` are taken to have a `TRUE` (i.e. > 1), or `FALSE` (0) value; one of them could be invoked on the command line when the compiler is run, or alternately be given `TRUE` or `FALSE` values with a define statement inside the program itself. Note that with many systems, invoking the compiler by a command such as CC (used on UNIX and some other systems) would in fact have the effect of first automatically invoking the preprocessor, which in turn would then supply the processed source code to the compiler proper for translation into machine code.

3.16 C PAST AND PRESENT, AND THE STANDARD LIBRARY

Early versions of C were notable for some of their rather informal definitions in areas such as structure definitions. The use of the language has spread in recent years beyond its original field of application. This leads to certain enhancements becoming desirable, and a tightening up on some of the earlier usages. In 1983 the American National Standards Institute (ANSI), set up a committee to produce a definition of the language that was machine independent, removed earlier ambiguities, and yet did not detract from the advantages in the language so valued by C programmers. It might be noted in passing that other languages, notably Ada, are large and complex to learn, and it is this danger that the ANSI standard attempts to avoid. All the changes introduced by the ANSI standard are enumerated, in summary, in Appendix C of Kernighan and Richie (1988). One of the most visible changes is the new kind of function definition, which allows better cross-checking. The function definitions used in this chapter follow the ANSI standard. A definition used in Sec. 3.11 such as:

```
demo_funct2 ( int x )
{
. . .
. . .
}
```

would in early versions of C have been written as:

```
demo_funct2 ( x )
int x ;
{
. . .
. . .
}
```

The first edition of Kernighan and Richie (1988) should be consulted for details of the earlier version of the language. This might be a useful reference if code written prior to the introduction of the ANSI version needs to be dealt with.

Contained in the ANSI standard is a clear specification of the standard library. The standard library contains a large number of definitions concerned with handling I/O, memory management, string manipulation and a variety of other tasks. The complete range of functions carried out by the standard library is described in Appendix B of Kernighan and Richie (1988). The functions, type definitions, constants and macros that comprise the standard library are declared in a number of header files, which are accessed by an include statement, as in:

```
#include stdio.h
```

The particular example header file quoted in the line above contains a variety of definitions relating to input/output, and the functions and other objects that it defines represent about one-third of the entire standard library.

3.17 EXERCISES AND REVIEW QUESTIONS

1. Write a C program which reads an 8-bit quantity (using `scanf`, say), and displays its value on the screen in binary form. (A simple but useful illustration of the use of the bitwise operators is involved here.)
2. Distinguish between global and auto type variables.
3. Write a function `swap` which interchanges the values of two arguments of type `int` which it is passed.
4. Write a function which returns the length of a character string. The function is passed by a pointer to the string. Use pointer arithmetic, as in the program earlier in the chapter, which copies one string on to another. Make the function return a value -1 if the string length exceeds some value `MAX_LENGTH`.
5. Define the term function prototype.

REFERENCES AND FURTHER READING

Brown, T. D. (1991) *C for FORTRAN Programmers*, Prentice-Hall, Englewood Cliffs, New Jersey.

Hutchinson, R. and Just, S. (1988) *Programming Using the Language C*, McGraw-Hill, New York.

Kernighan, B. W. and Richie, D. M. (1988) *The C Programming Language*, 2nd ed., Prentice-Hall, Englewood Cliffs, New Jersey.

Rojiani, K. B. (1992) *Programming in C for Engineers*, Prentice-Hall, Englewood Cliffs, New Jersey.

Schildt, H. (1990) *Turbo C/C++—The Complete Reference*, Osborne/McGraw-Hill, New York.

4

FORTRAN

4.1 INTRODUCTION AND BACKGROUND

FORTRAN (= FORmula TRANslating system), is one of the earliest high-level computer languages, having been introduced in the 1950s. Programming at assembler level was the norm in those days, so the introduction of FORTRAN represented an important advance in programming productivity. More recently, FORTRAN has been much criticized in relation to later languages because of some of its primitive features, but in its defence it has to be said that FORTRAN is one of a handful of early languages including COBOL and BASIC that have survived to the present day, and, in spite of any perceived shortcomings, is still used widely in electronic engineering circles and elsewhere. There are several reasons for this:

1. There is a vast amount of code that has been written in FORTRAN over the years, which represents a considerable financial investment not to be lightly discarded. The widespread popularity of FORTRAN is reflected in the availability of compilers which can run on almost any hardware platform from personal computers to supercomputers.
2. FORTRAN was originally conceived with the objective of achieving high code efficiency, and this is still true of later versions of the language. Efficiency was important when the language was introduced because it was seen as competing with assembler, which is inherently the most efficient way to program. Code efficiency makes FORTRAN attractive in 'number crunching' type applications such as the numerical solution of linear algebraic equations.
3. Unlike some other languages, FORTRAN conforms to carefully designed standards for each of the relatively small number of versions of the language that have developed over the years since its introduction; the importance of this latter aspect cannot be

overemphasized, especially in relation to the portability of programs between different computers.

This chapter aims to give an overview of FORTRAN, emphasizing its distinctive features so that its position *vis-à-vis* other languages can be seen more clearly, and also to point to applications for which it is best suited. This follows similar objectives to those of Chapter 3 dealing with C. No attempt is made to give a comprehensive treatment of FORTRAN, there being many good references available for those readers who actually need to carry out serious programming in the language (Balfour and Marwick, 1979; Hahn, 1987). The bulk of the chapter will deal with the FORTRAN 77 version of the language, but recently FORTRAN 90 has been introduced to modernize the language, and rectify some of the shortcomings of earlier versions. The enhancements forming part of FORTRAN 90, and other earlier versions are compared in Sec. 4.8.

4.2 INTRODUCTORY PROGRAMS

As in Chapter 3, we include some minimal programs as a means of giving an idea of what some FORTRAN code actually looks like. The first program simply displays a message on the standard output device:

```
C----- simple introductory FORTRAN program
       PROGRAM  HELLO
       PRINT *, 'Hello World'
       END
```

The program is almost self-explanatory, but note that the statements cannot start on columns 1 to 6, these six columns being reserved for either statement labels or alternatively the letter 'C', to indicate that the entire line is a comment—as in the first line of this program in fact. This 'C' should appear in column 1. Statements, in this particular program at least, are separated by virtue of being on successive lines. Some programs may have statements that cannot fit on to one line, these can be divided between several lines indicated by putting a **continuation character** in column 6 on the second and any other further lines needed for the statement, these can be any ASCII character except '0' or a space character. In Pascal, C and some other languages, statements are terminated by a ';' character, with carriage return being regarded purely as a white space character, and as such ignored by a compiler. The '*' in the PRINT statement has the effect of using the standard output device, typically a terminal screen, to display the message. Other I/O statements can direct output to or from files, and can format data in a wide variety of ways to be described later.

The second example is roughly comparable to the C program given in Sec. 3.8. Note that there is no 'while do' construct in FORTRAN 77, so that, if in fact one is required, the WHILE DO needs to be manufactured using FORTRAN's version of a conditional statement (IF) in conjunction with a jump statement (called GO TO) instead. The Boolean expression in the IF statement reads literally 'COUNT Greater Than 0'. The GO TO is an unconditional jump to a statement identified by a label that needs to appear between columns 1 to 5 in the relevant line.

```
C-----decrement a number down to zero, displaying its value
      PROGRAM COUNTDOWN
      INTEGER   COUNT
      READ *, COUNT
1     IF(COUNT.GT.0) THEN
         PRINT *, COUNT
         COUNT = COUNT - 1
         GO TO 1
      END IF
      PRINT * , 'Blast Off!'
      END
```

The loop that features in this example illustrates some of the more primitive features in the language (rectified in FORTRAN 90), which, unless care is taken, can lead to badly structured code. One of the tenets of structured programming is that programs are built up as a set of nested blocks with no transfer of control from one level of nesting to another.[1] In languages that need to make heavy use of jump statements, as here, the nested structure can be violated if care is not taken, resulting in code that is difficult to debug or maintain. Provided loops and other constructs are constructed in a disciplined way, however, it is possible to apply structured programming methods as with languages like Pascal, albeit with more responsibility being placed on the shoulders of the programmer. One could make similar comments in relation to assembler, it being closer in structure to FORTRAN than languages that have features supporting structured programming methods as it were 'built in'.

4.3 DATA TYPES, VARIABLES AND CONSTANTS

FORTRAN 77 permits six different data types. These are: INTEGER, REAL, DOUBLE PRECISION, COMPLEX, LOGICAL and CHARACTER. Pascal programmers will be disappointed to discover that there is no provision for user-defined types as in Pascal, nor records, nor enumerated types. There is no provision for pointers either, making the implementation of data structures such as linked lists[2] difficult. Naturally, as with almost all languages, arrays can be used, and this will be described later. The provision of a DOUBLE PRECISION type reflects the common use of FORTRAN in precise floating point computation, while the COMPLEX type is particularly useful to electronic engineers in view of their special interest in the analysis of a.c. circuits and other types of systems for which complex arithmetic is required. Examples of the use of this type will be given later. Variables are declared, as in the second program, with TYPE[3] statements, using one of the key words above as required to identify the type of the variable(s). Some typical declara-

---

[1] See Chapter 5 for more on this.

[2] See Chapter 5 for more on linked lists.

[3] The TYPE statement should not be confused with the use of the reserved word TYPE in Pascal. There, a *data type* is being defined, as distinct from variables. Variables in Pascal can be subsequently defined in terms of predefined types listed in TYPE declarations. In FORTRAN the TYPE statement is used *directly* to define the type of variables.

tions of variables are given below. In the declaration of CHARACTER variables `MESSG*4` would mean that the variable MESSG contains four characters. Other more complex ways of declaring variables of type CHARACTER are available (Appleby, 1991).

```
COMPLEX    VOLT1,VOLT2,Z1,Z2,I1
LOGICAL    XOR1,NAND1
CHARACTER MESSG*4
```

4.3.1 Implicit naming convention

One of the arguably less desirable features of FORTRAN is the way variables of type INTEGER and REAL need no explicit type declaration at all, but are implicitly assigned a type dependent on the initial letter in the variable name. Variables starting with the letter I, J, K, L, M or N are assumed to be INTEGER, while variables starting with other letters are assumed to be of type REAL. However, explicit type declarations can be used to override this form of typing, which can be regarded as a legacy from the early versions of FORTRAN and should perhaps be avoided in the interests of good programming style. Implicit naming is in fact the very opposite to the strong typing principle embodied in Pascal and many other modern languages.

4.3.2 Constants

Constants for all of the predefined types can be used in programs, and some examples are quoted below. The precision of integer and floating point numbers is dependent on the computer hardware. Thus integers could be (signed) 32-bit or 16-bit numbers, while floating point numbers could be commonly 32-bit, or 64-bit for double-precision quantities. This diversity of precision arising out of differing word lengths for various machines could lead to difficulties in transferring FORTRAN programs from one machine to another.

- **Integer constants**

  3456
  −99

- **Real constants**

  34.789
  −88.6
  −5.5E-6 {or written in conventional form -5.5×10^{-6}}

- **Double-precision constants**

  5.7D19 {note the exponent is signified by a 'D'}

- **Complex constants**

  (2,3) {equivalent to 2 + j3}
  (−4.7,7.1E3) {equivalent to $-4.7 + j7.1 \times 10^3$

- **Logical constants**—the two values of logical quantities are written as:

.TRUE.
.FALSE.

- **Character constants**—(in many other languages these would be called *strings*) are a list of characters enclosed in single quote characters as in:

'An Introduction to FORTRAN'
'An Intro   to FORTRAN'

Note that space characters are significant inside character constants.

4.3.3 The PARAMETER statement

The PARAMETER statement is used when programmers wish to specify constants in the form of symbolic names rather than quoting the constants in terms of their actual values. The provision is similar to CONST declarations in Pascal. The advantages are that, firstly, the meaning of code is clearer and, secondly, if a given constant is to occur a number of times in a program, it is only necessary to edit the (single) line of code where the constant is declared. For example a constant may be machine dependent, in which case a single definition will make it more convenient to adapt the program for use on different computers. It might be noted that PARAMETER statements are only effective within a program or subprogram in which associated constants are defined. The parameter statement has the general form:

```
PARAMETER(<name1>=<value1>,<name2>=<value2>,...)
```

name1, etc., are the symbolic names to be used in quoting the various constants, while value1, etc., are constant expressions defining the actual value of the constant. An example of a PARAMETER statement together with relevant TYPE statements is:

```
REAL        MAXVOLT
COMPLEX     ZMAX
CHARACTER PCBOARD*8
PARAMETER(MAXVOLT=415,ZMAX=(100.0,250.0),PCBOARD='MVME133')
```

4.3.4 The DATA statement

The DATA statement is a nonexecutable statement used to give initial values to variables. For instance it can be very conveniently used to initialize array elements, where the only other alternative, and clumsy, option open would have been to use a separate assignment for each array element to be initialized. The general form of a data statement is:

```
DATA var1/const1/,var2/const2/,....
```

var1, etc., are each lists of one or more variable names, or array names, array element names.[4] const1, etc., are each lists of constant values to be assigned to the variable list just before the preceding '/' character. An example of a typical DATA statement is given below:

---

[4]There are other possibilities, but these will not be discussed here (Appleby, 1991).

```
REAL    STEP,INPUTV
REAL    VOLT(10,10)
        PARAMETER (PI=3.1415926536)
DATA    STEP,INPUTV/10.0,
      ! 230.0/,OMEGA/2.*PI*50./,VOLT(1,1)/1.0/
```

4.3.5 Arrays

Arrays of any of the data types described previously are allowed in FORTRAN, and multi-dimensional arrays up to seven dimensions can be used (in FORTRAN 77). Bounds on array indices may have any valid integer value, including negative values. Declaration of arrays can take a number of alternative forms: they can appear in lists of variables in type declarations, or else in a DIMENSION statement or a COMMON statement. (COMMON statements and their purpose will be considered later.) Some typical array declarations are:

```
DOUBLE PRECISION POWER, OMEGA
DIMENSION    POWER(-4:5), OMEGA(1:10,1:10)
CHARACTER    NAMES(100)*10
COMPLEX      ZA(0:10,0:10)
```

In these examples, POWER is a one-dimensional array in which the index has a lower bound of -4 and upper bound of 5. OMEGA is a two-dimensional array. Note the way the *type* of POWER and OMEGA is declared. Without the type declaration used, both arrays would have been of type REAL in view of the implicit naming convention described above. NAMES would be an array of variables of type CHARACTER, each element of which in this case would comprise ten characters. Finally, ZA would be a two-dimensional array of type COMPLEX—with lower bound 0 and upper bound 10 for both dimensions.

Array elements can be manipulated much as in Pascal or C. Array indices can be any expression that yields a valid integer value. Thus:

```
X = POWER(I*(Y+8)) + 10.9
```

would be valid, provided Y is of type INTEGER.

4.4 EXPRESSIONS

Expressions in FORTRAN are not very different from expressions in many other languages. The range of operators provided is more restricted than that of C described in the previous chapter, and are: '+' '−' '/' '*' and '**' for exponentiation. Parentheses can be used freely to force a particular order of precedence in evaluating expressions. Care must be taken in dealing with mixed expressions that involve integers and real quantities. In this situation, if two quantities are being multiplied, added, etc., then, if one quantity is real and the other integer, the integer is converted to real form and real arithmetic used to produce a real result. The value assigned to a variable following evaluation of such an expression depends on the variable's type, however. Thus if we carried out the assignment $I=X*J$, with X (REAL) having a value 5.7, and J (INTEGER) having a value 2, the expression would evaluate to 11.4, but the value subsequently assigned to I (INTEGER) would be 11. Many FORTRAN programmers seem to employ the kind of mixed mode arithmetic used in this example.

Logical expressions Logical expressions are used in control structures such as IF THEN to be discussed later. They can be formed by variables including arrays of type LOGICAL, or relational expressions, i.e. expressions which are .TRUE. or .FALSE. depending on the results of some test involving a comparison between two arithmetic expressions.

Operators in logical expressions are as follows:

- **.AND.**—e.g. VAR1.AND.VAR2 gives a .TRUE. value if both VAR1 and VAR2 have .TRUE. values.
- **.OR.**—e.g. VAR1.OR.VAR2 gives a .TRUE. value if VAR1 or VAR2 have .TRUE. values.
- **.NOT.**—e.g. .NOT.VAR gives a .TRUE. value if VAR is .FALSE. and vice versa.
- **.EQV.**—e.g. VAR1.EQV.VAR2 gives a .TRUE. value if VAR1 and VAR2 are the same.
- **.NEQV.**—e.g. VAR1.NEQV.VAR2 gives a .TRUE. value if VAR1 and VAR2 are *different*.

Complex logical expressions can be formed using the above operators, possibly using parentheses to enforce the order of precedence required. The order of precedence is .NOT. (highest), .EQV., .EQUV. (lowest with .AND. and .OR. in between and both of equal order of precedence. Operators of equal precedence are applied from left to right, always subject to the order of precedence forced by any parentheses. A further complication arises in the case of relational expressions that also involve arithmetic operators. In these cases all arithmetic operators take precedence over logical operators. Finally, note that although in many programs the most common use for logical expressions is in a Boolean test with a control statement, it is also possible to have assignments involving logical expressions, subject to the above rules.

Complex expressions Variables of type COMPLEX can in most respects be manipulated in expressions as would expressions that involve integer, real or double-precision quantities. The operators '*', '/', '−', '+' and '**' may be used freely in expressions which can also make use of parentheses. This is one of the more attractive features in the language as far as electronic engineers are concerned. Additionally, however, there are further operations which are only appropriate for complex quantities, such as extracting the real part of a number. Such operations are dealt with by the provided functions listed below. Some familiarity with complex quantities is assumed in the description that follows. The functions produce a result that can be assigned to a variable of appropriate type as indicated in Table 4.1.

Table 4.1 Descriptions of functions

Function name	Description	Type of argument	Type of function
CSQRT	Square root	COMPLEX	COMPLEX
CEXP	Exponential	COMPLEX	COMPLEX
CLOG	Natural logarithm	COMPLEX	COMPLEX
CSIN	Sine	COMPLEX	COMPLEX
CCOS	Cosine	COMPLEX	COMPLEX
CMPLX	Convert to complex	*INTEGER	COMPLEX
		*REAL	COMPLEX
		*DOUBLE	COMPLEX
		COMPLEX	COMPLEX
INT	Conversion to integer	COMPLEX	INTEGER
REAL	Extract real part	COMPLEX	REAL
DBLE	Convert to double precision	COMPLEX	DOUBLE
CABS	Modulus	COMPLEX	REAL
AIMAJ	Extract imaginary part	COMPLEX	REAL
CONJ	Complex conjugate	COMPLEX	COMPLEX

* These functions have two arguments, all others have one.

The code fragments below demonstrate the application of some of these functions:

```fortran
COMPLEX   W,X,Y,Z
REAL      A,B,C
INTEGER   R,S,T
   ...

   ...
A = 10.3
B = 30.6
R = 20
S = 44


C     Y given the value 10.3+j30.6
Y = CMPLX(A,B)


C given the value of the modulus of A+jB i.e 40.6
C = CABS(CMPLX(A,B))


C     X given value of complex conjugate of A+jB
      ie 10.3-j30.6
X = CONJG(CMPLX(A,B))


C     D given the value that is the imaginary part of Y
      ie 30.6
D = AIMAG(Y)


C     W given the value that is the square root of 10+j10
W = CSQRT(10,10)
```

In constructing expressions that include complex quantities, implicit type conversion occurs when an operand of type REAL or INTEGER is combined with an operand of type COMPLEX using one of the standard arithmetic operators.

An analysis of the circuit shown in Fig. 4.1 will now be given to demonstrate the use of the complex type. The driving point impedance of the circuit is to be found for fixed values of the various impedances; the expression for impedance can easily be shown to be given

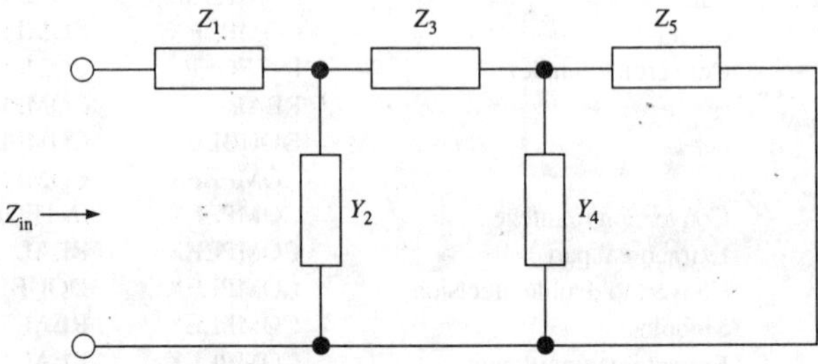

Figure 4.1 Ladder network.

by the expression below. Many references on circuit analysis could be consulted for a justification for the formula, e.g. Desoer and Kuh (1969). The program is written such that Z_1 is read in as data, while the other impedances are given as constant quantities.

One might note in passing that, perhaps more typically in a.c. circuit problems, the frequency response of a network would be required, in which case the calculation would be done repeatedly for a range of frequencies, not simply the single frequency implicit in this program. It is suggested that the reader adapt the program to enable this to be done with this network for any convenient set of component types and values.

$$Z_{in} = \frac{Z_1(1+Y_4 Z_5+Y_2 Z_5+Y_2 Z_3(1+Y_4 Z_5)) + Z_5 + Z_3(1+Y_4 Z_5)}{1 + Y_4 Z_5 + Y_2(Z_5+Z_3(1+Y_4 Z_5))}$$

```
PROGRAM LADDER
COMPLEX Z1,Y2,Z3,Y4,Z5,DENOM,NUMER,ZIN
REAL  R5,X5
PARAMETER(Z1=(1,1),Y2=(10,10),Z3=(2,2),Y4=(5,5))
PRINT *,'R5 AND X5?'
READ *, R5,X5
Z5 = CMPLX(R5,X5)
NUMER=Z1*(1+Y4*Z5+Y2*Z5+Y2*Z3*(1+Y4*Z5))+Z5+
! Z3*(1+Y4*Z5)
DENOM=1+Y4*Z5+Y2*(Z5+Z3*(1+Y4*Z5))
ZIN=NUMER/DENOM
PRINT *, 'DRIVING POINT IMPEDANCE'
PRINT *,'REAL PART IMAGINARY PART'
PRINT *, REAL(ZIN), AIMAG(ZIN)
END
```

4.5 CONTROL STRUCTURES

4.5.1 Conditionals

The range of control structures in FORTRAN 77 is somewhat limited compared to Pascal (but see also Sec. 4.8). Conditional statements, as used in the example in Sec. 4.2, are termed 'BLOCK IF' in that the IF and also the ELSE clauses can include *blocks* of statements, and are not just limited to a single statement in each clause. Other forms of conditional (not described here) are provided, but were features of early versions of the language; they survive only in the interests of making old FORTRAN code compatible with more modern FORTRAN standards. A typical BLOCK IF, somewhat more general than the example of Sec. 4.2, is given below. Here, the first two statements are executed if X is less than 100 otherwise the last two statements are executed. Indentation is used to clarify the range of control in both cases:

```
IF(X.LT.100) THEN
    VOLT = I*RES1
    FREQU = S + 10
ELSE
    VOLT = I*RES2
    FREQU = S + 25
ENDIF
```

In place of the simple statements used here, it would be possible to include other BLOCK IF statements or blocks of other types. A jump out of a BLOCK IF is allowed, but a jump from outside the statement is not. (Why would you think this a reasonable restriction?)

More complex forms of conditional statement are provided for, in the form of ELSE IF type logic, when decisions involve the choice between more than two alternate sets of statements. An ELSE IF precedes a block of statements with which it is to be identified, up to the ENDIF statement that was associated with the originating IF. Unfortunately there is no CASE type logic in FORTRAN 77, so somewhat complex forms of IF are required for what can be accomplished much more clearly in Pascal and some other languages. The example below gives an illustration of ELSE IF. Note again the use of indentation to clarify the IF statements structure.

```
IF(X.LT.100) THEN
    VOLT = I*RES1
    FREQU = S + 10
ELSE IF(X.LT.200) THEN
    VOLT = I*RES2
    FREQU = S + 100
ELSE
    VOLT = I*RES3
    FREQU = S + 1000
ENDIF
```

4.5.2 Looping constructs

A **counted loop** in FORTRAN, the counterpart to a Pascal **for** loop, is here called a **DO loop**. There are, however, more than minor syntactical differences between the DO loop and the Pascal for; notably, the loop control variable can be of type INTEGER, REAL or DOUBLE, *and* can start or end with either negative or positive values. The control variables can be incremented by any valid real, double or integer amount. A simple example of a DO loop is shown below:

```
      DO 103 I=1,9,2
         SUM = SUM + A(I)
         PROD = PROD*A(I)
  103    CONTINUE
```

CONTINUE is a dummy statement used here in conjunction with the label 103 to act as a terminal statement which ends the loop. CONTINUE can be compared to a NOP in assembler code; in other words it results in no computation as such. In the example, the action of the loop would be to execute the statements up to the terminal statement, each time increasing the control variable I by 2. Thus I would assume values 1, 3, 5, ..., 9, after which the loop would terminate. As with the BLOCK IF it is possible to have, in place of simple statements, further DO loops or BLOCK IFs forming nested control structures of various kinds.

4.5.3 The unconditional GO TO statement—uses and restrictions

The unconditional GO TO statement, of general form GO TO <statement label>, as featured in the example given in Sec. 4.2—incidentally a good illustration of its application—is a means of implementing that which in other languages would be termed **while do** type logic. Care must be exercised in using GO TO statements in conjunction with other control structures as we have already noted. As well as transfer of control from a GO TO into an IF block being prohibited, it is also prohibited to transfer control into an ELSE block, an ELSE IF block, or into a DO loop from outside that loop.

4.6 INPUT/OUTPUT IN FORTRAN

Input and Output (I/O) provision in FORTRAN can handle data in a wide variety of formats and types, and is particularly powerful in the way it can deal with information in data files. I/O *can* be simple to use, however, as in the case of the PRINT statement which featured in an earlier example. This might be convenient for relatively simple cases involving I/O to a terminal screen and keyboard, but users would need to be prepared to live with whatever default ways of reading and outputting data were provided. However, tighter control over specifying the formatting and forms of accessing data is possible, as will be shown. The discussion in this section should only be regarded as an overview of the I/O facilities available in FORTRAN, and a reference such as Balfour and Marwick (1979) should be consulted if you need to carry out extensive programming involving I/O.

4.6.1 Data, records and files in FORTRAN

Data in a FORTRAN program may originate from, or be sent to, a variety of devices such as a printer, a modem, a file on magnetic tape or a device such as a terminal, the latter often being the device accessible to programs by default. Whatever its source or destination, however, we consider data in the abstract as consisting of **files**, each of which has the information in it divided into **records**. Records could for instance correspond to a line of characters to be printed on a printer, but can take a wide variety of alternate forms. Records can be considered as divided further into the primitive data items themselves, each of which is contained in a so-called **field** in the case of data that is formatted. Thus an integer as printed on a page could be viewed as a set of digit characters contained in a field of (say) ten columns. Files that are associated with devices such as magnetic tape drives, disks, etc., are termed **external files**, and are accessed by so-called **external units**, this being the generic name for the I/O devices. Special provision must be made in programs to carry out operations such as opening or closing external files. Files stored in the computer's memory are called **internal files**, and may be looked on as a way of supporting data structures that lend themselves to this form of organization, comparable to the support for other data structures provided by constructs such as arrays in fact.

4.6.2 Formatted I/O

It is possible to view the data in a record as comprising a list of characters, and this needs to be converted to (or from) a form with which the computer can deal. Thus a row of integer numbers to be displayed on a screen is, in the final analysis, just a string of characters such as '1', '2', ..., '9', but in this example the internal representation could, say, be in the form of signed 32-bit words, perhaps as two's complement, which the computer's arithmetic unit can properly deal with. The conversion of records to/from internal and external forms is achieved by a process termed **formatting**.[5] In simple cases a programmer would not need to do anything special to format data. Thus in the statement `PRINT *, A`, where `A` is an integer, the internal representation of the variables is converted to a displayed integer character value in some predefined number of columns, perhaps left justified, in the field allocated to the number. However this would not be appropriate if we wanted to display, say, a row of 15 such numbers on one line. The field width would then need to be restricted to allow all the numbers to fit on a single line, and the programmer would then need to make use of the formatting provision in the language.

There are a variety of ways of specifying the format of records, one of which employs a special nonexecutable statement called FORMAT. This is used in conjunction with input or output statements and gives the format of relevant records in the data. As an alternative, FORTRAN 77 allows formatting information to be included as part of the input or output statement itself. Whichever alternative is used, however, there needs to be some form of specification of format information giving the type and layout of the data. This will now be described in simplified form.

---

[5] There is no suggestion that formatting is peculiar to FORTRAN of course, but we are here giving special emphasis to its treatment in FORTRAN in relation to features not always found in other languages.

4.6.3 Editing codes

Fields corresponding to individual data items are specified by **editing codes**. A list of data items would therefore need to be specified by a list of editing codes. The editing codes are different for the various types of data—integers, floating point numbers, characters, etc. Editing codes can control field width, and can also control the format of data in other ways. For example, it is possible to repeat groups of editing codes so that if we wanted to display five integer numbers each contained in a field 10 wide, a *single* editing code could be used. The following is a short list of some of the more heavily used editing codes together with a brief explanation of their action. A fuller discussion can be found in Balfour and Marwick (1979).

- Iw specifies an integer contained in w columns, right justified. Thus an editing code I7 would display the number 123 as bbbb123 where 'b' stands for a blank.
- $Fw.d$ specifies a fixed point real number occupying w columns and with d fractional digits, rounding the fractional part of a number to the number of digits allocated to it. A '.' and sign are included in the field width. F8.2 would display the number -12.679 as bb-12.68.
- $Ew.d$ specifies a floating point real number in mantissa/exponent form, where w is the field width and d is the number of digits in the mantissa. Thus E10.4 could display the number -12.679 as $-.1268E+02$. A more general form of this editing code Ew.dEe gives control over the exponent width e.
- wX skips columns leaving the next w columns blank; in other words any following data would be separated from the previous data by w columns.
- Aw prints an alphanumeric character string over a field width w, left justified. The w is optional and if omitted the A format can be used to print a literal string specified in the associated PRINT statement.

4.6.4 Simple use of the FORMAT statement

If we want to display the value of some integer J on the default output device in a field of seven characters, using the I editing code, this could be done as:

```
        PRINT 15, J
15      FORMAT(I7)
```

We could, if needed, display a sequence of data by having an appropriate list of editing codes separated by ',' as in the following example where the four numbers are displayed in four contiguous fields all on a single line.

```
        PRINT 20,J,K,FLT,X
20      FORMAT(I7,I12,E12.4,F10.2)
```

4.6.5 More advanced formatting

In many cases more flexible control over the way data is formatted is required. Repetition of editing codes is achieved by prepending the editing code by the number of repetitions required. Thus 3I10 would make provision for 3 fields, 10 columns wide for 3 integers. Editing codes can be repeated in groups as well as singly. The two FORMAT statements below thus have the same effect:

```
FORMAT(I5,I5,E12.4,I2,E12.2,I2)
FORMAT(2I5,2(E12.4,I2))
```

4.6.6 Dealing with files

Recall first that files in FORTRAN are either internal or external. The discussion that follows applies in the main to both types of file although differences are commented on as appropriate. External files are generally of greater importance as far as most users are concerned. I/O involving files is comparatively sophisticated in FORTRAN 77, with provision for a wide variety of data formats as already described, and also choices possible between direct and sequential forms of access. There is good provision to deal with end-of-file and error conditions, and also to carry out a variety of operations on external units, such as connection and disconnection. READ and WRITE statements are used for I/O involving files, possibly in conjunction with FORMAT statements. The input/output statements have provision for specifying the following:

- **Unit**—an integer number identifying hardware such as a disk unit that is currently associated with a particular file.
- **Format**—edit codes are used as already described, but note that with READ and WRITE statements there are a number of alternative ways of stating this information, either in a separate FORMAT statement or else as part of the input/output statement itself.
- **Error**—a label associated with a statement to be executed in the event of an error condition. Errors would include cases where data is inconsistent with edit codes in a format statement.
- **End-of-file**—a label associated with a statement to be executed in the event of an attempt to read beyond the end of a file.
- **Status**—an integer variable that is set to a zero value if no error condition or end-of-file condition has been detected, a positive value if an error condition has been detected, and a negative value if an end-of-file condition has been detected *and* there is no error condition. This provision allows for added flexibility in the treatment of error and end-of-file conditions.
- **Record**—a number identifying a particular record in a file: this only applies to files in which the direct mode of access is used. Records in direct access files are accessed in a way analogous to the way memory locations are accessed 'directly' by address. The record number can therefore be viewed in some ways as a counterpart to a memory address.

The examples below give illustrations of the application of these facilities:

- Read three real numbers from the file with unit identifier 20 and with format specified at line number 100, and jump to line 200 if an error condition has been detected, at which the programmer would no doubt have written some error-handling code, to display an error message; for example:

```
      READ(FMT=100,UNIT=10,ERR=200) A,B,C
100   FORMAT(3F12.4)
```

- Write three integers to the file with unit identifier 20 and with format specified at statement number 101, and set the value of the variable STATUS to zero, positive or negative values depending on whether an error condition or end-of-file condition was detected:

```
      WRITE(FMT=101,UNIT=20,IOSTAT=STATUS) I,J,K
101   FORMAT(3I10)
```

More on managing files Files are made accessible, as far as READ and WRITE statements are concerned, by the OPEN statement. This specifies whether the file name, the file status (specifying whether the file is new, i.e. about to be created, or existing), the unit number, the access type (sequential/direct), the form of data (unformatted/formatted), and, in the case of a direct access file, the record length. A statement called CLOSE can be used to 'disconnect' a file from a particular unit. After executing a suitable CLOSE statement, it would be possible by a subsequent OPEN to access a file in different ways inside the same program, although a CLOSE is implemented implicitly by default when a program terminates. An example of the use of the OPEN statement is given below. This relates to a *new* file called 'demo.dat', which is to be given sequential access. It is to be 'connected' to external unit number 5.

```
 OPEN(UNIT=5,FILE='demo.dat',ACCESS='SEQUENTIAL',
 ! STATUS='NEW', FORM='FORMATTED')
```

The second example relates to an existing file called 'oldfil.dat' which is to be given direct access. Each record has a 'length' of 512 characters.

```
 OPEN(UNIT=6,FILE='oldfil.dat',ACCESS='DIRECT',STATUS='OLD',
 ! RECL=512,FORM= 'FORMATTED' )
```

It might be added that abbreviated ways of writing open statements are allowed. The fuller versions of the statements have been used here in the interests of clarity.

Sequential access If a file is being accessed sequentially, a data transfer operation causes the *next* record in the file to be either written or read. Somewhat more flexible operations than this are possible by use of the BACKSPACE statement, which would reposition the file before the current record. For example BACKSPACE 5 would position the file connected to unit 5 at the record just prior to the current record. A REWIND operation is also provided: this positions the file at its *first* record.

Example of I/O involving sequential file access The two programs below illustrate the use of sequential access files and methods of handling error and end-of-file conditions. Note the way that the second program is written to handle files with any number of records up to the allotted maximum of 100, and the use of the error specifier to enable the program to stop looping. There will not be any question of an error condition being generated in the second program, if the data it reads was prepared by the first program, but obviously there are other situations where there will be error conditions.

```
C          PROGRAM TO WRITE SOME DATA TO A FILE 'DEMO.DAT'
           PROGRAM FILEWRITE
           INTEGER I, J, K, L
C          create and open a sequential access file on disk
           OPEN(5,FILE='demo.dat',ACCESS='SEQUENTIAL',STATUS='NEW')
           DO 1 I = 1 , 5
              J = I
              K = I*2
              L = I*3
              WRITE( 5, FMT=80 ) J, K, L
1          CONTINUE
80         FORMAT( 3I10 )
           END
```

Contents of the file demo.dat

1	2	3
2	4	6
3	6	9
4	8	12
5	10	15

```
C          PROGRAM TO READ 'DEMO.DAT' AND DISPLAY ITS CONTENTS
           PROGRAM FILEREAD
           INTEGER MAXRCD
C          upper limit on the number of records in the file
           PARAMETER(MAXRCD=100)
           INTEGER I, J, K, L
C          program opens 'demo.dat' for sequential access.
C          It must already exist!
           OPEN(5, FILE='demo.dat', ACCESS='SEQUENTIAL',STATUS='OLD')
           DO 2 I = 1 , MAXRCD
              READ( 5, FMT=80, END=99, ERR=100 ) J, K, L
C             echo data just read from current record on screen
              PRINT * , J, K, L
2          CONTINUE
           PRINT *, '?? number of records in file exceeds',MAXRCD
80         FORMAT( 3I10 )
           GO TO 101
```

```
99      PRINT *, 'end of file reached, normal termination'
C
C       Demonstration of BACKSPACE, the code below will enable
C       the LAST record to be accessed again
        BACKSPACE 5
C       reread last record
        READ( 5, FMT=80 ) J, K, L
        PRINT *, 'read the last record again'
        PRINT *, J, K, L
C
        GO TO 101
100     PRINT *, '?? data error on reading file'
101     END
```

4.7 SUBPROGRAMS—SUBROUTINES AND FUNCTIONS

Subprograms are available in FORTRAN, and are called **subroutines** or **functions**, the former being roughly comparable with a procedure in Pascal while the latter is comparable with a Pascal function. There are, however, some important differences in detail between the two languages, as will be seen. FORTRAN subprograms serve similar purposes to subprograms in many other languages, most importantly they are used as a means of breaking programs into manageable modules, and, since this aspect is dealt with in Chapter 5, it is not elaborated on further here.

4.7.1 Subroutines

Subroutines[6] are invoked by a CALL statement that quotes the name of the subroutine together with a list of arguments[7] corresponding to the data passed *to* the subroutine, and also data passed back *from* the subroutine to the calling program. Arguments used in the calling program are called **actual arguments**. The subroutine itself also has a line in which the subroutine name and the arguments appear. Parameters here are called **dummy arguments**.[8] The names used for corresponding actual and dummy arguments need not be the same, but all names must be consistent with any type declarations used inside or outside the subprogram, and must appear in corresponding order in the list of arguments. The names used for dummy arguments have a meaning that is local to the subroutine, so that identical names which might appear in other program modules could refer to different quantities. The exception in the case of variables declared in so-called COMMON blocks will be considered later. Subprograms can also make use of local variables declared inside

---

[6] The use of the term subprogram in the following sections will be taken to apply equally to subroutines or functions.

[7] **Argument** is the term in FORTRAN for which, in Pascal, C and some other languages, the term **parameter** would be employed.

[8] Equivalent to **formal parameters** in Pascal and some other languages.

the body of the subroutine, which would not have a meaning in any other module provided they did not appear in COMMON blocks.

A simple example of a subroutine is given below. It is the FORTRAN counterpart to the program in Sec. 3.11 which calculated the roots of a quadratic equation.

```
C      THIS IS A SAMPLE CALLING PROGRAM
       REAL   COEFFTA, COEFFTB, COEFFTC, ROOT1, ROOT2
       PARAMETER (COEFFTA=1,COEFFTB=16,COEFFTC=4)
C      COEFFTA etc are ACTUAL arguments
       CALL ROOT( COEFFTA, COEFFTB, COEFFTC, ROOT1, ROOT2 )
       PRINT *, ROOT1, ROOT2
       END


       SUBROUTINE ROOT( A, B, C, R1, R2 )
C      ROOTS OF A QUADRATIC, REAL ROOTS ASSUMED
C      A B C R1 and R2 are DUMMY arguments
C      TEMP1 AND TEMP2 ARE LOCAL VARIABLES
       REAL A, B, C, R1, R2, TEMP1, TEMP2
       TEMP1 = SQRT(B**2-(4.0*A*C))/(2.0*A)
       TEMP2 = -B/(2.0*A)
       R1 = TEMP2 - TEMP1
       R2 = TEMP2 + TEMP1
       END
```

The code should be largely self-explanatory; but note that unlike C and Pascal there is no distinction, as far as the syntax involved in the argument list is concerned, between arguments passed *to* or *from* subprograms. The two program modules above could be compiled as a single monolithic program, but alternatively it is possible to compile them separately, and subsequently have them combined in a linking stage.[9] This advantage—of being able to prepare and compile programs as a number of separate units—is shared with C, Modula-2 and Ada. The advantages of being able to prepare programs as a set of separate modules will be more fully explored in Chapter 5.

SAVE—a way of holding on to variable values between CALLs We saw in the chapter on C, that, by default, the values of local variables in functions would be lost between one invocation of the function and the next. This is also true, by default, in FORTRAN subprograms. In cases where this might prove inconvenient, a facility called SAVE allows such values to be stored between successive calls. This simply involves including a statement SAVE followed by a list of variables whose values are to be retained between calls.

RETURN—an alternative way to end a subprogram In the previous example, the subroutine was terminated when the END statement was reached. In other circumstances it might instead be desirable to return to a calling program at one of several alternate points:

---

[9] This is described in Sec. 7.5.

the RETURN statement is provided for this purpose. It can be regarded as a high-level counterpart to a RTS (return from subroutine) in 68000 assembler.[10] Note that RETURN has a somewhat different role from a return statement in C where it is there used in conjunction with data to be passed back to a calling program.

Recursion Pascal in common with many languages allows for recursive calls to subprograms, that is a subprogram may have within it a call to itself. Many elegant powerful algorithms can be devised for languages in which this is possible. The bad news is that recursion is not allowed in FORTRAN 77!

4.7.2 Functions

As in C and Pascal, a function can return a value when the function name is used in assignment statements and elsewhere. Subroutines can only return values via arguments. A function, as opposed to a subroutine, must have a *type* associated with it. Functions may be written by programmers in the interests of developing somewhat more elegant code than a subroutine that does a comparable task, but in addition **intrinsic functions** are available for commonly used purposes such as the implementation of trigonometric functions. Intrinsic functions are featured in the section dealing with complex expressions. Intrinsic functions represent a robust efficient set of functions that programmers can employ in programs, obviating the need to write their own counterparts. A complete list of intrinsic functions is given in Appleby (1991).

4.7.3 Library subprograms

In many programs it is convenient to make use of pre-prepared subprograms which can be 'plugged in' to a user's application program. This saves programming effort, and also has benefit in that it allows the use of robust, well-tested routines that can be relied on. Examples of external subprograms include the NAGLIB[11] library, used in numerical analysis work such as problems involving the solution of differential equations and linear algebraic equations: these are just two examples in what is a very comprehensive set of routines. The NAGLIB library happens to be written in FORTRAN—perhaps a commentary on the suitability of FORTRAN for this type of work—but, in principle, external subprograms written in any language could be incorporated into a FORTRAN program, provided data is communicated between the various modules consistently.

4.7.4 The COMMON statement—communicating global values

One way of communicating data between the various program units that constitute a program is via the subprogram arguments. As an alternative it is possible to declare variables to have a meaning that is global to several modules (by default, variable names have a meaning which is local to the module in which they are defined including those quoted as dummy arguments). Global variables can be defined using a COMMON declaration. There are well-known arguments against communicating data via global variables compared to data passed as arguments, mainly concerned with the dangers presented to the design of well-

---

[10] See Appendix 1 for the 68000 instruction set.

[11] The Numerical Algorithms Group (NAGLIB) can be contacted at the following address: Wilkinson House, Jordan Hill Road, Oxford OX2 8DP, UK.

structured and easily understood programs (Attikiouzel, 1988). A list of six good reasons why one should *not* use the COMMON facility is given in Balfour and Marwick (1979)! This reference gives full details of the syntax involved in the COMMON statement, and its application.

4.8 FORTRAN STANDARDS PAST AND PRESENT

The first widely used version of FORTRAN, FORTRAN IV was introduced in the mid-1960s. This standard acted as a unifying influence on the large number of partly incompatible FORTRAN variants that had developed from the inception of the language. FORTRAN IV allows for separate compilation of program units. Many FORTRAN programs make use of large mature libraries such as NAGLIB, so this facility was of major importance. FORTRAN IV added COMPLEX and DOUBLE PRECISION to its range of variable types; it provided for subroutine calls using a return jump mechanism (as opposed to the method used in other languages which allows for recursion). However there were numerous restrictive features such as not allowing more than six characters in identifiers, arrays of at most three dimensions, and ten characters per integer. I/O was restricted by factors such as the then widespread use of punched cards as a data medium. This meant that lines needed to fit on the 80 columns of a punched card. By degrees, later versions of FORTRAN have removed more and more restrictions like these.

The next important advance in standards was FORTRAN 77, which effectively replaced FORTRAN IV. However, it is fully compatible with the earlier standard, a necessary condition in view of the heavy investment represented by existing FORTRAN IV code. Many changes were introduced in the 77 standard, the following list summarizes some of the more significant improvements:

- Arrays up to seven dimensions
- The IF THEN ELSE conditional construct (i.e. as described earlier in the chapter, the earlier conditional statement was very crude by comparison)
- More powerful DO loop provision
- More sophisticated I/O
- Better facilities for dealing with information in character format

FORTRAN 77 represented a considerable advance on what went before, but still could be said to be relatively primitive in comparison with languages such as Pascal and Modula-2 which were designed from their inception with the needs of structured programming and software engineering in mind. The latest standard, FORTRAN 90 (Appleby, 1991), introduced in late 1991, attempts to address this issue, and has major extensions added to the language which include user-defined types, pointers and dynamic data storage, recursive use of subprograms, and modular data and subprogram definition. The control structures provided are extended to include case logic, as found in Pascal, and WHILE DO logic. DO loops can be terminated by an END DO statement, eliminating the need for numbered labels. Collectively the enhanced control structures are as powerful as those found in Pascal and other such languages. In the context of the discussion in the previous chapter on C involving bit-wise logic operations, it is interesting that such operations are provided in FORTRAN 90. There are a large number of detailed improvements in the language, and only the more significant are described here. Taken as a whole it seems that the improvements to the language will ensure that FORTRAN will remain very widely used for many years to come, particularly in applications where code efficiency is the important consideration. Finally, the needs of programmers who have written code in earlier versions of the language are not forgotten: any FORTRAN 77 or FORTRAN IV code they may have written will be compatible with the new standard.

4.9 EXERCISES AND REVIEW QUESTIONS

1. Give examples of improvements that have been made in relation to the provision of data types in the various FORTRAN standards over the years.
2. Write a FORTRAN 77 program that multiplies a two-dimensional array with N rows and M columns by a vector.
3. Compare the provision in FORTRAN 77 regarding *intrinsic functions* with comparable functions in Pascal. Which language would you say is superior in this respect?
4. Write a FORTRAN 77 program that calculates the frequency response of the circuit shown in Fig. 4.2. The circuit arguments are to be supplied as data, and the results displayed as columns showing the amplitude and phase of the output voltage for each frequency. You could initially write your program for linear increments in frequency, and then make it more useful by making it suitable for a graphic display of the amplitude in decibels versus a logarithmic frequency scale. If you are feeling more ambitious, and you have access to a compiler that comes complete with a graphics library, you might try adding a graphics routine actually to implement the graphics display (routines like this are often called **postprocessor** programs and are a feature of most commercial ECAD (= Electronic Computer-Aided Design) packages).

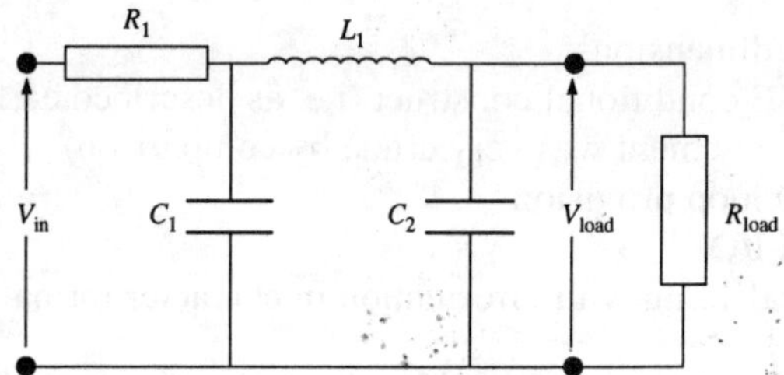

Figure 4.2 Low-pass filter circuit.

5. Distinguish between *actual* and *dummy* arguments in FORTRAN subprograms.
6. Noting that there is nothing in FORTRAN 77 comparable to the VAR form of declaration used for arguments in Pascal procedures, what possible problems could this give rise to?

REFERENCES AND FURTHER READING

Appleby, D. (1991) 'Classic languages—FORTRAN', *Byte*, September, pp. 147–151.
Attikiouzel, J. (1988) *Pascal for Electronic Engineers*, Van Nostrand, London.
Balfour, A. and Marwick, D. (1979) *Programming in Standard FORTRAN 77*, Heinemann Educational Books, London.

Desoer, C. A. and Kuh, E. S. (1969) *Basic Circuit Theory*, McGraw-Hill, New York.

Fidler, D. E. and Nightingale, R. (1978) *Computer Aided Design*, Nelson, London.

Ford, B., Hague, S. J., Prentice, J. A. and Taylor, D. B. (1973) 'The development and maintenance of multi-machine software in the NAG project', *Proc. Conference on Software for Numerical Mathematics*, Academic Press.

Hahn, B. D. (1987) *Problem Solving with FORTRAN 77*, Edward Arnold, London.

Press, W. H., Flannery, R. P., Teukolsky, S. A. and Vetterling, W. T. (1986) *Numerical Recipes—The Heart of Scientific Computing*, Cambridge University Press, Cambridge.

5

DATA ABSTRACTION, DATA STRUCTURES AND ALGORITHMS

5.1 INTRODUCTION

Later chapters in the book concerned with software engineering examine the problem of producing software; very briefly this involves a specification of what the software is to do, its design, subsequent implementation and finally its updating (maintenance). The entire cycle of producing software constitutes what has come to be known as **software engineering** (Ince, 1989). The topics of data structures and algorithms, discussed here, are elements that underpin software engineering, so an understanding of the material in this chapter can be seen as a prerequisite to the later discussion of software engineering. Specifically, we will be concerned with modelling the real-life objects which programs have to deal with, using appropriate programming constructs such as arrays or records. The chapter also supports material on compilers and operating systems in later chapters. A selection of commonly used algorithms of various kinds will also be discussed in this chapter, sorting algorithms being a notable example in view of their practical importance.

Choice and implementation of a data structure is conditioned by a program's specification. Consider the following specification:

- The program is to read a list of lines of text, and to produce as output a list consisting of integers that are the number of characters in each of the words.

Descriptions such as this are divorced from any method that would subsequently be used in an implementation, and in fact it should be possible for the implementation to be changed, independently of the specification. Design approaches may alter, perhaps in the interests of run-time efficiency for instance, but an implementation will be satisfactory, i.e. **correct,** provided the design's specification is met. A specification describes in abstract form actions that the program will eventually need to make concrete. Abstraction is in fact a problem-solving method that can simplify programming, separating as it does the specification and design phases—a *divide and conquer* approach.

5.2 ABSTRACTION AND THE ROLE OF DATA TYPES

Stepwise refinement is a familiar method of decomposing a problem by stages into increasingly simpler units, usually structured as procedures, and we are there progressing from a relatively high level of abstraction to increasingly lower and lower levels of abstraction until it is possible eventually to deal in the code. This process could be termed **procedural abstraction**. It is also possible to deal with *data* at various levels of abstraction as well as procedures. Data at a very low level would involve objects such as bytes accessed by their physical address. High-level languages deal with data at a higher level of abstraction using constructed types to deal directly with objects such as floating point numbers, and at a higher level still structured objects such as arrays or records. This provision in a language makes possible a closer approach to the problem domain. Data types provided in a language will seldom be capable of directly modelling many types of problem however; it is then advantageous for programmers to build on the range of data types provided by the language to form further, higher-level data types. Such (constructed) data types carry the process of abstraction further and so are often referred to as **abstract data types**. Creation and use of higher-level data types brings with it several objectives: firstly, to identify the useful range of data types in attacking the range of problems required; secondly, to define a range of primitive operations that manipulate objects; finally, there is the problem of actually *implementing* the data type, which may bring with it issues of efficiency in choices between alternative methods, a matter which is frequently of great importance in real-time systems where performance in terms of code efficiency is critical.

5.3 THE QUEUE

The queue is a data structure used widely in this book; notably in some of the material on operating systems, and so will prove a convenient vehicle to demonstrate the principles involved in the use of abstract data types. A possible definition of a queue is:

- an ordered list of items of data of similar type, additions to which may be made at one end—often called the **tail** (or **rear**)—and deletions from and access to which may be made at the other end—often called the **head** (or **front**).

It is, in principle, the same as the common idea of a queue. Queues are sometimes called **first in first out** data structures, or **FIFO**s for short, because of the way data is added or removed from them. Figure 5.1 shows some events in the life of a queue where items called Q1, Q2, Q3 are added to the front of an initially empty queue and then an item from the rear of the queue is extracted. This would need to be Q1 according to the definition of a queue

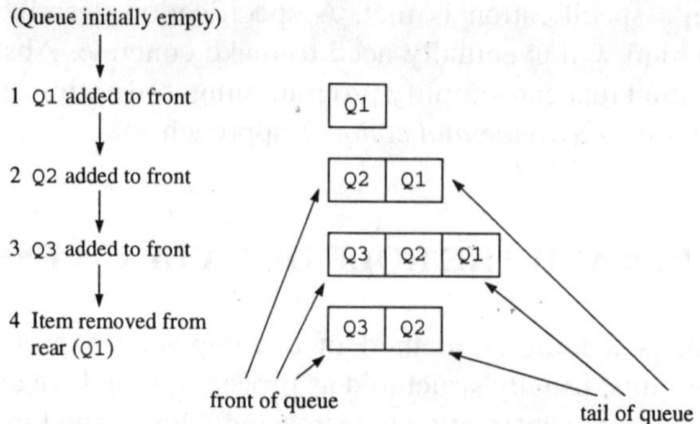

Figure 5.1 The queue.

given. Q1 etc. could be bytes, strings, records, etc., depending on the use to which the queue is being put.

The different ways in which queues need to be operated on now need to be examined. Most obviously, methods of adding data to, and removing data from, a queue are required. In addition however, it clearly makes no sense to try to remove data from a queue that contains no elements, so there needs to be a way of checking for empty queues. Finally, it may be that limitations of storage require some upper limit on the size of a queue. Summarizing, four possible operations on queues have been described. Depending on the range of applications envisaged, further operations might be required. Some general-purpose procedures (often called **access procedures**) are needed to implement whatever operations are needed. The procedures would be written to access a given named queue (we may want to have several queues in a program), and, where relevant, the data to be added or removed needs to be specified as a parameter to the relevant procedure. Decisions concerning the implementation of the queue follow definitions of the ways the queue access procedures need to be used. The names and operations of possible access procedures are given below:

- `CreateQ( Q )` creates a new queue.
- `InsertQ( Q, qdata )` adds some data `qdata` to the queue called `Q`.
- `RemoveQ( Q, qdata )` removes some data `qdata` from the queue `Q`.
- `FrontQ( Q, qdata )` returns the front queue element of the queue `qdata`, but returns the queue unchanged.
- `LengthQ( Q )` is a function which returns an integer in the form of the number of elements (i.e. the length) of the queue `Q`. This can then be tested against the maximum number permitted if required.
- `EmptyQ( Q )` is a Boolean function returning TRUE if the queue length is zero, FALSE otherwise.

5.3.1 Array implementation of queue access procedures

Having established a list of access procedures (with an eye to questions such as the performance needs of likely prospective applications), possible implementations can be investigated. Definitions of `Q` and `qdata` need appropriate *type* definitions, which raises the question of appropriate data structures to represent them. Decisions will depend to some degree on the language used; and choices may be restricted in a language such as FORTRAN where there are no constructs such as the pointers or records which feature in Pascal, C and some other languages. Use of the array to represent a queue might seem an obvious first choice and is comparatively simple to explain. Arrays certainly *could* be used for this purpose in FORTRAN programs.[1] One could envisage the elements at the head and tail of the queue being identified by two array indices used as pointers. Suitable checks on these indices could be made to test for empty queues. The `Length` function would be implemented simply as the difference between the head and tail pointers. If, for simplicity, the elements in the queue are taken as integers, the following data structure suggests itself:

```
CONST
   MaxQLength = 500 ;


TYPE
   QueueElement = INTEGER ; { sample queue data type }


   QueueStatus  = ( Normal, Empty, Full ) ;


   Queue          = record
      Contents      : ARRAY[0..MaxQLength] OF QueueElement ;
      Head, Tail    : 0..MaxQLength ;
   end ;
```

Variables of type `Queue` can be used in the queue access procedures and elsewhere. Notice that, for convenience, the actual data that is used with the queue has been defined separately from the record that defines the queue. This will change from one application to another, and is chosen to be of type INTEGER here by way of illustration. Other aspects of the TYPE definition should remain invariant from one application to another. The type `QueueStatus` is included, so that access procedures can test a queue to establish whether or not it contains any elements, or whether the length exceeds some maximum allowed value; an indication of the queue's status could be passed back to a calling program which would then be able to respond appropriately.

Queue access procedures (see also Exercise 7 in Sec. 5.9)


```
procedure CreateQ( VAR Q : Queue ) ;
begin
Q.Head := 0 ;
Q.Tail := 0
end ;
```

---

[1] The alternatives to the use of arrays to represent queues will be examined shortly.

```
function EmptyQ( Q : Queue ) : BOOLEAN ;
begin
if( Q.Tail = Q.Head ) then
    EmptyQ := TRUE
else
    EmptyQ := FALSE
end ;


procedure InsertQ( VAR Q: Queue; qdata : QueueElement ) ;
begin
Q.Contents[Q.Tail] := qdata ;
Q.Tail := Q.Tail + 1
end ;


function LengthQ( Q : Queue ) : INTEGER ;
VAR LengthBuf : INTEGER ;
begin
Lengthbuf := Q.Tail - Q.Head ;
if (LengthBuf >= 0) then
    LengthQ := LengthBuf
else
    LengthQ := 0   { 0 signals queue is empty }
end ;
procedure RemoveQ( VAR Q        : Queue ;
                   VAR qdat     : QueueElement ;
                   VAR Status : QueueStatus ) ;
begin
if (Q.Tail = Q.Head) then
    Status := Empty
else if LengthQ( Q ) >= MaxQLength then
    Status := Full
else
    begin
    qdat := Q.Contents[Q.Head] ;
    Q.Head := Q.Head + 1 ;
    Status := Normal
    end
end ;
procedure HeadQ( Q                 : Queue ;
                 VAR HeadOfQueue : QueueElement ;
                 VAR Status      : QueueStatus   ) ;
begin
if (Q.Tail = Q.Head) then
    Status := Empty
else if LengthQ( Q ) >= MaxQLength then
    Status := Full
```

```
    else
      begin
      HeadOfQueue := Q.Contents[Q.Head] ;
      Status := Normal
      end
  end ;
```

Note the use of the variable `Status` in the procedures `HeadQ` and `RemoveQ` to give an indication of an empty queue, or a queue whose length exceeds some upper limit; an application program needs to contain some code to handle such conditions. Any abstract data type should in fact have provision for dealing with exceptional cases,[2] and constraints such as limits on the maximum length of queues should form part of the specification. The approach adopted here is one of several possible alternatives; another idea would be only to invoke an access procedure under certain conditions, for example to invoke `EmptyQ` before deciding whether to invoke `RemoveQ`. Another possibility is to arrange for the access procedure to pass back a key value in the data to indicate its status. The following short demonstration program shows how the various access procedures could be used.

```
PROGRAM QueueTest( Input, Output ) ;
{ CONST, TYPE and Procedure declarations here, as above }

VAR
    i      : INTEGER ;
    Q      : Queue ;
    Qdat   : QueueElement ;
    Status : QueueStatus ;
begin
CreateQ( Q ) ;
{ insert 4 elements in the queue }
for i := 1 TO 4 DO
    begin
    Write( '? ' ) ;
    ReadLn( Qdat ) ;
    InsertQ( Q, Qdat )
    end ;
{ flush queue to the screen }
while NOT EmptyQ( Q ) DO
    begin
    RemoveQ( Q, Qdat, Status ) ;
    WriteLn( Qdat )
    end
end.
```

5.4 LINKED LISTS

One of the problems in implementing lists and for that matter other data structures with

---

[2]Exception handling is covered in Chapter 6.

arrays is that a fixed amount of store needs to be specified to hold the array at compile time, irrespective of how much store is actually required when the program runs. Linked lists[3] represent an alternative that overcomes this and other problems in representing data structures such as queues. Instead of storing list elements as contiguous array elements, each item, in addition to holding the data, contains a pointer to the next item in the list. Pointers in this context could be thought of in general terms as a kind of 'address' for the next element in the list. Chapter 3 discussed the support for pointers in C, and support for pointers also exists in Pascal, Modula-2, Ada and some other languages. We first examine the use of linked lists in a general way, divorced from the syntax of any particular language, and later give some details of methods of implementing linked lists in Pascal and C. The queue abstract data type will be revisited, using a linked list implementation in place of the earlier array.

5.4.1 Insertion and removal of elements from a linked list

If we consider the list depicted in Fig. 5.2, to which the element X is to be added, the pointer in element A needs to be changed so as to now point to the new element, while the pointer associated with the new element needs to reference what was originally the second element. The utility of this lies in the fact that these are the *only* adjustments required, unlike the comparable situation with the array implementation, where the position of *all* elements needs to be adjusted. Much the same applies to the deletion of an element from a list.

5.5 POINTERS IN PASCAL

It is appropriate at this point to review the provision in Pascal for pointers, before discussing implementation of linked lists in this language. Where appropriate, comparisons will be made between C and Pascal realizations of linked lists. Pointers are generally associated with objects for which storage space is allocated dynamically (while the program executes); this is different to some other types of variable, where storage space is

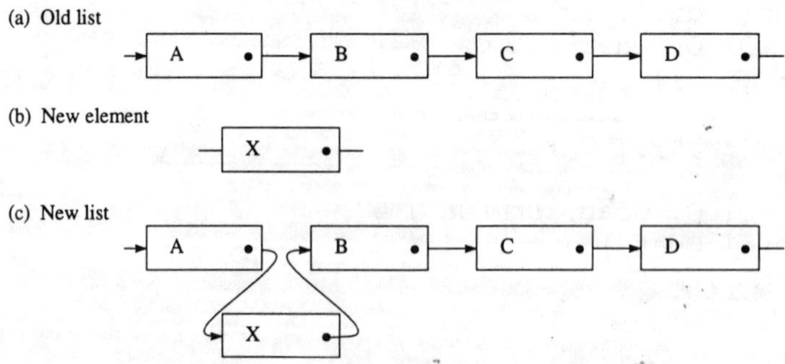

Figure 5.2 Insertion of a new node into a linked list.

---

[3]This section makes reference to pointers as implemented in C, so a review of Sec. 3.10 might prove useful here.

specified (by a suitable VAR declaration) at compile time. Dynamically allocated storage does not have a variable name associated with it, and can only be referred to by means of a pointer to the variable type. Note that auto-type variables in C, and comparable types of variables in other languages, are stored on a stack which naturally changes in size at run time. The consequence is that some estimate of the amount of store needed by objects such as arrays must be made when the program is written, and allowing for some margin of error it is clear that some store will be unused and thus a source of inefficiency. Part of the function of a compiler is to set the amount of store needed for the code, data and the stack.[4] It is one of the advantages of dynamic data structures that storage space for them can grow or contract during execution. Special procedures are available to programmers to allocate and deallocate store for use by dynamic data structures.

To set up a linked list in Pascal, a notation for defining pointers is needed. The following type definition defines a pointer `Ptr` that points to a structure called `Node`. It is important to distinguish between the pointer and the object being pointed to. A variable being pointed to by some pointer `MyPtr` would be written as `MyPtr^`.

```
TYPE
      Ptr = ^Node ;
```

In the case of a linked list, `Node` would most likely be a record comprising fields defining the data, and a pointer to the next element in the linked list, as in the example below where the data is simply an integer:

```
TYPE
   Node  =  record
    Data : INTEGER ;
    Next : Ptr
   end ;
```

The field `Next` points to the next element of the linked list. If a variable `LList` of type `Ptr` was defined, then, to give an element in our linked list a value, we could write code such as `Llist^.Data := 122 ;` i.e. `Llist` is the pointer and `LList^` the object being pointed to. Naturally the field `Next` would also need to be suitably defined. A linked list is accessed using a pointer to its first node. Successive nodes are then accessed via the `Next` field of the preceding node. This raises the question of how the end of the list is to be indicated. One method is to use a special pointer value called NIL. Thus the `Next` field in the last node in the list must be made to point to NIL. Pointers in Pascal can be assigned, and compared using the comparison operators '=' and '<>'. Pointers may also be passed as parameters to procedures.

Storage for nodes in a linked list is acquired by the procedure NEW, so that, for the variable `LList`, the code `NEW(LList)` would acquire enough storage to hold an integer and also the pointer forming the remaining field in the record. If NEW is used repeatedly, however, the point may eventually be reached where no more store is available, and so another procedure called `DISPOSE` is provided to return any store used by unwanted nodes, or other variables associated with pointers.

---

[4]The provision in C was discussed briefly in Sec. 3.13.

The example below shows how a simple linked list with *N* nodes can be built. The last node points to NIL, this being the first node in the list, while the list is accessed using the pointer `ListHead`. It is important to recognize that there is no other way the list can be accessed, in other words it is not possible to directly address the internal nodes of the list in the form it is written here, which implies that if it is necessary to search for a particular node, this must be done one node at a time starting with the node at the head of the list. Clearly this could prove to be a source of inefficiency if the list is of any length. No such problem exists in the case of a list implemented as an array, since it is possible to address any element of an array directly in terms of its index. Note the use of the pointer `Temp` to buffer the current node being built.

```
VAR
    Temp, ListHead, LList : Ptr ;
    . . .
    . . .


    NEW( Temp ) ;
    Temp. Next := NIL ;
    Read( Temp.Data ) ;
    LList := Temp ;
    for i := 0 TO N DO
       begin
       NEW( Temp ) ;
       Read( Temp^.Data ) ;
       Temp^.Next := LList ;
       LList := Temp
       end ;
    ListHead := Temp ;
    . . .
    . . .
```

The code fragment below demonstrates how the list pointed to by `ListHead` could be used in this case simply to count the number of nodes in the list.

```
VAR
    Number          : INTEGER ;
    Temp, ListHead : Ptr ;


    . . .
    . . .
    Number := 0 ;
    Temp := ListHead ;
```

```
while Temp <> NIL DO
   begin
   Number := Number + 1 ;
   Temp := Temp^.Next
   end ;


...
...
```

5.5.1 A queue implementation using pointers

It is possible to implement a queue in terms of pointers, as an alternative to arrays; and a comparison of the relative merits of the two different approaches is illuminating. The queue would be viewed as a linked list of records, each record comprising the data in the queue together with a link to the next node. Before looking in detail at the implementation, note that we deal with the same specification as before, and as far as an application program is concerned a similar set of access procedures appear to be used; only the procedure bodies and type definitions will change. A possible set of type definitions is given below. Notice the way `Queue` is defined so that it could be used as a name in the data type implemented before as an array, as well as in the present approach with pointers. It would only be necessary to use the type definition names `QElement` (here of type INTEGER by way of example), and `Queue`. The linkage of one element of the queue to the next using the type `QueueNode` would be hidden as far as an application program is concerned, except for the two nodes at the tail and the front of the queue respectively. However, the queue's type definition is accessible in the program that calls the access procedures. This is a defect in Pascal, rectified in Modula-2.

Type definitions using pointers

```
TYPE
   QueueElement = INTEGER ; { sample queue data type }
   QueueStatus  = ( Normal, Empty, Full ) ;


   QueueNodePtr = ^QueueNode ;


   QueueNode    = record
      Contents : QueueElement ;
      Next     : QueueNodePtr
   end ;


   Queue        = record
      Head     : QueueNodePtr ;
      Tail     : QueueNodePtr
end ;
```

5.5.2 Implementation of queue access procedures using pointers

```
procedure CreateQ( VAR Q : Queue ) ;
begin
Q.Head := NIL ;
Q.Tail := NIL
end ;


function EmptyQ( VAR Q : Queue ) : BOOLEAN ;
begin
   EmptyQ := (Q.Head = NIL)
end ;


procedure InsertQ( VAR Q : Queue; qdat : QueueElement ) ;
VAR
   NewNode : QueueNodePtr ;
begin
NEW( NewNode ) ;
NewNode^.Contents := qdat ;
NewNode^.Next := NIL ;
if EmptyQ( Q ) then
    Q.Head := NewNode
else
    Q.Tail^.Next := NewNode ;
Q.Tail := NewNode
end ;


function LengthQ( Q : Queue ) : INTEGER ;
VAR
   LengthBuf : INTEGER ;
   Temp      : QueueNodePtr ;
begin
if Q.Head = NIL then
   LengthQ := 0
else
   begin
   LengthBuf := 1 ;
   Temp := Q.Head ;
   while Temp^.Next <> NIL DO
      begin
      LengthBuf := LengthBuf + 1 ;
      Temp := Temp^.Next
      end ;
   LengthQ := LengthBuf
   end
end ;
```

```pascal
procedure RemoveQ( VAR Q       : Queue;
                   VAR qdat    : OldNode^.Contents ;
                   VAR Status : QueueStatus    ) ;
VAR
   OldNode : QueueNodePtr ;
begin
with Q DO
   begin
   OldNode := Head ;
   if (Head = NIL) then { dont remove from empty queue }
      Status := Empty
   else if LengthQ( Q ) >= MaxQLength then
      Status := Full
   else
      begin
      Qdat := OldNode^.Contents ;
      Head := Head^.Next ;
      if Head = NIL then
         Tail := NIL ;
      DISPOSE( OldNode ) ;
      Status := Normal
      end ;
   end
end ;




procedure HeadQ( Q          : Queue;
                 VAR HeadOfQ : QueueElement;
                 VAR Status  : QueueStatus ) ;
begin
with Q DO
   if Head = NIL then
      Status := Empty
   else if LengthQ( Q ) >= MaxQLength then
      Status := Full
   else
      begin
      HeadOfQ := Head^.Contents ;
      Status := Normal
      end
end ;
```

The same approach to handling constraints as that used in the previous implementation with arrays has been retained. Notice, however, that checking a queue to see if it exceeds some maximum length involves stepping through the entire list, which could lead to an excessive time penalty in some cases, possibly requiring some other implementation. The 'array' implementation was much faster in this respect.

To see the meaning of the access procedures, and the role of pointers in particular, you should examine a sequence of events leading to a queue being constructed, and then the effect of deleting a node from the queue.

5.5.3 Implementation of queue access procedure in C

The access procedures for `emptyq` and `insertq` coded in C are shown below, together with suitable type definitions. It may be useful to review the material in Chapter 3 on pointers in C before trying to understand the code. Where possible, the capabilities and idioms of C have been exploited, but the structure of the code has been kept close to that of the earlier Pascal version. If you understand the two access procedures below it should be reasonably straightforward to develop the remaining procedures.

```c
enum      q_status{ NORMAL,  EMPTY,  FULL } ;


typedef   int             q_data ;
typedef   enum q_status   queue_status ;
typedef   unsigned char   boolean ;
typedef   struct q_node{
              q_data            contents ;
              struct q_node   *next ;
} queue_node ;


typedef struct{
    queue_node   *head ;
    queue_node   *tail ;
}queue ;


#define      NULL_PTR      (queue_node *)NULL




/*-------------------------------------------------------*/
boolean empty_q( queue q )
{
    if(q.head == NULL_PTR)
        return TRUE ;
    else
        return FALSE ;
}
```

```c
/*--------------------------------------------------*/
queue *insert_q( queue *q1, q_data new_data )
{
    queue_node      *new_node ;
    extern   void *malloc() ;

    if((new_node = (queue_node *)malloc(sizeof(queue_node)))
                        == NULL_PTR){
        fprintf( stderr, "no more store\n" ) ;
        return NULL_PTR ;
    }
    new_node->contents = new_data ;
    new_node->next = NULL_PTR ;
    if( empty_q( *q1 ) )
        q1->head = new_node ;
    else
        q1->tail->next = new_node ;
    q1->tail = new_node ;
    return q1 ;
}
```

It is possible to return a pointer to a function in C, and this is used in `insert_q` to return the new queue by making the function a pointer to `queue`. A null pointer is returned in the case where there is no more store available for the new queue node.

The library function `malloc` is used to return an amount of store for the new node in the queue. It in turn is given the number of bytes needed in terms of the library function `sizeof`, which returns the number of bytes needed by its argument, the struct `queue_node`. `Malloc` is being used here in the same way as `NEW` in a comparable Pascal program to obtain some storage. Note the use of the `fprintf` to indicate that there is no available store. This is the preferred way of showing error messages in C. The code fragment below shows a typical way of using `insert_q`:

```c
queue  *q1 ;
q_data data ;
...
...
if( (q1 = insert_q(q1, data)) == NULL_PTR )
    /* insert_q must have said out of store,
       so handle exception case here
    */
else
    /* otherwise proceed normally */
...
...
```

5.6 STACKS

A stack, like a queue, is a form of list; and useful contrasts between these two types of data structure can be drawn. With a stack, however, insertions and deletions occur at the same end of the list, not at the head and tail of the list, as with queues. Stacks are called **last in first out** data structures in contrast to queues which as we have noted are called first in first out.

Stacks, at the level of machine code, are a familiar way of storing procedure return addresses and for passing parameters between procedures, for which purpose hardware support is usual, in view of the heavy use of a stack.[5] A simple analogy of a stack is of a pile of plates in a cafeteria. Plates are picked up from the top of the pile, and clean plates are added to the pile of plates. The last item to be placed on the stack is the first to be subsequently removed. The two operations on stacks are called respectively **pushing** and **popping**. Stacks are stored in memory in a reserved block of locations, one of which, the top of the stack, is pointed to by a register termed the stack pointer. The push and pop operations are associated with store and load operations together, in conjunction with adjustments to the stack pointer such that it always points to the top of the stack. For a push, the stack pointer is decremented and the data stored at the address then pointed to. In the case of a pop, the data at the location pointed to by the stack pointer is loaded into the CPU, and *then* the stack pointer is incremented. This description assumes that a stack will 'grow' downwards in memory as data is pushed on to it from some initial top of stack position, which is common but not mandatory. The implementation details would be dependent on the design of the hardware.

Evidently data is pushed on to a stack before popping it off again subsequently, so it follows that care must be taken to avoid popping data from an empty stack, since the data would then be some garbage value stored in memory. Note that stacks change in size dynamically, so some estimate of their size needs to be made so that no encroachment is made on other areas of memory which might be holding code, or other data.

After this illustration of one use of stacks, it will be useful to discuss the stack as a data structure in a more general sense. Stacks may be used as a programming tool just like any other data structure; the requirement is for some general specification of stacks devoid of specific implementation details, together with operations to be carried out on them. This would form the basis for developing an abstract data type to access any stack, no matter how it is to be implemented eventually. Stated more formally, the operations described in the previous paragraph may be summarized in terms of the access procedures below:

- `CreateStack( S )` performs initialization, returning an empty stack. This can be compared to the process of giving a value to a stack pointer.
- `Push( S,Sdata )` returns the stack `S` with the data `Sdata` placed at the top.
- `Pop( S, Sdata )` takes the stack `S`, and returns the stack with `Sdata` removed from the top.
- `EmptyStack( S )` is a Boolean function which returns TRUE if the stack contains no data, or FALSE otherwise.
- `StackTop( S, Sdata )` returns the data at the top of the stack `S`, without removing the data from the stack.

---

[5] This aspect of stacks is discussed further in Chapter 7 in connection with compilers.

The question of implementation can now be approached. The type of stack discussed earlier was effectively implemented as an array accompanied with low-level hardware support in order to obtain satisfactory run-time efficiency. We can also choose to implement a stack as an array using the high-level support available in Pascal as was done for queues, or alternatively a linked list could be used. Whichever implementation is adopted however, the definition of the abstract data type should be such as to make the behaviour of a program that uses it quite independent of any implementation details. The example below uses linked lists; it is left as an exercise to the reader to develop an alternative implementation using arrays.

```
TYPE
    StackElement   =   INTEGER ; {Sample data type for the stack}

    StackStatus    =   ( Normal, Empty ) ;

    Stack          =   ^StackNode ;

    StackNode      =   record
       Contents : StackElement ;
       Next      : Stack
    end ;

procedure CreateStack( VAR S : Stack ) ;
begin
S := NIL
end ;

function IsEmptyStack( StackPointer : Stack ) : BOOLEAN ;
begin
if ( StackPointer = NIL ) then
   IsEmptyStack := TRUE
else
   IsEmptyStack := FALSE
end ;

procedure StackTop(     StackPointer : Stack;
                    VAR TopOfStack   : StackElement;
                    VAR Status       : StatusStatus ) ;
begin
if IsEmptyStack( StackPointer ) then
   Status := EMPTY
else
   begin
   TopOfStack := StackPointer^.Contents ;
   Status := Normal
   end
end ;
```

```
procedure Push( VAR StackPointer : Stack;
                    TopOfStack   : StackElement ) ;
VAR
   Temp : Stack ;
begin
NEW( Temp ) ;
Temp^.Contents := TopOfStack ;
Temp^.Next := StackPointer ;
StackPointer := Temp
end ;



procedure Pop( VAR StackPointer : Stack;
               VAR StackData     : StackElement ;
               VAR Status        : StackStatus ) ;
VAR
   OldNode : Stack ;
begin
if IsEmptyStack( StackPointer ) then
   Status := Empty
else
   begin
   OldNode := StackPointer ;
   StackData := StackPointer^.Contents ;
   StackPointer := StackPointer^.Next ;
   DISPOSE( OldNode );
   Status := Normal
   end
end ;
```

5.7 SORTING ALGORITHMS

Methods of sorting are very commonly encountered in programming (Knuth, 1973). A simple example is sorting a list of names into alphabetical order, needed in a situation where a database containing names needs to be updated when a new name is added. Sorting is applied to data that consists of what will be called here records (not to be confused with a Pascal type record), where each record is associated with various kinds of data and also a **key**, which is a subfield of each record. For instance, the data could consist of an array, where each array element is a (Pascal) record, with elements consisting of a string (names perhaps), and an integer for the key. Many other types of data are possible, however. Keys are used as a basis of sorting records. In the example just cited, an unsorted array might appear as shown below:

key	name
5	'fred'
6	'betty'
3	'bert'
4	'john'
2	'mary'
1	'alf'

After sorting, the array would be:

key	name
1	'alf'
2	'mary'
3	'bert'
4	'john'
5	'fred'
6	'betty'

There are a large number of sorting algorithms of varying efficiency and complexity, and it is a matter of judgement whether the time spent devising a sophisticated sorting program is justified in a particular case. For simple problems, it might be better to use a relatively inefficient but quick-to-develop sort program. Efficiency is judged in terms of computation time, and also storage space, although the former is usually the dominant consideration. Data can after all be stored on disk if needed. The time taken by a particular algorithm can be assessed by measuring its efficiency for a wide representative sample of data. Data that is nearly sorted can, with some algorithms at least, require less time than other data. A sort program in Sec. 5.7.1 exercises the particular algorithm used for a list of integers that are initially in reverse order. The sort time here is proportional to the square of the number of elements (Tenenbaum and Augenstein, 1989). Had the list been nearly sorted the time needed would have been proportional to the number of elements. The reader is invited to experiment with the program for this and other lists, and to make time comparisons.

5.7.1 Example of quicksort

This is a very commonly used method of sorting, and is relatively simple but efficient. Essentially the algorithm is as follows: an element is selected from the array to be sorted. This element is repositioned in the array so that now the elements from the first-array element up to the repositioned element are less than or equal to the repositioned element, and the remaining elements are greater than or equal to the repositioned element. It follows that the two parts of the array are sorted with respect to the repositioned element, which does not need to change its position. This process is now repeated for the two subarrays, which will be increasingly divided themselves as the solution proceeds. Eventually the array will have been fully sorted.

Later a program that uses the quicksort algorithm will be described, but first a simple illustration demonstrates the main principle involved. Consider the array of integers below (the key and the 'data' for the record elements here happen to be one and the same, this would not necessarily be so in other cases):

```
10  23  89  5  55  3
```

The first element is selected as the one to be repositioned. The repositioning of the element 10 gives:

```
(5  3)  10  (23  89  55)
```

The brackets indicate the two as yet unsorted subgroups. In further iterations, the 10 value will not change its position of third from the start of the list. The second operation of the algorithm leaves the first subgroup (5 3) fully sorted as 3 5, while the second subgroup gives 23 (89 55), writing out the entire array we get:

```
3  5  10  23  (89  55)
```

Finally it remains to sort the only unsorted subarray (89 55) to give a completely sorted array:

```
3  5  10  23  55  89
```

As an exercise, the reader is invited to carry out a similar procedure to sort the following array:

```
1  2  100  56  23  7  200  56  78  60
```

The Pascal program below demonstrates the application of quicksort to arrays of integers, although it would be relatively simple to extend its use to more complex data such as strings. The core of the program is the procedure QuickSort, and it will be noted that this procedure is used recursively until all sublists are fully sorted.

```pascal
PROGRAM QuickSort( Input, Output ) ;

CONST
   MaxListSize = 2000;

TYPE
   List          = ARRAY[1..MaxListSize] OF INTEGER ;

VAR
   X             : List ;
   Index         : INTEGER ;
```

```pascal
procedure Swap( VAR SList : List; i, j : INTEGER ) ;
VAR
    temp : INTEGER ;
begin
temp := SList[i] ;
SList[i] := SList[j] ;
SList[j] := temp
end;


procedure QuickSort ( VAR SList : List; left, right : INTEGER) ;
VAR
    i, last : INTEGER ;
begin
if( left < right ) then
   begin
   Swap( SList, left, (left+right) DIV 2 ) ;
   last := left ;
   for i := left + 1 TO right DO
      if ( SList[i] < SList[left] ) then
         begin
         last := last + 1 ;
         Swap( SList, last, i ) ;
         end ;
   Swap( SList, left, last ) ;
   QuickSort( SList, left, last-1 ) ;
   QuickSort( SList, last+1, right )
   end ;
end;



begin
{ demonstration array to be sorted by QuickSort Procedure }
{ list elements are in reverse order - worst case example }
for Index := 1 TO 1000 DO
   X[Index] := 1000 - Index ;
QuickSort( X, 1, 1000 ) ;
{ display sorted array, 10 numbers to a line }
for Index := 1 TO 1000 DO
   begin
   Write( X[Index] : 6 ) ;
   if ( (Index MOD 10) = 0 ) then {new line every 10 numbers}
      WriteLn
   end
end.
```

5.8 SEARCH ALGORITHMS

As with sorting, search algorithms are used in connection with data organized as records, each of which is associated with a key. In this case the object is to find a particular record as identified by its key. Search algorithms are often structured so as to add a record in cases where there is no record corresponding to the key provided; the algorithm would then be more appropriately termed a **search and insertion algorithm**. Other variants of search algorithms might allow for deletion of records.

The table constituting all records can be organized in a number of ways such as an array or a linked list. Linked lists have the usual advantages compared with arrays, such as simpler deletion of records and provision for the size of the table being determined dynamically, rather than at compile time, as would be necessary if an array structure were used. Choice of data structure for the table is also bound up with the question of time taken to find a certain record, and it will often be the case that this is determined by the search technique used.

5.8.1 Sequential search

This is the simplest search method, and involves stepping through the table of records until the record with the required key is found. The method is slow, and we examine it here as a simple illustration of search methods rather than as a practical technique. The somewhat simple Pascal program below shows the application of sequential search for the case where the data table is organized as an array of records, the keys being integers, while the data are characters. Provision is made for inserting records at the end of the table, in cases where the search proved unsuccessful. The code should be self-explanatory. It should be quite simple to write a comparable program where the data table is organized as a linked list instead of as an array as here.

```
PROGRAM Search1( Input, Output ) ;


CONST
   MaxTableSize = 20 ;


TYPE
   RecordDataType = CHAR ;
   TableType = record
      Data  : RecordDataType ;
      Key   : INTEGER
   end ;
```

```
VAR
    Table      :   ARRAY[1..MaxTableSize] OF TableType ;
    Index      :   INTEGER ;
    SearchKey  :   INTEGER ;
    Data       :   RecordDataType ;
    Found      :   BOOLEAN ;
    TableSize  :   INTEGER ;
begin
{ construct a table with sample data and keys }
TableSize := 0 ; { Table initially contains no records }
writeln( ' enter a key(-1 to end), and data value ' ) ;
repeat
    TableSize := TableSize + 1 ;
    Read( Table[TableSize].Key, Table[TableSize].Data ) ;
until (Table[TableSize].Key = -1) ;


{ get search key value }
WriteLn( ' search key?? ' ) ;
Read( SearchKey ) ;
{ search for record }
Index := 1 ;
Found := FALSE ;
while( ( Index <= TableSize ) AND NOT Found )
    begin
    if( SearchKey = Table[Index].Key ) then
        begin
        Writeln( ' record with data - ', Table[Index].Data,
                '- Found' ) ;
        Found := TRUE
        end
    else Index := Index + 1 ;


{ add new record, if not found in original table }
if NOT Found then
    WriteLn( ' record not found, new record added to table..' ) ;
    TableSize := TableSize + 1 ; { Insert new record at end of table }
        Writeln( 'new record data?' ) ;
        Read( Table[TableSize].Data ) ;
        Table[TableSize].Key := SearchKey ;
        end ;
end.
```

The efficiency of sequential search depends on the position of the required record in relation to the first record. The worst case would be if the record were the last, involving, for a table with n records, n comparisons, while obviously only a single comparison would be needed if the record happened to be the first in the table. On average one might think that a time proportional to $n/2$ would be required over a large enough number of searches, but in fact this would not be the case where particular records were being accessed more heavily than others. A possibility then is that the algorithm could arrange to place the more heavily used records near the beginning of the table.

5.8.2 Getting greater efficiency—example of the hashing approach

For many applications, the sequential search method is far too slow. Bornat (1979), reports that compilation time for early compilers was ten times as long compared to that for more recent compilers. This is because the sequential search method was used in the search procedures carried out as part of compilation, instead of more efficient algorithms. Many other search algorithms of varying efficiency and suitability for different types of data are available. Only one further technique, called **hashing**, will be examined here. A more comprehensive treatment of various search algorithms is given in Tenenbaum and Augenstein (1989), Knuth (1973) and Gries (1971).

The problem with sequential search is the repeated comparisons made involving the key. Clearly, for greater efficiency, the aim should be to reduce the number of comparisons, and ideally we would like no comparisons to be necessary at all. In principle this could be possible if the table were organized as an array, and keys were integers used to index into that array. Comparing the sequential search algorithm on which the Pascal program in Sec. 5.8.1 was based, we would now be using the counterpart of the variable `SearchKey` as an array index. Additionally, however, there may be cases where there is no existing record corresponding to some of the key values. This situation could be accommodated by making the data for an as yet nonexistent record some unique value, distinct from 'normal' data.

The disadvantage of the algorithm just described is that it becomes necessary to make provision for the maximum number of records in the data table at compile time, irrespective of the actual number of records that the table may ultimately contain. This, although potentially wasteful of storage space, might be acceptable for relatively small tables. Often, however, the identification of record keys with a limited range of array index values might prove too restrictive. There are many cases where we need to use record keys that have a large range of values in relation to the number of records. For example, keys can be in the form of character strings, and if for instance we allowed for strings of up to six characters, the number of distinct values for the strings amount to many millions. Thus on the basis of the simple direct access algorithm described above, this would entail declaring a possibly very large array, many or most of whose elements moreover could correspond to empty records. With the hashing approach shortly to be described, this problem is dealt with by making certain key values share a common record. Hashing algorithms aim to convert a key value to an integer of limited range to be used to index into the array of records. The transformation is achieved by a **hash function**.

The hashing method of inserting records in a table, or searching for them, can perhaps best be demonstrated by an example. Consider a table consisting of at most 1000 records, but where the key values are integers ranging between 0 and 10^6. For the sake of illustration the data is assumed to be in the form of characters as in the program in Sec. 5.8.1. A special key value of -1 is to denote an empty record. A typical table is shown in Table 5.1.

Table 5.1 Inserting records

Array index	Key	Record data
0	555000	'a'
1	444001	't'
2	−1	
.		
.		
801	996801	'y'
802	412802	'%'
803	−1	
804	100804	'!'
805	−1	
.		
.		
999	756999	'+'

The hash function used here to generate an array index from the key is the last three digits of the key, so that a key 996801 gives an index of 801 as shown in Table 5.1. It might be added here that this is simply one illustrative example of a hash function, others are possible. Expressed more formally, if we denote the present hash function as `Hash(Key)`, then in this case we can write `Hash(Key) = Key MOD 1000`. If we use this approach to insert new records with given keys into the table, a problem is that a hash function may generate the same array index as that corresponding to an existing record. Consider the record with key value 412802 in Table 5.1; if a new record with key value 333802 is to be inserted, it clearly cannot occupy the same position in the array as the existing record. This situation is called a **hash collision**. One possible way of dealing with hash collisions is to insert a new record at the first unoccupied position following the position where a collision occurred. For example, in the case of insertion of the record with key 412802, the position at 803 is unoccupied, and the new record could be placed there. If, in the process of attempting to insert records in this way, the last element in the array were reached, the process could continue at the start of the array.

The procedure just described for inserting records into a table is in some respects somewhat similar to the procedure for searching for a record. A given key to be used to search for the given record is hashed, and the resulting index used to access a record. If the record is empty it would have a special indicator such as a key of −1 as in the example to indicate the fact. If the key for the record is different to the search key, a sequential search is commenced starting from the record just accessed. In many cases, however, the first access will give two matching keys, in which case further sequential searches would not be required. An algorithm, in pseudocode form, to carry out the search is given below:

```
record index i = Hash( search key ) ;
Record_Found := TRUE ;
initial index := i ;
```

```
while search key <> key for record i AND Record_found DO
begin
if key for record i <> special empty record key then
    begin
    if i = Maximum Number of Records then
        i := 0
    else
        i := i + 1 ;
    if i = initial index then
        Record_Found:= FALSE ;
    end ;
else
    Record_Found := FALSE ;
end
```

5.9 EXERCISES AND REVIEW QUESTIONS

1. Discuss and contrast procedural and data abstraction as approaches to the design of software.
2. A company manufactures electronic circuit boards which are to be described by a computer program. The attributes of the boards can, in the interests of simplicity, be assumed to be as listed:
   (a) A PC board type identified by one of the following strings 'PC1', 'PC2', 'PC3'.
   (b) Resistors, in the set of preferred values, and of tolerances 5 per cent, 10 per cent and 20 per cent. The preferred values and tolerances are to be represented as enumerated types, rather than numerical values.
   (c) Integrated circuits.
   No board can contain more than 40 resistors or 75 integrated circuits, and there will be no more than 2000 boards.

   Give suitable data structures describing a number of such circuit boards. Assume all integrated circuits have already been predefined by some type declaration IC. As a further exercise you might write comparable declarations in C. (Preferred resistor values are as follows: 1, 1.5, 1.8, 2.2, 2.7, 3.3, 3.9, 4.7, 5.6, 6.8, 8.2.)
3. A **dequeue** is a set of data from which elements can be deleted at either end, and into which elements can also be inserted at either end. Specify a set of access procedures for inserting and deleting elements together with other access procedures. Use the access procedures described for queues to gain insight into possible approaches.
4. List the advantages and disadvantages of using arrays and linked lists in the representation of queues and other forms of list.
5. Write the code for the access procedures for the dequeue described in Qu. 3 using linked lists in either Pascal, C or both!
6. The Pascal quicksort program in Sec. 5.7.1 sorts an array of integers. Modify the program so that it sorts records for data in the form of strings and record keys in the form of integers. Further modify the program so that it sorts records that consist of addresses together with keys in the form of names. Sorting is to be done in alphabetical order of names.

7. The queue access procedures in Sec. 5.3.1 were designed to demonstrate a principle in a simple way. However, they have a flaw in cases where the pointer to the rear of the queue exceeds `MaxQLength`, even when the queue is not full. State possible solutions to this (Tenenbaum and Augenstein, 1989).

REFERENCES AND FURTHER READING

Attikiouzel, J. (1988) *Pascal for Electronic Engineers*, Van Nostrand, London.

Bornat, R. (1979) *Understanding and Writing Compilers*, Macmillan, London.

Findlay, W. and Watt, D. A. (1988) *Pascal—An Introduction to Methodical Programming*, Pitman, London.

Gries, D. (1971) *Compiler Construction for Digital Computers*, John Wiley, New York.

Ince, D. C. (1989) *Software Engineering*, Chapman and Hall, London.

Knuth, D. E. (1973) *Art of Computer Programming*, vol. 3, *Sorting and Searching*, Addison-Wesley, Reading, Mass.

Prospero Software (1987) *Prospero Pascal Manual*, London.

Tenenbaum, A. M. and Augenstein, M. J. (1989) *Data Structures Using Pascal*, Prentice-Hall, London.

Thomas, P., Robinson, H. and Emms, J. (1988) *Abstract Data Types*, Oxford University Press, Oxford.

Welsh, J. and Elder, J. (1987) *Introduction to Modula-2*, Prentice-Hall, London.

Wirth, N. (1976) *Algorithms + Data Structures = Programs*, Prentice-Hall, Englewood Cliffs, New Jersey.

6

LANGUAGES—CHOICES AND CHARACTERISTICS

6.1 INTRODUCTION

Choice of programming language can seem to be one of the more complex and contentious decisions to be made in using computers. This to some degree is a reflection of the diversity of problems solved on computers, each with specialized needs. Wilson and Clark (1988) report that over 1000 different programming languages have been developed, many for specialized applications—only a handful of these are in widespread use. One should say also, in connection with choosing a language, that many applications can be reasonably well served by a number of languages, often perhaps with no obvious 'best choice' between several candidate languages.

Arguments often extend beyond considerations based on the purely technical merits of a language and its design. For example, what one might call 'inertia' often plays a role in choosing a language. Many programmers display an understandable reluctance to change from a language they know well, in spite of advantages that another language might seem to offer for their application. Learning a new language represents a substantial commitment in time and there is also the problem of existing code written in other languages (including assembler) which also represents an investment of another kind. Any reason for changing from one language to another must be compelling. In some cases programmers have no choice of language, and customers purchasing software may insist that software vendors write code in one specified language. An important case in point here is the language Ada, which some customers, especially the military, insist be used to write software they purchase. FORTRAN has been used for many years, and will continue to be used in the future in spite of the fact that it is an early second-generation language with primitive features in comparison with later languages such as Pascal. It has to be said in its defence that FORTRAN has particular characteristics which make its use attractive for certain scientific/engineering applications, as we saw from the discussion in Chapter 4. It has also been updated to make it compete more effectively with more modern languages.

Many classes of applications in computing fall outside the range covered in this book, and we restrict the discussion to languages for those applications likely to be of interest to electronic engineers. For example, languages such as PROLOG and LISP, used in areas such as artificial intelligence, and COBOL, used in data processing, will not be discussed here.[1] One of the main thrusts of the book is in the area of real-time systems, so considerable emphasis will be given to languages that are used in this field.

The material in this chapter should be seen as complementing the discussion in Chapters 3 and 4 on C and FORTRAN. It was felt that the importance of these two languages justified devoting a chapter to each of them, but clearly it is not possible to discuss other languages to the same depth, and where other languages are involved the treatment will be to illustrate some general point. The emphasis is on comparisons between languages rather than the detailed programming techniques.

The main issues addressed in this chapter can be summarized as follows:

- The needs of real-time programming alluded to above. This often involves systems in which several sequential programs execute in parallel. There is a clear need for this in cases where there may be several processors, but we will see that it is also true in many instances in systems with a single CPU. Getting the various programs to work consistently with each other is the issue here. Robustness is crucial in real-time systems which naturally is particularly true of the software used. Careful consideration needs to be paid to matters such as procedures that allow for proper recovery from errors without causing catastrophic system failure. The languages Ada and Modula-2 will be used to illustrate some of the concepts involved.
- Discussion of some general features of languages in outline, in terms of aspects relating to strong typing, structured features, and orthogonality, with a review of some of the languages commonly used by electronic engineers.
- Requirements in a language when it is to be used as a method of developing large programs, real-time or otherwise. Programs in excess of 100 000 lines of code are not uncommon, and such programs would invariably be prepared by a team of programmers rather than one individual, bringing with it the problems of software management and maintenance referred to briefly in Chapter 5, and which will be discussed in detail in the chapters dealing with software engineering. Modula-2 will be used to illustrate ideas here, although it should be mentioned that one of the strong points in Ada is the provision in the language to facilitate the production of large-scale programs.

In summary, the present chapter examines the support required of a language viewed as a tool in the production of robust large-scale software, and comparisons between languages are made largely on this basis. It can be viewed in conjunction with some of the material in Chapter 5 as providing support for later discussion on software engineering. Some of the chapter will be a review, assuming some knowledge of 68000 assembler in particular. The chapter could be read in conjunction with Chapter 7 which includes material on compilers.

---

[1] A full discussion of most commonly used languages can be found in Wilson and Clark (1988).

6.2 ASSEMBLERS

An assembler is an example of a **translator** program; that is a program that accepts as input a program written in one language, and produces as output a program in another language, commonly the machine language in the form of binary codes that a computer can actually execute. As with any translator, an assembler improves the efficiency with which programmers can produce code. The most primitive method of programming operates at the level of the binary codes used by the computer, an almost impossibly long and error-prone process for all but the smallest of programs. Programming at the level of machine code was used in the early days of computing when software development tools were rare or nonexistent, but it is now confined to cases where there is a need to patch short test routines directly into store.[2]

Programming using an assembler allows the use of mnemonic forms of machine instructions, and is the next step up from programming at machine code level. Most assemblers permit symbolic names for jump addresses, labels and constants, as well as instruction op codes. In addition, there is usually provision for management of the code and data areas of store. Further improvement in programming efficiency is achieved with a high level language, and this aspect will be discussed shortly. What we do here is to examine the various types of assembler package, and give some idea of the cases when programming in assembler is to be preferred to a high-level language. No discussion will be given of the somewhat specialized question of how to actually *write* an assembler program.

6.2.1 High-level languages and assembler compared

Before comparing one high-level language against another, it is appropriate first, perhaps, to reflect on the reasons for using a high-level language as an alternative to assembly language. Essentially, programming in assembler or machine code involves the programmer in managing the resources of a computer at the level of registers and physical store locations, and this implies considerable knowledge of the computer's architecture. A compiler will automate many of the functions carried out by an assembler programmer, making the nature of the detailed machine architecture largely transparent at the programming level, and many high-level language programs can be used on a variety of hardware platforms. Programming in assembler in fact is only done if the factors discussed a little later are of importance.

Productivity in terms of code produced in a given time is comparatively low with assembler. Commonly it will take at least ten times as long to code and test a program written in assembler in comparison to an equivalent high-level language program. Just as important is the fact that a program written in a high-level language is much easier to understand than a comparable assembler program. Since with most programs the bulk of the programming is carried out after the program is first installed, i.e. when the program is being maintained, this latter point is vital, particularly so since the programmer maintaining a program will likely be different from the individual who initially wrote it.

The advantages of programming at assembler level can be summarized as follows:

---

[2] Chapter 14 contains a discussion on testing methods.

- Access to hardware registers, ports and physical store locations becomes possible, enabling a freer choice to be made in the storage of data of various kinds. For instance a variable can be stored in a register rather than in memory in the interests of run-time efficiency. As we have noted in Chapter 3, C is an exception, where there is in fact provision for using registers to store variables.
- Closer control over the interrupt facilities. If it is obligatory to carry out operations such as disabling interrupts, then, with a program written in a high-level language, it would be necessary to make provision to embed a machine code routine for this purpose in the rest of the code. PL/M is a notable exception having special instructions to deal with interrupts. (PL/M is briefly discussed later.)
- Control over operand type and size. One can opt to use say 8-, 16- or 32-bit integers, signed or unsigned, depending on the needs of the application, whereas with some high-level languages signed 32-bit integers may be the only type of integer allowed. Similar control is possible for other types of operand. However, note the remarks on operand type and size relating to C in Chapter 3.
- Choice of address mode is possible with assembler, but seldom with compilers. In the interests of run-time efficiency it may be desirable to choose a particular address mode different from that which a compiler would have selected for a comparable problem.

These factors can be of considerable practical importance in some applications, but must be balanced against the obvious advantages of high-level languages in terms of the much greater productivity in writing code. Although it is difficult to give a general rule, assembler should usually be reserved for short routines where one or more of the features listed above is of importance. Note that many high-level languages make provision for the inclusion of assembler routines, and this feature can be of particular importance in real-time programming in code such as interrupt service routines, where run-time efficiency is critical. Operating systems contain code such as this and usually make use of code written in assembler to some degree.

6.3 SOME SPECIFIC LANGUAGES—A BRIEF SURVEY

6.3.1 BASIC

BASIC (Borland, 1988) is a language that was first introduced in 1965 at Dartmouth College, New Hampshire when it was envisaged that it would be used mainly for teaching programming, rather than as a production language. When it was first introduced there were few other high-level languages available, FORTRAN being the only other alternative of comparable power. One of the main attractions of the language is that it is very easy to learn, and is readily available on personal computers. As a language it could be said to be primitive, with no built-in support for structured programming as found in languages such as Pascal. BASIC is available in interpretive form, where translation occurs at run-time and is thus comparatively slow, as well as in compiled form. Later versions of BASIC have attempted to enhance the language, notably to give it more structured features (Borland, 1988). Standard versions of BASIC lacked features such as labels and recursive use of subroutines. Standardization is not a strong feature of BASIC, there are literally hundreds of versions, making for extreme difficulty in porting programs from one machine to another. Hewlett-Packard have introduced a version of the language called 'Real-time Basic' which has facilities for interrupts, timers and many other very useful features for use in real-time work.

6.3.2 Ada

Ada (Barnes, 1989) is a language designed originally for use by the US Department of Defense (DOD). It was intended for use by the DOD mainly, although not exclusively, as a language for real-time embedded systems, and intended to replace the plethora of languages used for such work. It has now gained widespread acceptance both in the USA and elsewhere, notably in the military sphere, although its use has spread outside its initial intended application area (Myers, 1987). Ada has been designed from the ground up as a language which embodies the needs of software engineering, with good support for structured programming techniques, strong typing, features to enable recovery from run-time errors, and the ability to produce code in the form of separately compiled modules. There is, as there has to be for work with real-time systems, built-in support for concurrent programming, and this is discussed in some detail later.

6.3.3 Modula-2

Some readers may have graduated to this book from a background of Modula-2. However, for the benefit of those that have not, and might instead be familiar with Pascal, Modula-2 can be viewed essentially as an improved version of Pascal. Nicklaus Wirth, the inventor of Pascal, intended the language as a teaching tool rather than as a production language, although subsequent usage after its introduction in 1968 led to its widespread use in commerce and industry. This can, to some degree, also be said about BASIC. Nevertheless there are certain shortcomings in the language, and Wirth addressed these problems when he started on the design of Modula-2 in 1977. Currently Modula-2 is gaining widespread use in educational institutions, largely at the expense of Pascal. The syntax of the two languages is very similar, and a Pascal programmer should find it easy to adapt to Modula-2. A good introduction to Modula-2 is to be found in Welsh and Elder (1987).

Programmers moving from Pascal to Modula-2 would note a number of useful improvements in the detailed syntax of the language, but perhaps the most striking feature is the support it offers for producing large-scale software. As with Ada, programs can be divided into separately compiled *modules*—hence the name Modula-2. A specific example illustrating the module concept is given later. Concurrency is also supported in Modula-2, and definitions of types that support primitive operations involving concurrency are provided in a library module.[3] The operations provided allow parallel tasks to be started and executed, and also to exchange data securely.

6.3.4 FORTH

FORTH (Brodie, 1986; Winfield, 1983) is a language used mainly in the development of microprocessor software. The design requirement was for a language that could produce extremely efficient code and could moreover produce code that was compact, this being of

---

[3] Concurrency issues are also discussed in Chapter 9.

importance in small microcontroller applications where memory space is limited. This in fact is the kind of situation where assembler might be thought necessary, and FORTH could be regarded as a language that combines the efficiency of assembler with some of the advantages of high-level languages in terms of effective program development. Like BASIC, FORTH can be used in interpretive mode, which is an important feature in program debugging. When used in compiled mode, it is possible to compile portions of a program rather than the complete program, and this enables the programmer to compile and debug individual parts of a program, with further parts perhaps being tested in interpreted mode, with no need to recompile or relink the remaining tested code. Program development can thus be much more interactive than with other languages.

6.3.5 PL/M (= Programming Language/Microprocessors)

PL/M (McCracken, 1978) is a language developed by Intel specifically for use with their 8080 and 8086 families of microprocessors. (There are two variants of PL/M for these two different chip families.) PL/M has the structured programming constructs and data typing one might expect in a high-level language, but is relatively low-level compared to languages such as Pascal. It has support for the interrupt facilities on the 8080 or 8086 microprocessors, which makes it unusual for a high-level language, and is the advantage gained by being targeted at a specific computer architecture.

6.4 LANGUAGE DESIGN

We examine here attributes of languages that are needed to support the aims of properly engineered software, including a brief review of the principles of structured programming and strong typing. Important areas are the topics of information hiding, real-time provision and exception processing, which are discussed in Secs 6.5 to 6.7.

6.4.1 The needs of structured programming

To solve any problem there must be some way of enabling the three basic operations of sequencing, looping and alternation. A language that supports structured programming principles will provide considerable assistance to programmers in the production of robust and maintainable code, while with primitive languages (notably assembler) the programmer has more responsibility for structuring a program, leading to greater time and effort on his or her part and greater opportunity for the introduction of errors.

In the syntax of Pascal, the three operation types described above can be written as:

1. Sequencing

```
begin
statement1 ;
statement2 ;
...
end ;
```

2. Looping

```
while <boolean condition> do
   begin
   statement1 ;
   statement2 ;
   ...
   end ;
```

3. Alternation

```
case <expression> of
   <value1> : <statement1> ;
   <value2> : <statement2> ;
   ...
end ;
```

The three operations contain statements which themselves would be structured in one of these three different forms; and a principle of structured programming is that a complete program can be built up of one or more nested constructs of this type, with no transfer of control being made from one level of nesting to another. The concept applies in a way that is independent of language so that, in principle, structured programming can be applied to assembler as well as to languages like Pascal. Clearly, when programming in assembler, machine code instructions in the form of unconditional and conditional jumps would be used. With structured programming they would be used in a highly disciplined way so as to obtain the overall nested structure described.

6.4.2 Data typing

A computer program needs to model real-world data. This varies widely, and can be in the form of floating-point numbers, signed integers, characters, strings and other types. The facility of a language to deal with data in these various forms will determine the ease with which a problem can be mapped into a program that represents it. Assembler languages deal with data in the form of information in registers and store locations such as bytes or words—in other words signed or unsigned integers. Assembler programmers are only too well aware of the difficulty in translating data at this level into (say) a floating point word. All high-level languages provide some support for describing data at a level higher than the machine code level. Integers, floating point numbers and characters are supported by most languages, but data at even this level can be regarded as primitive by some standards. For instance, if we need to have a program respond to decisions based on the day of the week, in a FORTRAN program we would be reduced to having some kind of alternation construct in which days were represented (say) by numerical values leading to tricky, error prone and difficult-to-read code. Strongly typed languages such as Pascal aim to enable programmers to devise data of types suited to the problem in hand, possibly using the primitive types provided as part of the language. In the specific example just quoted, an enumerated type would be appropriate:

```
VAR
    Day = ( sun,mon,tues,wed,thurs,fri,sat ) ;
```

The variable `Day` could be used in code such as:

```
Day := mon ;
case day of
   mon  : <action for monday> ;
   tues : <action for tuesday> ;
    ..
    ..
end ;
```

Values of `Day` other than those defined in the declaration would not be recognized, and would be interpreted by the compiler as a bug. Had we been using some other representation of 'days' such as integers, the compiler would not have been able to detect invalid data in this way.

6.5 MODULES AND INFORMATION HIDING

In the implementation of an abstract data type for queues in Chapter 5, two different approaches were examined. Comparisons between implementations may be made in terms of factors such as code efficiency, but the key point which should have emerged by now is that both have been designed to meet the same specification; the interface to an application program in terms of the procedure headings is identical. The internal composition of the access procedure bodies is not apparent to a program that uses the access procedures as 'black boxes'. This approach to program design, called **information hiding** (Parnas, 1972), allows changes to access procedures to be made with no observable effect on a program that uses them, and is of great value in the maintenance phase of an application. Provision of a *correct*, *robust* and *properly tested* set of access procedures simplifies program design and gives greater assurance of reliability. The advantages of information hiding become weakened if it is possible for programmers writing an application program to be able to make changes to access procedures that have already been carefully tested to meet their specification. The ideal is, therefore, to have provision in a language to allow a program to be prepared as a collection of separately prepared and compiled modules, one of which could comprise a set of access procedures, such that application programmers only have access to the source code they themselves produce. Pascal falls some way short in this respect since it is possible to view the source code for the access procedures. One version of Pascal allows for the compilation of programs as a set of separate modules (Prospero Software, 1987) but this does not form part of the latest ANSI Pascal standard. In any event, even with this particular compiler, the structure of the abstract data type is not completely hidden since the associated TYPE declarations need to be included in a program and are thus visible. What is needed is that provision for information hiding is built into a language. Modula-2 and Ada are examples of languages that have just such a provision.

6.5.1 Information hiding in Modula-2

In Modula-2, programs are prepared as a set of **Modules**, which can be written and compiled independently of each other. Use of variables, declarations and procedures involving different modules can be placed under tight control by statements called IMPORT and EXPORT. EXPORT lists those objects that are available for use by modules other than the particular module that defines them. IMPORT lists those objects that a particular module needs, and which are defined inside other modules. Thus in the case of the access procedures discussed in Chapter 4, one module would include the procedure bodies, and would list the names of the access procedures with an EXPORT statement. Modules that wanted to use one or more of the access procedures would include an IMPORT statement listing the access procedures required. Objects not listed by IMPORT or EXPORT statements would not be *visible* to other modules. Modules are split into two parts, one, the **implementation module**, gives procedure bodies and CONST, TYPE and VAR declarations. This module is the one to be compiled and forms one of possibly several library routines to be later combined with the object code generated by other modules that need to use them. The other part, called the **definition module**, is a list of the names of objects defined in the associated implementation module. In the case of the earlier abstract data type for queues, the implementation module might appear as shown below. The strong resemblance between the syntax of Modula-2 and Pascal will be noticed. There are many differences in the two languages, but these will not be elaborated on here, other than those concerned with the module concept. However, note that a FUNCTION in Pascal is referred to in Modula-2 as a procedure.

```
DEFINITION MODULE QueueADT;

    CONST
        MaxQLength ;
    TYPE
        QueueStatus ;
        QueueNode ;
        Queue ;

procedure CreateQ( VAR Q : Queue ) ;
procedure  EmptyQ( VAR Q : Queue ) : BOOLEAN ;
procedure InsertQ( VAR Q : Queue; qdat : QueueElement ) ;
procedure LengthQ( Q : Queue ) : INTEGER ;
procedure RemoveQ( VAR Q         : Queue;
                   VAR qdat       : QueueElement ;
                   VAR Status     : QueueStatus ) ;
procedure FrontQ( Q              : Queue;
                  VAR FrontOfQ   : QueueElement;
                  VAR Status     : QueueStatus ) ;
end QueueADT.
```

A module that needed to utilize the queue access procedures would employ the information in the definition module properly to interpret procedure parameters, this being the mechanism by which the position and nature of parameters is made to agree between the implementation module and other modules using the access procedures.

The definition of the type `QueueElement` would best appear in the module associated with the application program and not in the present module, which should be as general as possible in its scope. `QueueADT` would therefore need to import this definition. A possible implementation module is:

```
(*
   Implementation Module for queue abstract data
   type using arrays
*)
MODULE  QueueADT ;
   IMPORT   QueueElement ;
   EXPORT   CreateQ, FrontQ, EmptyQ, LengthQ, RemoveQ ;
CONST
   MaxQLength = 500 ;
TYPE
   QueueStatus = ( Normal, Empty, Full ) ;
   Queue       = RECORD
      Contents    : ARRAY[0..MaxQLength] OF QueueElement ;
      Front, Tail : 0..MaxQLength
   end ;
(*
 Modula-2 code corresponding to the Pascal code
  for the access procedure bodies would go here
*)
begin
(* empty! *)
end QueueADT.
```

The structure of a module needing to use the queue access procedures would be as follows:

```
MODULE QueueUser;
FROM QueueADT IMPORT Creatq, EmptyQ, InsertQ, RemoveQ, FrontQ,
                     QueueStatus, QueueNode, Queue, MaxQLength;
  (* other IMPORT and EXPORT statements here *)
  (* CONST, TYPE, VAR and procedure Declarations of objects
     local to the module here *)
begin
(* Module code here *)
end QueueUser.
```

The list following **EXPORT** in the implementation module comprises the names of the procedure headings that are associated with the access procedures that appear inside the module, and to which access is to be permitted by other modules. Other modules would be able to use them by including an appropriate **IMPORT** statement.

6.5.2 Information hiding in Ada

The concept of dividing a module into two parts as in Modula-2 is also a feature in Ada (Barnes, 1989). The term **package** is used in Ada for what would have been a module in Modula-2. The interface to the outside world is called the **package specification** while implementation details are in a **package body**. An **Ada program** consists of package specifications and package bodies. One of the package specification units (the main master program) is in the form of a procedure. Although there are detailed differences between Modula-2 and Ada in relation to producing programs in modular form, the general philosophy of the language design in this respect is similar for both.

6.6 REAL-TIME PROGRAMMING LANGUAGES

6.6.1 Introduction

The programming needs of real-time programming have already been referred to in Sec. 6.1. To recall: the main aspects are concerned with the need to support concurrently executing programs together with tackling the problem of generating code that satisfies run-time constraints. Similar concepts feature in Chapter 9 in relation to multiprogrammed operating systems, and before reading this section it is suggested that Chapter 9 is read first. It might be added that the discussion here encompasses real-time systems in more general terms, operating systems being but one instance of a real-time system. Languages that support concurrency are used in uniprocessor machines, but much of the performance gain possible in concurrent systems is obtained in multiprocessor systems. Embedded type systems are the most important application area here. The increasing use of systems has resulted in the more active interest in concurrent languages.

Many specialized languages have been developed for real-time work, and no attempt is made here to give a comprehensive discussion of real-time languages. We concentrate on Modula-2 and Ada, which were both designed with the needs of real-time systems in mind from their inception. The same cannot be said for some other so-called real-time languages. Some languages provide concurrency with each task structured as a separate program. The complete assembly of programs would then execute concurrently with the aid of a multi-tasking operating system. This approach entailed relatively low-level system calls into the operating system. The whole approach is error prone since the true concurrent nature of the programs is not known at compile time, so the compiler is not capable of carrying out checks on aspects relating to intertask communication and synchronization.

6.6.2 Concurrency in Ada

The rendezvous as a task communication/synchronization mechanism In Ada, the concept of a task is used to specify a program unit that can execute concurrently with other tasks (Barnes, 1989). Tasks can be completely and independently executing programs that run concurrently (or pseudoconcurrently) without interacting with each other. More commonly, however, tasks need to interchange data and synchronize with each other. Ada has two types of provision for this. One method is by the use of shared global variables, the other by a technique called **Rendezvous** which encompasses the requirements of mutual exclusion, task synchronization and intertask communication. Rendezvous can in some respects be compared to the message passing technique for intertask communication discussed in Chapter 9. A rendezvous involves a **caller** task and a **server** task. The caller communicates with a server by an **entry** statement—a form of procedure call. The server uses an **accept** call to 'accept' the call from the sender. If the entry occurs before the corresponding accept, the caller task blocks until the occurrence of the accept. On the other hand, if the accept occurs before the corresponding entry, the server suspends itself until the occurrence of the entry. When a call has been accepted the rendezvous has occurred. Data can be transferred from a caller to a server following a rendezvous, and also data can be returned from the server to the caller at this time, the rendezvous primitives having provision for input and output parameters as in Pascal. There is no buffering between the two tasks involved with the rendezvous mechanism, so they will run in lockstep with each other. The server can accept calls from more than one caller, which is useful in situations where there needs to be provision for tasks such as print servers to be accessed from several sources. Only one caller can rendezvous with a given server at any given time thus ensuring proper mutual exclusion. After a rendezvous with one caller, other callers need to wait, and are serviced on a first-come-first-served basis.

Declaration of tasks in Ada Tasks in Ada must be declared inside a procedure or other program unit. The actual code for individual tasks is split into two parts: the specification and the body, both of which should be inside the parent program unit. The specification will identify the task name and the declaration of entries used to communicate with the task if any. The task body will comprise local declarations followed by the actual executable code. The overall structure can thus be written as in this (skeleton) example (note that lines starting with '--' are comments):

```
task DEMO_TASK is
   -- declaration of entries to DEMO_TASK
   entry  DEMO_ENTRIES( <parameter list> ) ;
end  DEMO_TASK ;

task body DEMO_TASK is
   -- local declarations here (not visible to other program
        units)
begin
  -- imperative part of task here
     accept( <parameter list> ) ;
end DEMO_TASK ;
```

A procedure could include declarations for several tasks declared in this way, and would have the typical structure given below:

```
procedure  DO_TASKS is
   -- list if task declarations, declared as above in DEMO_TASK
   -- list of task bodies as quoted above in DEMO_TASK
begin
   -- imperative part of main procedure (tasks are now active)
end DO_TASKS ;
```

The tasks would all become active together, when the program reaches the imperative part of the parent procedure, i.e. following the `begin`. The code for the main procedure and the tasks then runs concurrently until the program is allowed to reach the `end` statement for the main program. Note that the procedure would wait until *all* tasks had terminated before *it* can terminate, and a return be made to whichever program unit called the procedure.

Illustration of the application of rendezvous A simple example will illustrate the use of the rendezvous. Three tasks, `TASK_A`, `TASK_B` and `SERV_AB`, are to run concurrently. `TASK_A` and `TASK_B` produce data which `SERV_AB` uses; following a rendezvous `TASK_A` or `TASK_B` then proceed to do further computation. `SERV_AB` clearly must not start until `TASK_A` or `TASK_B` has made some data available for it. This process is to occur continuously, a situation typical of many real-time systems. The timing involved is illustrated in Fig. 6.1, where the solid horizontal lines correspond to a task carrying out some processing, while the broken lines correspond to the period for which a task has finished doing some computation and is waiting to rendezvous with another task. Note the way `TASK_B` and `TASK_A` are queued when both are waiting to rendezvous with `SERV_AB`, waiting for it to reach the accept.

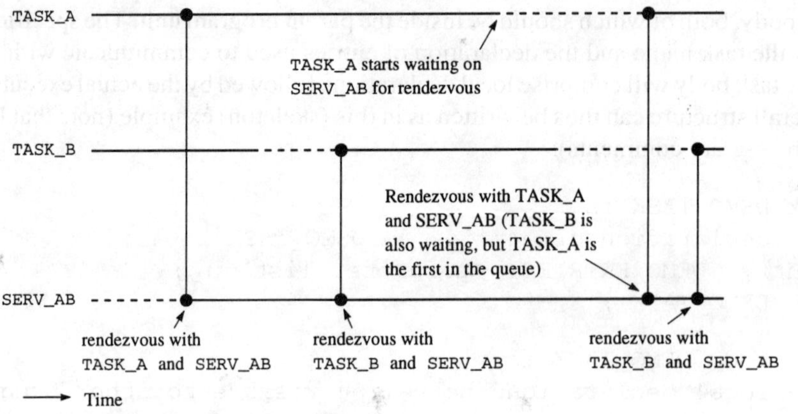

Figure 6.1 Demonstration of rendezvous of two tasks.

The code below just gives the structure of the program in outline. Additionally, further declarations would be required; notably for `DAT_TYP` which corresponds to the data being communicated between the various tasks. Note also that, in Ada, input and output formal parameters for entry and accept calls are distinguished by the key words in and out.

```
procedure SYNCH_DEMO is

    task TASK_A ;

    task TASK_B ;
```

```
    task SERV_AB is
        entry GEN_DAT( DAT : in DAT_TYP ) ;
    end SERV_AB ;

    task body TASK_A is
        loop
            -- generate data DAT for SERV_AB
            GET_DAT( DAT ) ;
            -- do some more processing, but only after rendezvous
        endloop ;
    end TASK_A ;

    task body TASK_B is
        loop
            -- generate data DAT for SERV_AB
            GET_DAT( DAT ) ;
            -- do some more processing, but only after rendezvous
            endloop ;
    task body SERV_AB is
        loop
            -- wait here for data to arrive from TASK_A or TASK_B
            accept GEN_DAT( DAT : in DAT_TYP ) do
                -- process data from TASK_A or TASK_B
            end GEN_DAT ;
        endloop ;
    end SERV_AB ;


begin
    -- main procedure together with SERV_AB, TASK_A & TASK_B all
    -- start up together for first time when this point reached
end SYNCH_DEMO ;
```

This example shows the way Ada treats concurrent tasks and the rendezvous principle in a comparatively simple form. There is provision, however, for more versatile handling of concurrent tasks. In the example, TASK_A and TASK_B are 'serviced' by SERV_AB in the order in which they occur in time. However, any real-time system needs provision for other aspects of task management, notably control over priorities between tasks. We have noted that, in the form described, Ada treats all tasks equally. It is possible in fact, by using a special form of task declaration, to give tasks any required priority, in which case a higher-priority task will always pre-empt a task of lower priority. Other provisions in Ada allow for suspending a task for some set interval, and allowing a task to select from one of several possible rendezvous. There is also provision for dynamic creation and termination of tasks, where it is not convenient to specify some fixed number of tasks at compile time. A comprehensive discussion of those aspects of Ada relating to tasking can be found in Barnes (1989), and this section should be regarded as a somewhat brief introduction.

6.7 DEALING WITH EXCEPTIONAL EVENTS

It is a common experience in executing programs that some exceptional events will occur, such as an attempt to divide by zero, making a normal continuation of the program impossible. Other examples of exceptions are variables going outside allowed bounds set for them, or an attempt to use more than the maximum amount of store allocated to a program. Some programs make no attempt to recover from an exception and, in cases like the divide by zero example quoted, the net effect would be for the program to terminate, simply displaying the reason for the program terminating with an appropriate message. Evidently some support is required here from the operating system. In many instances, however, the programmer may choose to detect certain types of exception and attempt a recovery procedure of some kind, for example typing invalid data in response to a prompt to input data to a program code to check the data and prompting for re-entry in the case of bad data. However, many types of exception would be difficult or even impossible to detect and recover from in this way. In the case of real-time systems, mechanisms for recovery from exceptions is essential if the system is to be prevented from failing, possibly with catastrophic, even life-threatening, consequences. It is this last aspect particularly which has led to the provision of exception handling as an integral part of some languages, Ada being a notable example.

In languages with provision for exception handling, an exception event will have the effect of transferring control to a routine called an **exception handler**. In Ada, exceptions are associated with program fragments such as that shown below. These contain statements that might give rise to some exception, together with, following the key word EXCEPTION, code to respond to any exceptions. The occurrence (raising) of an exception has the effect of setting predefined flags such as NUMERIC_ERROR. Note the similarity to a CASE statement, although there are important differences.

```
begin
  statement1 ;
  statement2 ;
EXCEPTION
  WHEN NUMERIC_ERROR =>
  -- statements to respond to numeric errors ;
  WHEN CONSTRAINT_ERROR =>
  -- statements to respond to constraint errors ;
end ;
```

Exceptions can be user defined, or predefined. Examples of commonly used types of predefined exceptions are:

- **CONSTRAINT_ERROR**—raised when a variable goes out of range. For example, a variable MyVAR defined by VAR MyVar : 1..10; would raise CONSTRAINT_ERROR if it were given a value greater than ten or less than zero. As with other types of exception, this is an instance of a problem that cannot be detected at compile time.

- **NUMERIC_ERROR**—an attempt to divide by zero would be a typical example of this type of exception.
- **STORAGE_ERROR**—raised if a program attempts to run out of storage space.
- **DATA_ERROR**—raised if a letter character is typed in response to `get (x)` when a number character is expected. The procedure `get` reads characters, it is roughly similar to its Pascal counterpart.
- **END_ERROR**—raised when an attempt is made to read past the end of a file.

This discussion of exception processing is, of necessity, rather brief, and Barnes (1989) should be consulted for a more comprehensive treatment. A discussion of some novel exception handling methods in C is given in Amsterdam (1991)—it should be mentioned that, in the case of C, provision for exception handling does not form a strong integral part of the language as it does in Ada, so that programmers need to put more time and effort into writing exception handling code.

6.8 EXERCISES AND REVIEW QUESTIONS

1. List any applications you know for which the coding needs to be done partly or entirely in assembler.
2. Summarize the improvements made in Modula-2, in comparison with Pascal, relating to information hiding. How would you rate FORTRAN in this respect?
3. Describe the respective roles of *caller* and *server* in the context of a rendezvous in Ada.
4. Extend the example used to demonstrate the rendezvous principle in Sec. 6.6.2 as follows: `SERV_AB` is to receive data from `TASK_A` and `TASK_B` as before, but, following the initial rendezvous, is now to rendezvous with `TASK_A` and return some data to it. `TASK_A` is not to transmit further data to `SERV_AB` until this has occurred.
5. Write a Modula-2 definition module suitable for the access procedures that define a stack as described in Sec. 5.6. As a further exercise you might try adapting the access procedures themselves such that they can be used as a Modula-2 implementation module. This would require some background reading in Modula-2 because there are differences in syntax compared to Pascal.

REFERENCES AND FURTHER READING

Amsterdam, J. (1991) 'Taking exception to C', *Byte*, vol. 18, August.
Barnes, J. G. P. (1989) *Programming in Ada*, 3rd ed., Addison-Wesley, Wokingham.
Baron, N. S. (1986) *Computer Languages—A Guide for the Perplexed,* Penguin Books, Harmondsworth, Middlesex.
Borland (1988) *Turbo BASIC Owners Handbook.*
Brodie, L. (1986) *Starting Forth*, 2nd ed., Prentice-Hall, Englewood Cliffs, New Jersey.
McCracken, D. D. (1978) *A Guide to PL/M Programming for Microcomputer Applications*, Addison-Wesley, Reading, Mass.
Myers, W. (1987) 'Ada: first users—pleased; prospective users still hesitant', *IEEE Computer*, March, pp. 68–73.

Parnas, D. L. (1972) 'On criteria to be used in partitioning systems into modules', *CACM*, vol. 14, no. 1, April, pp. 221–227.
Prospero Software (1987), *Prospero Pascal Manual*, London.
Welsh, J. and Elder, J. (1987) *Introduction to Modula-2*, Prentice-Hall, London.
Wilson, L. B. and Clark, R. G. (1988) *Comparative Programming Languages*, Addison-Wesley, Reading, Mass.
Winfield, A. (1983) *The Complete Forth*, Sigma Press.

Part III

Systems software

7

TRANSLATORS, LINKERS AND LOADERS

7.1 INTRODUCTION

Chapter 6 gave a comparison between various kinds of languages, the emphasis there being on choice of language to suit a particular application. This chapter examines more closely the mechanics of the processing of programs: from the stage following writing the source code, to the point at which a program is in a form suitable for loading into a computer's memory and then actually executing (see Fig. 7.1). This includes the process of compilation (or assembly), and relevant features of compilers and assemblers are in fact the main topics discussed. Compilation (or assembly) is only the first of several transformations that the source code for a program needs to go through before it is actually executed. The second stage involves combining **run-time library modules**, such as standard mathematical routines or I/O routines, with the binary module derived from the user source code in a process called **linking**. The resulting module is finally processed by the program responsible for loading the program into the computer's memory. Further transformation may be needed at this stage since, with some computers, the loader program has to fit programs into the space available, and will need to modify program addresses accordingly. Where necessary, concepts are illustrated using programming examples, which for convenience will be in C rather than Pascal, which is used in most other chapters in the book.

7.2 ASSEMBLERS

Recall from Chapter 5 that assemblers are translator programs that accept, as input, source code comprising instructions in mnemonic form, and transform it into the binary codes that the CPU actually executes. Most assemblers go much further than a simple substitution of a

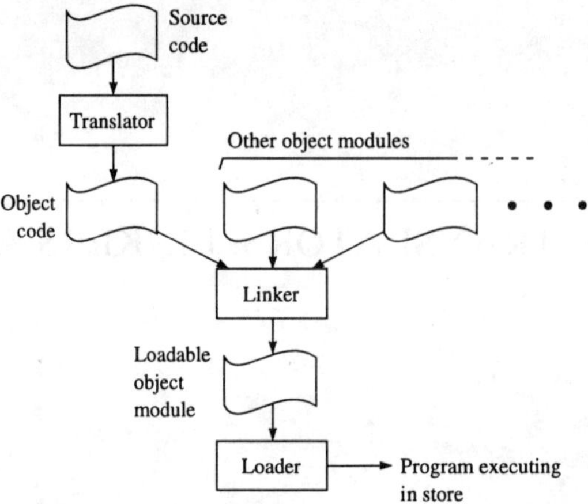

Figure 7.1 Processing source code for execution.

mnemonic name for a numerical code for instructions, however, and the general aim is to approach, albeit to a limited degree, the ease of use and gain in productivity characteristic of a high-level language. Symbolic names for jump addresses, labels and constants are the norm in all but the simplest assemblers; having to calculate numerical values for jump addresses for instance would be particularly annoying in programs of any length. Additionally, there is usually provision for management of the code and data areas of store. This is useful when programs are being prepared as a number of modules which eventually need to be combined; there is need to make use of store consistently, making sure that the memory space for code and data does not overlap for example. A variety of different types of assembler are available which are enumerated in this section and critically compared. No attempt will be made to discuss the somewhat specialized question of how to actually *write* an assembler however.

7.2.1 General features of assemblers

Assembler instructions are generally formatted in four fields as in the 68000 fragment below:

label field	mnemonic field	operand field	comment field
	. . .		
	. . .		
LOOP	MOVE.L	(A0),D0	GET NEXT NUMBER FROM STORE
	ADD.L	D0,(A0)+	DOUBLE IT
	SUBQ	#1,D1	DECREMENT LOOP COUNTER
	BNE.W	LOOP	LOOP UNTIL COUNTER ZERO
	. . .		
	. . .		

The label field for the first instruction here contains the label LOOP, but later we will see that it is possible for different information to appear in the 'label' field. The mnemonic field contains the instruction op code in mnemonic form. The operand field contains either data or addresses specified in one of the 68000's address modes. The comment field is optional, but very necessary in practice if a program is to be properly documented! Assemblers differ as to how the various fields are distinguished; some insist that a particular field must start at a specified column, others allow for a delimiting character such as a space or tab between the fields, some use a ';' ahead of the comment field, and some use a '*' character before a complete line of comment. In the interests of legibility, it is good practice to align the first column of the four fields as shown above.

7.2.2 Assembler directives

Assembler directives—sometimes called **pseudo ops**—are instructions to the assembler, and do not in themselves generate object code. An obvious need for a directive is in providing some means of specifying the start address of a program. The directive called ORG in many assemblers is used for this, for example the line ORG $80000 sets the address of the instruction in the following line to $80000, (ORG appears in the mnemonic field). Other directives are also used in specifying constants, reserving areas of store for data, and a number of formatting and housekeeping functions. Some of the more commonly used directives are listed below. Names for directives and other conventions differ from one assembler to another, and the directives described here apply to the FORCE 68000XPC macro cross-assembler (Flight Electronics, 1990):

- **DC** (= 'Define Constant') defines a data constant to be stored in specified memory locations. The constants themselves may be quoted as decimal, hexadecimal, ASCII characters, or as ASCII strings. One line can define a number of constants in which case they would be separated by ',' characters. In 68000 assembler constants may be aligned on a byte, word or long word boundary and the directive has provision for specifying the size of the constant, with either .B or .W or .L appended to the DC directive. The examples below should make the usage of the DC directive clear, note that labels can be used as part of the directive—this is useful in accessing data such as tables:

```
TABL    DC.W        $1200,$3400,$5600,$00
NUM     DC.L        $20000000
LOGON   DC.B        'hit carriage return to start','0'
```

- **DS** (= 'Define Storage') reserves blocks of store, typically being used to delineate workspace for data generated by a program. There are three variants corresponding to bytes, words, or long words: either .B or .W or .L are appended to the directive as for the **DC** directive. The first example below reserves $10 bytes of storage starting at location MSG. The second example reserves $1000 long words starting at WSPCE:

```
MSG     DS.B        $10
WSPCE   DS.L        $1000
```

- **EQU** (think of this as 'EQUate') allows symbols to be used in place of numerical values in code in the interests of clarifying its meaning, and has a comparable role to a CONST declaration in Pascal or a define declaration in C. Values may be specified in decimal, hexadecimal, binary or ASCII format with most assemblers. The symbol defined by an EQU directive appears in the label field as in the examples below:

```
COUNT   EQU        10
ADDRS1  EQU        $80000
```

7.2.3 Assembler expressions

It is possible to use varying forms of expression in the operand field of instructions, in the interests of clarifying the meaning of code. The expressions would be evaluated to numerical values by the assembler. In the 68000XPC assembler: arithmetic, shift and Boolean operators are allowed in expressions, and operands can be expressed in decimal, octal, hexadecimal, binary or ASCII formats. The examples below give some uses of various types of expression:

1. `MOVE.B   BASE+OFFSET,D0` moves the contents of the location found as `BASE+OFFSET` into D0. `BASE` and `OFFSET` would need to be defined, either by an EQU directive or by virtue of being a label.
2. `NUMA    EQU    'A'&00001111B` EQUates the constant NUMA to the logical AND of the ASCII code for 'A', i.e. 1.

7.2.4 Example of a complete 68000 assembler program

This short program demonstrates the use of the assembler directives discussed and other aspects of assembler code. The program simply displays a message on the screen with a time delay between consecutive characters. The message is stored in memory. The program ends when the message has been completely displayed. The program is written in such a way that it could be directly loaded into a Flight 68K single board computer (Flight Electronics, 1986). In particular it makes use of a monitor routine for displaying an ASCII character. It is assumed that provision has been made to return control to the monitor program when the program has finished. What would happen were this *not* done?

```
DTIME   EQU        10000         LOOP COUNT FOR TIME DELAY
        ORG        $400400
        SYM                      GENERATE SYMBOL TABLE IN LISTING
        LEA        #MESSAG,A6     POINT TO START OF MESSAGE
LOOP    MOVE.B     (A6)+,D0       GET NEXT CHARACTER FROM TABLE
        CMPI.B     #0,D0          TEST FOR TERMINATOR (NULL)
        BEQ        EXIT
        TRAP       #11            CALL MONITOR DISPLAY ROUTINE
        DC.W       6              (DISPLAYS CHARACTER IN D0.B)
        MOVE.L     #DTIME,D0      DELAY BEFORE DISPLAYING
                                  NEXT CHARACTER
```

```
DELAY  SUBI        #1,D0
       BNE         DELAY
       BRA         LOOP        PROCESS NEXT CHARACTER
EXIT   NOP
MESSAG DC.B        'HELLO WORLD',0
       END
```

The code below is a listing file produced by the assembler. This comprises the original source code, together with the address and object code for each instruction in hexadecimal—to the right of the line numbers. Notice that a list of all symbols together with their values is generated by the assembler; the directive SYM will make the assembler do this in the present case.

```
PAGE  001  book1.asm                     Fri Sep 13 20:02:39 1991

  1 00002710                    dtime   equ    10000          loop count for delay
  2 00400400                            org    $400400
  3                                     sym                   generate symbol table in listing
  4 00400400 2C7C00400424                lea    #messag,a6    point to start of table
  5 00400406 101E               loop    move.b  (a6)+,d0      get next character from table
  6 00400408 0C400000                   cmpi.b  #0,d0         test for terminator (NULL)
  7 0040040C 67000014                   beq    exit
  8 00400410 4E4B                        trap   #11           call monitor display routine
  9 00400412 0006                        dc.w   6             (displays ASCII in D0.B)
 10 00400414 203C00002710               move.l  #dtime,d0     delay before displaying
 11                                                           next character
 12 0040041A 04400001          delay    subi   #1,d0
 13 0040041E 66FA                        bne    delay
 14 00400420 60E4                        bra    loop          process next character
 15 00400422 4E71              exit      nop
 16 00400424 48656C6C6F20776F726C messag  dc.b   'Hello World',0
    0040042E 6400
 17                                      end
Assembly complete
Total bytes generated: 48
0 errors
0 warnings
Symbol table:

delay   0040041A  dtime   00002710  exit    00400422  loop    00400406
messag  00400424
```
5 symbols.

7.2.5 Assembler types

Two-pass/one-pass assemblers   One of the tasks of an assembler is to translate symbols, such as labels or constants defined by EQU directives, into numerical values which an object program can use. The information needed to do this is contained in the **symbol table**, which comprises all information relating to the symbols used by a program, and will contain fields for the following: the string describing the symbol, its type (label, EQU defined constant, etc.), and another field defining its numerical value, this is an address in the case of a label. **Two-Pass assemblers** make two passes in translating a program, in effect reading the complete program twice. In the first pass, the assembler examines the symbol table for the presence of a symbol in the current instruction being processed, and if it is not already present it follows that the symbol must be new, so the assembler will then insert it into the symbol table.[1] It is in the second pass that the source program is actually translated, and this is accompanied by a translation of symbols into numerical values such as jump addresses. Why take two passes to do the translation rather than one? The answer is to do with cases where there is a **forward reference** to a symbol, as in the 68000 program listed below:

```
        ORG     $80000
        . . .
        . . .
        BEQ     FORWD
        . . .
FORWD . . .
        . . .
```

Here the assembler will try to determine the third field in the symbol table, i.e. the numerical value associated with the symbol, as it processes the source code line by line, but in the case of the label FORWD this will not be possible in the first pass since the point in the program at which the label is located has not yet been reached. The symbol table entry for FORWD will, therefore, be only partially complete and is filled in at the second pass, at which time the value of the jump address can be established. The problem of forward references could also occur with symbols other than labels.

Some assemblers carry out translation in a single pass. In spite of what has just been said, it *is* possible to handle the problem of forward references in a one-pass assembler, but there are restrictions that make the use of one-pass assemblers inconvenient (Calingaert, 1979). **Single-line** assemblers, usually of a somewhat primitive kind, are sometimes included as part of the firmware on some kinds of microcomputer boards used in colleges to teach the principles of microprocessors. Single-line assemblers require code to be entered *directly* into the memory of the computer, translation occurring one instruction at a time. The program starts to execute following a monitor command, which would need to specify the start address. The programmer often has the task of *manually* calculating jump addresses. Assemblers of this type are really only suitable for small programs, typically short pieces of test code to be patched directly into store.

Native mode/cross assemblers **Native mode** assemblers produce object code that can execute on the same type of computer as that on which the assembler program itself is running. All computers need to be provided with native mode assemblers to carry out programming where high-level languages are unsuitable for some reason. Cross assemblers by contrast produce object code for a computer different from that of the computer on which the assembler program runs. Microprocessor development systems[2] use cross assemblers: here a host computer is used for program development for target systems such as single-board computers. Code produced by a cross assembler is downloaded into the

---

[1] The discussion in Chapter 5 on sorting is relevant here.

[2] Microprocessor development systems are discussed in Chapter 14.

target system's memory, or more commonly an EPROM programmer which is then used to prepare the EPROM or other nonvolatile type of memory in which the target system's firmware will eventually reside.

Absolute/relocatable assemblers **Absolute** assemblers read a complete source program, and produce object code in a form that is capable of executing directly after being loaded into store. *All* addresses are determined at assembly time. Preparing code in this way is usually only convenient for producing relatively small programs—debug code for example. The small demonstration program discussed in Sec. 7.2.1 was written in such a form as to be suitable for processing by an absolute assembler, the starting address being specified by the ORG directive, i.e. set at assembly time rather than run time.

It was noted in Chapter 6 that programs of any complexity are invariably developed as a number of modules, often by a team of programmers. Each module is assembled separately, and a complete working program formed in a later **linking** stage,[3] when the binary code of individual modules is combined to form a loadable program. The problem in writing assembler programs in this way is that the absolute addresses of code in an individual module cannot be determined until they are linked. In addition, the problem of allowing modules to exchange data must be addressed. **Relocatable assemblers** are used in these circumstances. This type of assembler produces object code that does not need to contain the final addresses of instructions; calculation of addresses and resolution of symbols global to several modules is deferred until link time. To do this, a package, sometimes called a **linkage editor**, takes as input the object code for all the modules generated by the assembler, and produces as output the code that is the final loadable program.

7.2.6 Macro assembly

There is a common need in most programs for the use of a particular code sequence at several points. The procedure/subroutine is a commonly employed method of avoiding the repetition of the code at each point of use with both high-level languages and assembler, and in addition giving greater clarity to a program. **Macros** are another tool, similar in some respects to procedures, which achieve a related goal. A macro allows the invocation of a code sequence in a program using a **macro call**, this being somewhat comparable to a procedure call. Access to a procedure, however, is by a transfer of control to the procedure body, followed by a return to the statement after the procedure call. There is only one copy of the procedure and control is transferred to it dynamically. This is different from a macro, where each occurrence of a macro call in a program causes the assembler to insert the instructions associated with the macro name. Thus it is possible to have multiple copies of the body of a particular macro in the same program. Macros can be associated with arguments to give control over their action, comparable in some ways to procedure arguments in a high-level language. Macros represent another attempt to give some of the advantages of a high-level language, in terms of providing a means for generating more readable and maintainable source code.

Additional directives are needed with macros, and, in the 68000XPC assembler for example, the directive MACR is used to precede a macro body, with the directive ENDM signifying its end. In the macro body, dummy arguments \0, \1, \2, etc., replace arguments

---

[3]Linkers are discussed further in Sec. 7.5.

in the macro call. The arguments themselves are registers or memory locations. In 68000XPC the macro definition must precede the point at which it is called. The example below is a simple example of the use of a macro, which adds the contents of three memory locations together, returning the result in a register:

```
        ORG   $400400

 * add3 macro definition: adds $400600,$400601,$400602
        add3MACR
            MOVE.B          $400600,D1
            MOVE.B          $400601,D2
            ADD.B           D1,D2
            MOVE.B          $400602,D1
            ADD.B           D1,D2
            MOVE.B          D2,D0       sum returned in parameter 0
            ENDM

 * Typical macro call
        add3D1
        ...
        ...
```

7.2.7 Conditional assembly

Conditional assembly allows sections of code to be included or excluded in assembly of a program depending on directives that are embedded in the code. There are a number of reasons for wanting to do this:

- It might be useful to include debug code in a program while it is being developed, but such code needs to be excluded in a production version of a program.
- Many programs need to be tailored to one of several possible uses, requiring one of several code sequences to appear in the final program, depending on the specialized needs of the application. In particular, I/O requirements commonly differ from one application of a program to another, requiring a selection of one of several different I/O routines. Replacement of a peripheral chip by one of a different type will require changes of this kind for example.

With the 68000XPC assembler, directives controlling conditional assembly include the following:

```
IFEQ    <Expression> = 0
IFNE    <Expression> <> 0
IFGE    <Expression> >= 0
IFGT    <Expression> > 0
IFLE    <Expression> <= 0
IFLT    <Expression> < 0
IFC     <Character String 1> == <Character String 2>
IFNC    <Character String 1> <> <Character String 2>
```

The following is a simple example of the use of one of these directives:

```
PORTA   EQU     1
PORTB   EQU     0
PORT    SET     PORTB
        ORG     $400400
        IFEQ    PORT
        MOVE.B  $800000,D0  READ PORT A
        ENDC
        IFNE    PORT
        MOVE.B  $800002,D0  READ PORT B
        ENDC
```

Hypothetically, this code could, say, be selecting one of two I/O ports accessed at addresses $80000 and $80002, depending on the value of the symbol PORT. The line PORT SET PORTB would then be changed by the programmer, depending on which alternative was required, the remainder of the code remaining invariant. The file listings below corresponding to the previous program shows that on line 11 the object code has been generated from the first of the two MOVE instructions.

```
PAGE  001  COND1.asm                                Sun Sep 15 13:45:25 1991

                 *CONDITIONALS DEMO


    1                                TTL     *CONDITIONALS DEMO
    2                                LIST
    3                                SYM
    4                                UNA
    5                                CLIST
    6 00000000                PORTA  EQU     0
    7 00000001                PORTB  EQU     1
    8 00000000                PORT   SET     PORTA       EDIT TO SELECT PORT A OR PORT B
    9 00400400                       ORG     $400400
   10 00000000                       IFEQ    PORT
   11 00400400 103900800011          MOVE.B  $800011,D0  READ PORT A
   12                                ENDC
   13 00000000                       IFNE    PORT
   14                                MOVE.B  $800013,D0  READ PORT B
   15                         ENDC
   16                                END
Assembly complete
Total bytes generated: 6
0 errors
0 warnings


Symbol table:


PORT   00000000  PORTA  00000000  PORTB  00000001
3 symbols.
```

7.3 COMPILERS

7.3.1 Introduction

A **compiler**, like an assembler, translates a program written in one language to a program expressed in another language. In the case of a compiler the language in which the object program is expressed can be machine code, assembler code or even another high-level language. If in fact the object code is in the form of assembler, then a further assembly stage is needed to produce code that can be processed by the linker. This section examines in some detail the various functions of a compiler. There are different types of compiler just as there are different types of assembler, and these are discussed.

As translator programs, compilers might be expected to share some common features with assemblers, and in fact the material in Sec. 7.2 should prove helpful in understanding some of the principles of compilers. Assemblers are relatively low-level translators; however, compilers are much more sophisticated programs, reflecting the added complexity of the languages they are translating. A crucial difference between an assembler and a compiler is that with an assembler, there is a one to one correspondence between a line of source code and the corresponding object code in the form of the associated (single) machine code instruction. This is not true of compilers, and the object code derived from a given source program will differ from one compiler to another. The implication is that there may be differences in the quality of the resulting machine code judged in terms of size or run-time efficiency. This is clearly going to be an important consideration for anyone needing to have criteria for selecting one compiler in preference to another.

7.3.2 Another role of a compiler—checking for errors

Although the primary role of a compiler might be thought to be translating the source code, another function is the task of checking the program for bugs or, put more precisely, checking the validity of fragments of the source code against the grammatical rules of the language. Moreover the nature of any problems found must be recognized in sufficient detail to enable the programmer to identify the bug. This checking function is for practical purposes essential, since even programs written by experienced programmers have a high probability of containing errors. The checking involves reading and interpreting the source code in a process termed **parsing**.

7.3.3 The syntax of a language

It will prove helpful at this point to digress somewhat and discuss the syntax of a computer language in a very general way. To express the structure of a program we need to develop rules for the language it is written in—generally called its **grammar**. Such rules can be compared to those that apply for *any* language, including English for example. One definition of a language is that it is a set of **sentences**, where sentences consist of a list of strings of various kinds (Bornat, 1984). Valid sentences are developed as a list of certain allowed strings in a proper order. For instance, in the sentence 'A black dog ate the biscuit', we break the sentence into constituent parts and test (parse) the sentence to see if the grammatical rules (i.e. syntax) of the English language are satisfied. The sentence can be conveniently described as a so-called syntax tree, as shown in Fig. 7.2.

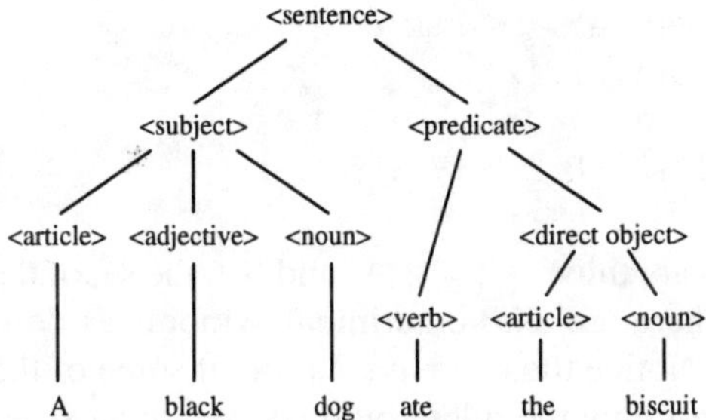

Figure 7.2 Typical syntax tree.

Entities such as *article* are all enclosed in '<' and '>' characters. The grammar states the various types of syntactic object that the language recognizes and the order in which they are allowed to appear. Thus a sentence must consist of a *subject* followed by a *predicate*. Definitions for objects such as *subject* must also be defined, possibly in terms of other syntactic objects, but eventually when the leaves of the tree are reached, in the actual words from which the sentence are constructed. These must satisfy the grammar so that 'black' is appearing correctly between an *article* and a *noun*, but a *verb* at this point would be grammatically incorrect.

More formally, a grammar can be defined in terms of several different types of object: **terminals**, **nonterminals** and a set of **production rules**. Terminals are quantities such as the words of the sentence above. Nonterminals are everything else, i.e. <sentence>, <adjective>, etc. Production rules define how lists of terminals and nonterminals can be related. A variety of ways to express the production rules are available. One, the **Backus normal form** (**BNF**), writes nonterminals in angle brackets as above, and terminal strings by themselves. The nonterminal defined by a particular production rule is written to the left of a list of nonterminals and terminals in the specific order in which the grammar allows them to appear, and are separated by a '::='. Thus in BNF form the production rule for <sentence> would be written:

```
<sentence> ::= <subject> <predicate>
```

Literally, this would read as: 'A sentence is defined as a subject followed by a predicate'. Other rules for subject, predicate and other nonterminals would be needed, forming a complete set of production rules. This approach can also be used to develop a grammar for a programming language. What this involves in the case of simple expressions is shown in the set of production rules below. To make life simple, parentheses and exponentiation are not included, and variables are only allowed to be letters. Note that the ' | ' symbol is used as a form of 'OR' operator when the production rule for some nonterminal gives it as one of several possible alternatives, and thus ' | ' does not stand for itself any more than '<' or '>'. It is not considered as one of the terminal symbols.

```
<expr>   ::= <term>|<expr>+<term>|<expr>-<term>
<term>   ::= <factor>|<term>*<factor>|<term>/<factor>
```

```
<factor>  ::=  <letter>|<number>
<letter>  ::=  a|b|c|...|z
<number>  ::=  <digit><number>
<digit>   ::=  0|1|2|3|4|5|6|7|8|9
```

The terminal symbols are thus: '*', '/', '+' and '–', the set of decimal digits, and the set of lower-case letters. There are six nonterminal symbols as defined by their respective production rules above. Notice the recursive nature of some of the production rules.

As a simple example of how the rules could be used to parse expressions, consider the following (correct) expression: x + 9. To parse this, we start at the left and step through the expression until the last terminal has been processed. So using the production for <expr>, a possible alternative form of <expr> is a <term>. Looking at the production for <term>, this can only be a <factor> here (there are no '*' or '/' terminals). A <factor> in turn can be a <letter> and since 'x' is a letter, the expression is correctly parsed at this point. Next, using the production for <expr> again, an expression can be an <expr> followed by a '+' followed by another <expr>. The first <expr> here is 'x', which has just been parsed, and the nonterminal following the (correctly placed) '+' evidently needs to be an <expr>, which is true since 9 is a <factor>, a <term> can be a <factor> and an <expr> can be a <term>. The **syntax tree** in Fig. 7.3 shows the parsing process for the expression more clearly.

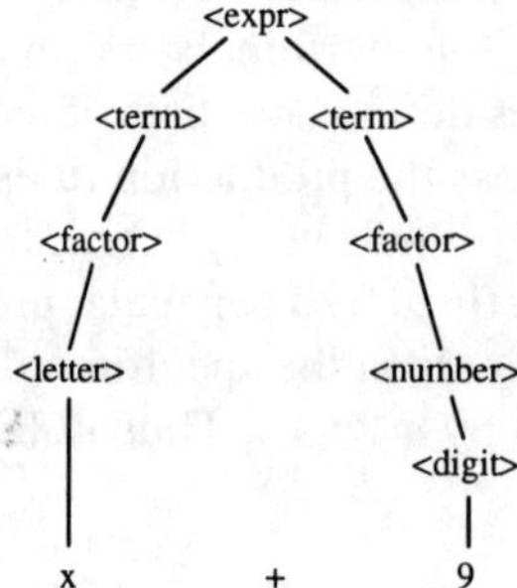

Figure 7.3 Syntax tree for a simple expression.

This somewhat lengthy and convoluted description might give some indication of the complexity of the steps that compilers need to go through in parsing a program. The grammar above is only a very small subset of that needed for realistic computer languages, including, as it must, provision for declarative parts, statement parts, conditional, looping, procedure and other constructs, all of which require production rules. A set of production rules for a Pascal subset is given in Aho and Ullman (1988). Many books on programming give the syntax of the languages with which they deal, often in the form of **syntax diagrams**, however, rather than the BNF form used here.

7.3.4 Simple demonstration of code produced by a compiler

Eventually, the structure of a compiler will be considered, but, to kick off with, it might be instructive to examine the complete process of compilation in terms of the end product in

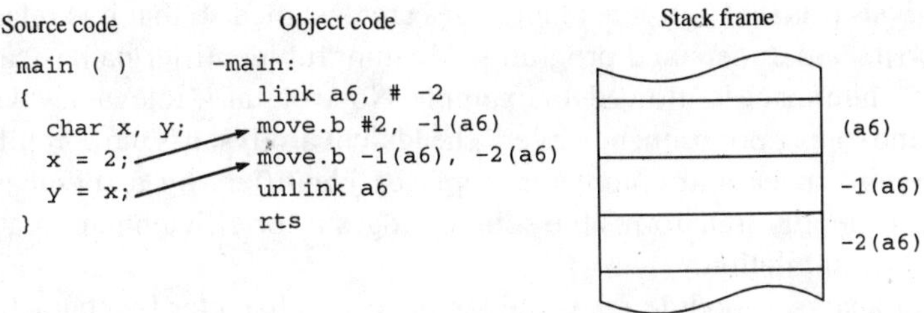

Figure 7.4 Example of compiler produced code.

the form of the object code for some simple C programs. The object code here is 68000 assembler. The code shown in Fig. 7.4 will not be unique to the original program; as we have already noted, a variety of possible assembler instructions could be developed depending on the compiler used.[4]

Notes:
- It will be remembered that the translation is done fragment by fragment, so that translation of each of the assembler statements into assembler code is done in isolation from the translation of the other fragment. This is not the approach adopted with some compilers, as we will see in Sec. 7.3.6 on optimizations.
- The compiler designates workspace for all variables in store with this particular program and compiler, although, as we noted in Chapter 3, in C there is provision to hold variables in registers. Some compilers will try to do this where possible in the interests of efficiency, as we will see later. Workspace is allocated after interpreting the information in the variable declarations. Local variables such as the '2' in the program are stored in an area that is pointed to by address register A6. **Auto** variables like these come into existence when the program starts to run, and the locations they occupy after the program terminates will likely be overwritten by data from a program that runs subsequently.
- The LINK and UNLINK instructions here are used to allocate and deallocate data areas on the system stack. These two instructions would be used for the same purpose had the program employed functions. A6 is here employed as a frame pointer, being the start address of the data area for local variables used by main.
- Even in a very small program like this,[5] the object code is less efficient than would be the case had the programmer used assembler: part of the price for the convenience of using a high-level language. You might speculate on a possible assembler code that is more efficient than that given.

7.3.5 Compiler structure

Compilers carry out translation in a number of sequential stages. The stage of actually developing the object code is simply the last step of translation. Remember that parsing and error reporting as well as translation are involved in the compilation process.

---

[4] The use of the LINK and UNLINK instructions in this context is explained later in the section dealing with run-time store allocation.

[5] Note that the program is rather *too* short to be useful; notably it does not do any I/O!

Lexical analysis phase High-level languages are designed so that it is relatively easy for humans to write and understand programs. Meaningful identifier names can be used and liberal use of white space is allowed for example. None of this is relevant when a *compiler* is interpreting the source program, however. The **lexical analyser** is the compiler's front end; it removes white space and comments, replaces identifiers by a **token** which is more compact and generally transforms the source program for efficient automatic analysis by later phases of compilation.

The lexical analyser module reads the source code character by character, and, where possible, the compiler tries to make sense of strings representing objects such as identifiers, special symbols (i.e. '=', '<', etc.) and reserved words. For example if the string 'main ' were encountered, the compiler would read the two spaces and ignore them. Next it would read the 'm' followed by the other characters, and it would check the complete string against a list of reserved words and insert a token corresponding to 'main' at the end of the list of tokens already assembled from previously read objects. Note that had 'main' been stored in its raw form rather than as a token, 4 bytes of storage would have been required, an important consideration if compilation speed is of interest, since the token will likely need to be accessed a large number of times at various stages of the compilation process.

If strings other than special symbols or reserved words are encountered, it is assumed that they are identifiers. The lexical analyser would then add the identifier to a table in memory reserved for identifiers. This is the **symbol table**, and is similar in some respects to a symbol table used by assemblers.

Lexical analysis is now examined in more detail using the earlier C program as an example. Each symbol that could appear in a program needs a token value, and a selected list of tokens for the symbols that appear in the program could be as follows:

Symbol	Token value
`main`	1
`<identifier>`	2
`+`	3
`{`	4
`;`	5
`=`	6
`}`	7
`char`	8
`(`	9
`)`	10
`<constant>`	11
`,`	12

The program would be used to generate a list of tokens, according to the values in the table above, and in addition there would be a field in the list that was a pointer to entries in the symbol table (see Table 7.1) in the case of the two identifiers and the constant 1. The symbol table would have three entries, which would at this stage have fields for the names x, y and 1, together with pointers to the place in the table where they occur. Other fields in the symbol table are filled in during subsequent stages of the compilation. It is assumed that entries to the symbol table are made in the order in which objects such as identifiers are added to it.

Table 7.1 Symbol table

Name	Pointer	Other fields...
x	1	
y	2	
1	3	

Finally the token list, including pointers to the symbol table, would be as follows:

Object	Token	Pointer to symbol table
main	1	
(	9	
)	10	
{	4	
char	8	
x	2	1
,	12	
y	2	2
;	5	
x	2	1
=	6	
2	11	3
;	5	
y	2	2
=	6	
x	1	1
;	5	
}		

Analysis and translation Following lexical analysis, the remainder of the compilation process involves first an analysis of the program in which the program is parsed, i.e. checked against the grammatical rules of the language as outlined in Sec. 7.3.3. A representation of the program is built using the token list and the partially complete symbol table. This representation is often called a **parse tree**, and is a realization of the syntax tree discussed in Sec. 7.3.3. In addition, the symbol table has further information added to it as more becomes known about objects such as identifiers, their type for example. Taking the earlier example of the C program to illustrate ideas, the syntax tree is built using the grammar for C, a very small subset of which is given below.[6] Note that the '|' character denotes separate alternatives, as described in Sec. 7.3.3:

---

[6] Some liberties have been taken in describing the grammar of C. Rigour has been sacrificed in the interests of clarity. A realistic grammar for the language is complex, and a full description would obscure the points being made here.

```
program          ::= <block>
block            ::= { <declarations> <statement list> }
declarations     ::= char <variable list> | int <variable list> ;
variable list    ::= <variable> , <variable list>
variable         ::= x | y
statement list   ::= <statement>; <statement list>
statement        ::= <variable> = <assignment> ;
assignment       ::= <var> + <var> ;
```

In the case of the program the syntax tree would be as in Fig. 7.5. The syntax tree needs to map on to a data structure that stores this representation of the program. This would comprise nodes, each of which would identify the nature of the fragment it describes (i.e. `<identifier>`, `<number>`, `<assignment>`, etc. The nodes would also include pointers to other nodes and, in the case of objects in the symbol table such as variables, pointers to the appropriate entry. The symbol table will by this time include the nature of the object (variable, constant, procedure name, etc.), its type (int, char, etc.), and the provision for storage). The tree structure corresponding to the statement part of the program is shown in Fig. 7.6.

Analysis is followed by translation into object code. Translation uses the two basic data structures generated during the analysis phase in the form of the parse tree and the symbol table. In its simplest form, code generation, as it is often termed, may be done on a fragment by fragment basis as was the case with the earlier 68000 example. In other words it is possible to identify a fragment of object code with a given fragment of source code such as an assignment with which it is associated. Other compilers do not translate on this simple basis, as is explained below.

7.3.6 Optimizations

Thus far we have discussed compilers in their most basic form, observing that the essential objectives a compiler has to achieve are translation (of source code into object code), detection and reporting on any errors found. Another objective, however, is to do with the quality of the code judged in terms of its run-time efficiency and also the overall size of the object

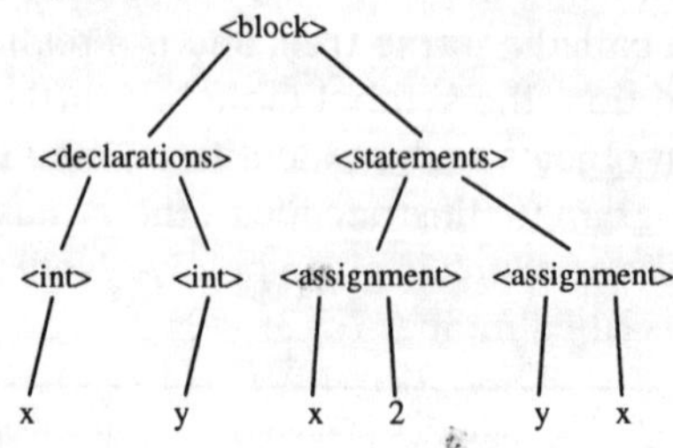

Figure 7.5 Parse tree for C program.

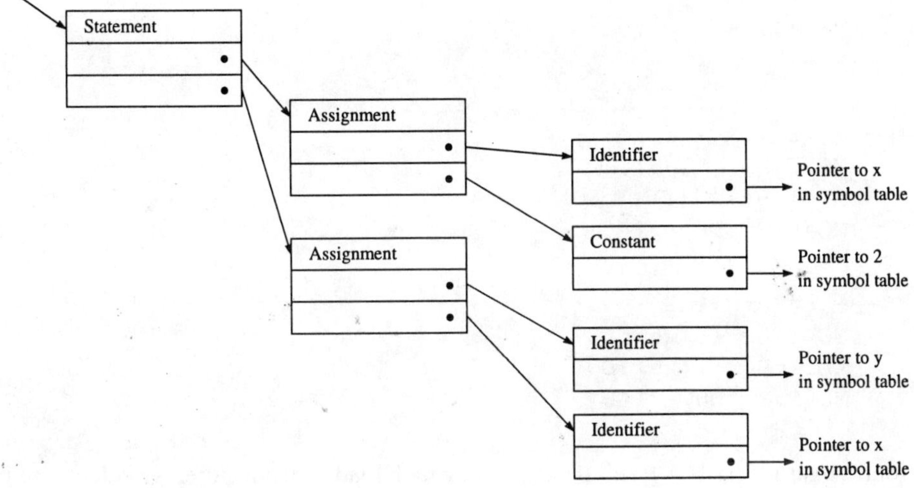

Figure 7.6 Data structure for part of parse tree.

program. **Optimization** refers to a range of techniques used by compilers to achieve these objectives. Run-time efficiency of the code produced by a compiler is of considerable interest to users, and is one of the most important measures of the quality of a compiler. The effectiveness or otherwise with which a compiler carries out optimizations could well be a reason for choosing one compiler in preference to another. Essentially we are involved in a variety of improvements to the code beyond those which the basic compilation process described would achieve. It used to be the case that optimizing compilers were only to be found on large mainframe computers, but in recent years they are being increasingly used on personal computers too, and thus are of interest to a very wide circle of users. It will probably be the case that in all but the most time-critical parts of a program, code produced by an optimizing compiler will approach the efficiency of that produced by writing in assembler. However, an optimizing compiler should not be looked on as a panacea, as a way of converting badly written source code into some optimal object program. Indifferent programming practices can easily defeat the most sophisticated optimizing compiler, while well-written code, with proper application of the rules of structured programming, will make life much easier for the compiler.

Two basic classes of optimizations exists: those made at the object-code level, and those that can be applied at the source-code level.

Optimizations applied at object-code level We have observed that simple compilation involves dividing a program into fragments such as assignment statements, and translating each fragment in isolation from the translation of other fragments. However the object program viewed as a whole, although correct, may contain sources of inefficiency such as redundant load and store instructions. Consider the 68000 fragment, corresponding to the following two statements:

```
        x=x+1;
        y=x+2;


  /* ---   x = x + 1 ; --- */
        MOVE.B   -6(A6),D0
        ADDI.B   #1,D0
        MOVE.B   D0,-6(A6)


  /* ---   y = x + 2 --- */
        MOVE.B   -6(A6),D0
        ADDI.B   -7(A6),D0
        MOVE.B   D0,-7(A6)
```

The fourth instruction, MOVE.B -6(A6),D0, which loads x from store, is evidently redundant, since x is already available in a register, and can be accessed more efficiently and using less code from there rather than from store. An optimizing compiler would keep a list of variables currently held in registers, and thus be able to eliminate any unnecessary load (or, in other cases, store) instructions.

Optimizations applied at source-code level

Constant folding Parts of expressions that are invariant at run time are best evaluated (once) at compile time, rather than carrying out identical evaluations repeatedly. Consider the instructions below:

```
  i = 100 ;
  y = x[i*10] ;
```

The array index used in the second instruction can be predicted at compile time and the instruction can then be processed as y = x[1000], avoiding a multiplication.

Register allocation for variables We noted earlier that some compilers will try to store variables in registers in preference to memory. Evidently the effectiveness of the technique depends on the number of free registers, and machines with large register files will show a greater improvement. The 68000 example below shows the compiler using registers to hold auto variables where possible, but since there are not enough registers there is a need to hold some variables in memory as well. Note that the compiler needs to save the prior contents of the registers on the stack, and at the end of the program restore the old values; naturally this partly negates the advantage of using registers, saving or restoring register values requires memory read or write cycles which can amount to a considerable overhead depending on the program involved.

As a final thought, it may be that some compilers would detect the fact that the program generates no output, and might eliminate the code completely! It pays to examine the compiler documentation to see just what sort of optimizations *will* be carried out.

Source code

```
main()
{
    int a,b,c,d,e,f,g;

    a=1;
    b=2;
    c=3;
    d=4;
    e=5;
    f=6;
    g=7;
}
```

Object code

```
main:
        link    a6,#-14
        /* --- save old register values on the stack --- */
        MOVEM.L D7/D6/D5/D4/D3,-(SP)
        /* --- use registers to hold a b c d and e --- */
        MOVE.L  #1,D7
        MOVE.L  #2,D6
        MOVE.L  #3,D5
        MOVE.L  #4,D4
        MOVE.L  #5,D3
        /* --- use stack frame to hold f and g --- */
        MOVE.W  #6,-12(A6)
        MOVE.W  #7,-14(A6)
        /* --- restore old register values --- */
        MOVEM.L (SP)+,D7/D6/D5/D4/D3
        UNLK    A6
        rts
```

Optimizations specifically involving loops It is well recognized that most programs spend
the bulk of their execution time inside loops, and optimizations which involve loops will in
many cases prove the greatest source of improvement in performance. This is particularly
true in the case of nested loops. For example, a loop that contains code that is invariant
during execution of the loop can be profitably moved outside the loop, as in the example
below:

```
for(i=0;i<10000;i++)
    for(j=0;j<10000;j++){
        b[i] = i*i ;
        a[i,j] = b[i];
    }
```

The assignment b[i] = i*i is invariant inside the inner loop, and can be moved outside as shown by the equivalent piece of source code below—never forget that we are thinking of the situation where the compiler would be carrying out a transformation like this rather than the programmer:

```
for(i=0;i<10000;i++){
   b[i] = i*i;
       for(j=0; j<10000;j++)
          a[i,j] = b[i] ;
   }
```

Loop unrolling Loops might be powerful programming tools, but the overhead involved in testing for conditions that terminate a loop can lead to a serious source of inefficiency in cases where tight, heavily used loops are involved. This is because of conditional branch and other instructions concerned with management of the loop. **Loop unrolling** aims, partially or completely, to eliminate such instructions by replicating the loop a number of times. A penalty is the increased code size, but frequently code size is deemed to be less important than run-time efficiency.

By way of example, consider the program fragment below, which features a for loop. It can be seen that, apart from the instruction ADD.W D7, D6, most of the remaining instructions are concerned with the management of the loop. The total number of clock cycles per iteration of the loop can be shown to be 38. A grand total of 380 clock cycles are thus needed for the ten iterations of the loop. Had loop unrolling been employed—with the loop *fully* unrolled—the number of clock cycles would be simply ten times the number of clock cycles for the ADD.W D7, D6 instruction, i.e. 80 clock cycles. This example is an extreme case of the inefficiency possible in implementing loops; clearly the improvement gained by unrolling would be less when there is more code inside the loop than just the single assignment statement here.

Source code

```
int   sum, i ;

. . .
. . .
sum = 0 ;
for( i=0;  i<10;  i++ )
   sum = sum + i ;
. . .
. . .
```

Object code

```
          . . .

          . . .
/* --- sum = 0 ; --- */
          CLR.W           D6
/* --- for( i=0; i<10; i++ ) --- */
          CLR.W           D7
I1_5:
          CMP             #10,D7
          BGE             I1_2
/* --- sum = sum + i ; --- */
          ADD.W           D7,D6
          ADD.W           #1,D7
          BRA             I1_5
I1_2:
          . . .
```

7.3.7 Run-time support

An important function of a compiler is to allocate storage space for the code and data that a program will require when it executes. The compiler will in some cases be able to predict the amount of store needed by some data—such as global variables. In other cases, however, data needs to be stored on a stack or else on a heap[7] so that storage requirements change dynamically. Some upper estimate for the maximum possible store requirement must be made, so that the operating system can provide suitable provision when the program is loaded. We saw in Chapter 3 that, in the case of C, storage space is divided into three parts for code and global variables, for the stack, and another so-called 'free' area which can be used as a heap if required by a program. In the case of C the stack will hold function return addresses, local variables, function parameters and possibly also act as a scratch area for temporary variables used in intermediate stages of expression evaluation. The stack will grow downwards, while, with requests for dynamic storage from routines such as `malloc`, the remaining area allocated for use as a heap will grow upwards toward the stack.

Methods for allocation of storage space for data vary, depending on the needs of the particular language. In block-structured languages such as C there must be support for recursion, and this means that there may be more than one activation of a given function at any time, so that provision for storing data for each activation must be made. Use of a stack in the way indicated previously makes this possible. In the case of FORTRAN, however, no recursion is allowed and it is possible to determine at compile time the space needed for storing all the data needed by a program. This simplifies the problem of run-time support considerably, but at the expense of the expressive power of the language since algorithms that use recursion cannot be used.

Run-time support with the 68000 Techniques involved in run-time support will now be illustrated using code produced for the 68000. This microprocessor has provision in its instruction set for support in handling procedure/function calls. Specifically, there are two

---

[7]This is described in more detail a little later.

instructions that are used to handle passing parameters between functions together with management of the stack. The instructions are called `LINK` and `UNLK`. A `LINK` instruction normally appears at the start of a function to make a part of the stack called the stack frame accessible to the function for the purpose of storing local variables. `UNLK` appears at the end of a function (just prior to an `RTS` instruction typically) to restore registers to the values held prior to the execution of the `LINK`, and to clean up the stack. Specifically, the `LINK` instruction causes three sequential actions:

1. The current contents of the address register specified in the instruction is pushed on to the system stack.
2. The updated contents of the stack pointer are loaded into the specified address register.
3. The sign extended two's complement displacement specified in the instruction is added to the stack pointer.

The first step saves the address register so that its original value may be later restored at the end of the function—normally by an `UNLK` instruction. This address register is to be used as a pointer to the **stack frame** (termed a **frame pointer** naturally enough) for storing variables local to the function. The second step makes the frame pointer point to the top of the area of memory to be used as a stack frame. The third step is needed to enable the stack to be used for subroutine calls and other purposes without overwriting the stack frame area. The stack pointer at the conclusion of the `LINK` instruction would point to the location just below the stack frame, the size of which is determined by the displacement.

Example 1 `LINK A6, #-2` creates a stack frame pointed to by the original value of the stack pointer minus 4, with `A6` being used as the frame pointer. The stack pointer will have decreased by 6 as the result of this instruction. The six locations on the stack are used to store a long word for `A6`, and two locations for the stack frame (see Fig. 7.7).

Example 2 Consider the following C program. The function `funct` is passed a parameter from a calling program. The function has one local variable and, as can be seen, this features in an assignment that uses the parameter.

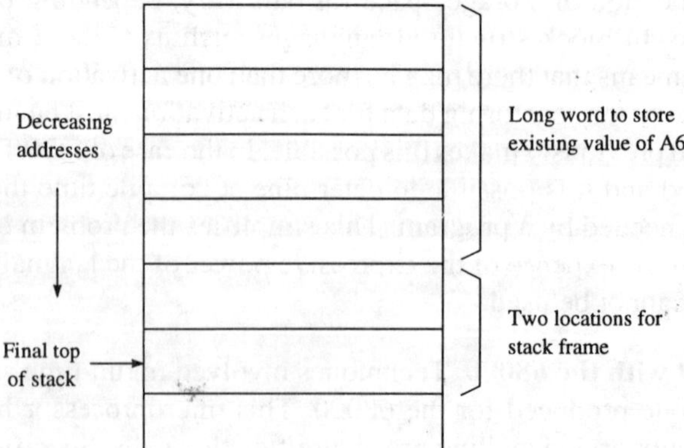

Figure 7.7 Stack usage for a LINK #−2 instruction.

```
main()
{
    short  int x ;

    x = 30  ;
    funct( x )  ;
    x = x*2  ;
}

funct( short int x )
{
    short int y ;   /* 16 bit */

    y = 15 + x ;
    y = y*2  ;
}
```

The object code generated by a compiler for this is shown below. Being well behaved, the compiler saves all registers used by the program, so that the initial values can be later restored when the program ends. The history of the stack usage is shown in Fig. 7.8, up to the point at which the call to funct has been made and the parameter x is being used by the function. The stack pointer has been assumed to be set to 100 initially. The reader is invited

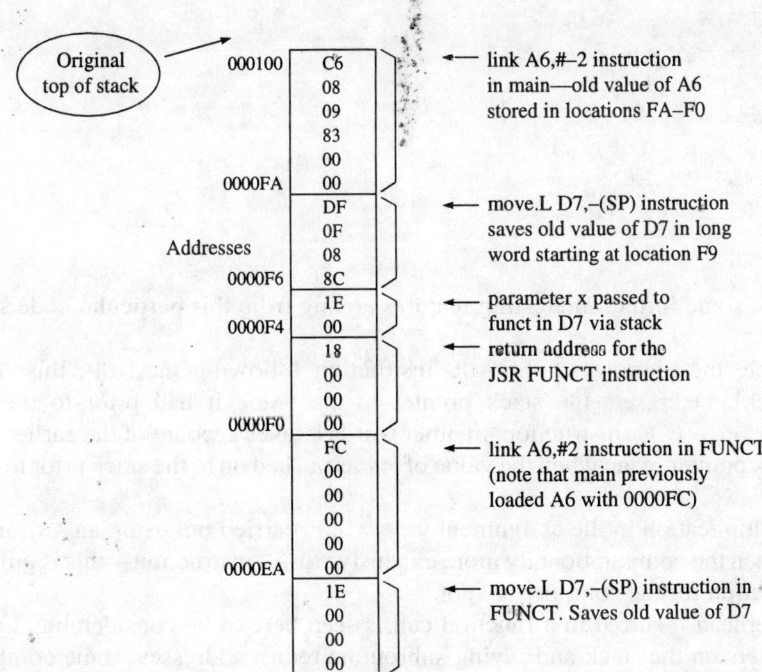

Figure 7.8 Typical history of stack usage.

to further examine the history of the stack up to the point when the program terminates. At this stage the stack pointer should have been restored to its original value of 100.

```
MAIN:
      LINK     A6,#-2
      MOVE.L   D7,-(SP)
      /* --- x = 30; --- */
      MOVE.L   #30,D7
      MOVE.W   D7,-(SP)
      JSR      FUNCT
      /* --- clean up stack --- */
      ADD.L    #2,SP
      /* --- x = x*2 ; --- */
      ASL.W    #1,D7
      MOVE.L   (SP)+,D7
      UNLK     A6
      RTS

FUNCT:
      LINK     A6,#-2
      MOVE.L   D7,-(SP)
      /* --- y = 15 + x ; --- */
      MOVE.L   #15,D2
      ADD.W    8(A6),D2
      /* --- y = y*2; --- */
      MOVE.W   D2,D7
      ASL.W    #1,D7
      MOVE.L   (SP)+,D7
      UNLK     A6
      RTS
```

There are some further interesting features arising from this particular code sequence:

- First note the purpose of the ADD instruction following the JSR; this 'cleans up the stack', i.e. resets the stack pointer to the value it had, prior to the previous MOVE.W D0,-(SP) instruction, in other words it takes account of the earlier change to the stack pointer made when the value of x was pushed on to the stack prior to the call to FUNCT.
- The multiplication in the assignment y = y*2; is carried out using an ASL instruction rather than the computationally more expensive MULT instruction—this is quite an easy optimization for the compiler to spot.
- The overhead incurred in a function call is seen here to be considerable, i.e. passing parameters on the stack and saving subroutine return addresses; some compilers will attempt an optimization in which function calls are replaced where possible by a functionally equivalent block of in-line code.

Heaps—allocation, de-allocation and fragmentation We have seen in the chapter on C, that heaps are a mechanism for dynamically creating space for data. This is different from the allocation of space for (say) some global variable, for which the compiler would allocate some *fixed* amount of storage. It is also different from the mechanism used to allocate space for data such as auto variables, a stack being a suitable form of data structure for storage in this case. Languages like C or Pascal, which make use of pointers to reference data created dynamically, have a need for this form of dynamic storage allocation. It can be shown that it is not practicable to use a stack form of organization to manage some types of data for which storage space is to be created dynamically (Bornat, 1984).

The area of store made available as a heap is divided into blocks, corresponding to the various requests for some stated amount of storage space by a routine such as `malloc` in the case of C. A natural consequence of allocation of store in this way is provision of some mechanism for releasing it when no longer required;[8] otherwise available space in a heap might become exhausted after repeated requests for more store. The further implication then is that the heap will consist of a number of blocks of free and allocated store. A problem that arises out of this is **fragmentation** of the store, since after a number of calls to request and subsequently free blocks in the heap, a relatively large number of interspersed small blocks of used and unused store will exist. Eventually it may well be that blocks of free store are so small that requests for a given amount of store cannot be satisfied, even in cases when the *total* amount of free store might be sufficient. A technique used on some machines is—from time to time—to move the blocks of allocated store so that they form a single contiguous block with only one hole. The method is called **memory compaction**, and is little used because of the amount of CPU time required and consequent effect on performance.

7.3.8 Bootstrapping a compiler

One of the more practical problems facing many people using compilers is **installation**. Compilers themselves are invariably written in a high-level language, and, unless the object code for the compiler is available, it is necessary to compile the compiler itself before it can be used on a particular machine. This process is often referred to as **bootstrapping**.

Possible forms of object code Before discussing the bootstrapping process itself, it will prove useful to comment on possible types of code that different types of compiler can produce. Much of the discussion of the different types of assembler applies to compilers in a general sense here too. The different types of code are:

- Native machine code, which is directly executable on a target machine after having been loaded in store.
- Relocatable machine code, which needs to be combined (linked) with other modules before being loaded. Most compilers produce relocatable rather than absolute code, since libraries of various kinds and other modules usually need to be included in the final (loadable) binary program.

---

[8] A function called `free` is available in C for this purpose.

- Assembler code, which is native to the target machine. In this case a separate pass through an assembler is necessary. Interestingly this could in principle be done on a machine different from that used for compilation.
- Machine code for a computer different from that on which the compiler is running. Directly comparable to cross assemblers discussed in Sec. 7.2, a cross compiler is used to prepare code for eventual use on a target system different from the host computer.
- Object code in the form of another high-level language.

To understand the bootstrapping process it is important to appreciate the role of the various languages involved. These are the language of the source code, the object code and the language in which the compiler itself is written. Bootstrapping involves compilation of the compiler program in the native machine code of the host machine (i.e. the machine on which the compiler is to run). In the case of (say) a FORTRAN compiler written in C, running on a VAX host computer, native VAX code would be produced from the compiler source code. As a footnote it is as well to observe that most software vendors do not sell the source code for their compilers. If only the object code for the compiler is available, evidently it can only be used on one type of computer.

7.4 INTERPRETERS

Compilers carry out translation into object code for a complete program prior to the time the program is loaded into store. Interpreters on the other hand defer translation of a program until run time. Naturally this makes interpreted code much slower than compiled code, typically by a factor of between 20 and 100; their use is attractive for machines such as home computers, which are often provided with BASIC interpreters. The interpretive mode of translation allows the editing of individual statements of a program, as opposed to the translation of a complete program necessary with compilation. This makes it more convenient for testing programs. Interpreters are available for a number of languages, including Pascal and FORTRAN, for a variety of machines.

The mechanism for actually interpreting a program may be illustrated by comparison with the mechanism by which a computer executes a program loaded into its memory. The sequence is: fetch the next instruction from store, increment the program counter in preparation for fetching the next instruction, decode the instruction and finally actually execute the instruction. This sequence is repeated until the program is finished. A software interpreter operates in much the same way, and we can think of a **statement counter**, identified with individual statements in the program, as pointing to the next instruction to be interpreted. The statement counter would correspond to a **line number** in the case of an interpreted BASIC program. The source code statements would be 'fetched' from the place in store identified by the statement, the statement counter would be incremented, the statement would be translated, and finally the statement implemented by executing the routines it needs.

7.5 LINKERS

In the sections on assemblers and compilers, the stages in translation that lead up to the production of object code were discussed. Invariably, a further stage is required to produce a program in a form suitable for being loaded into store and actually executed. Linkers were referred to in Sec. 7.2.5, in relation to relocating assemblers, which need to produce code suited for processing by a linkage editor. We now need to extend the discussion to see how code produced by compilers needs to be processed. Most programs make use of **run-time libraries** for functions such as standard mathematical routines, I/O routines or graphics routines. Libraries are compiled/assembled as separate modules, and bound with any user-produced modules to form a single binary program using a **linker** package.

7.5.1 Characteristics of linkers

Linkers need to be designed for use with object modules produced by a variety of different translators, including assemblers. The implication is that the translator packages must all produce code compatible with what the linker expects. Thus object modules must have information inserted in them by the translator program, specifically for later use by the linker package. Examples of programs prepared in this way are I/O routines written in assembler for use by (say) a program written in Pascal. Another example is the NAGLIB library of mathematical routines[9] which are written in FORTRAN but which can be used by a program written in another high-level language. A difference between code developed by a relocating assembler and a compiler is that assemblers of the former type often allow closer control over the final code produced by a linkage editor than is the case with code that is derived from a compiler. It is common to be able to specify the absolute start address of the (complete) program for example.

Object modules will invariably make reference to other modules. For instance a program may call a procedure that has been separately compiled (or assembled). The problem of mechanisms for resolving common objects such as starting address of the modules and common data clearly needs to be addressed. In the case of code produced by a relocating assembler, additional directives are needed to handle features specific to relocation and linking. For example directives are needed if symbols such as jump addresses are to be defined either internally or externally to a module. Provision for selected modules to start at an absolute address declared by a suitable ORG directive is usual; other relocatable modules will have their start address determined by the linkage editor. Special declarations are also required for high-level languages.

It is possible to develop C programs as a number of independently compiled modules, which are bound together using a linker. As we have seen in Chapter 3, it is possible to declare variables outside of functions and in so doing make them globally available to all functions in the file. If we want to prepare a program as separate modules, it is evidently necessary to make functions, variables and constants available to files other than that in which they were defined. Declarations for a variable of a function defined in one translation unit but needed in another translation unit must be prepended by the keyword extern. Thus if a variable elsewhere is defined in some file as a global variable, the form of the declaration would be:

```
int   elsewhere ;
```

---

[9] Referred to in Chapter 4.

If access is needed to the same variable `elsewhere` in *another* file, it is declared as:

```
extern  int  elsewhere ;
```

Without the keyword `extern`, `elsewhere` would refer to two different objects in the two files, in spite of sharing a common name.

In the case of Pascal there is no provision in the ANSI standard version of the language for separate compilation, although there are nonstandard implementations that do in fact allow external declarations. As we have noted in an earlier chapter, Modula-2 addresses precisely this weak feature of Pascal. For each Pascal procedure, the main program would contain a (single) statement with the procedure heading in place of the full procedure, as with the following example:

```
Procedure Fred( <parameter list> ) ; EXTERNAL ;
```

Special extensions to the language are needed for the compilation of the EXTERNAL procedure(s) as a separate module.

7.6 LOADERS

A loader is a program that takes as input a loadable module, usually prepared by a linker, and loads it into the computer's main memory ready for execution. In many cases, once the program has been loaded, the loader program will actually cause the program to execute by loading the computer's program counter with the starting address of the program.

7.6.1 Absolute binary loaders

An absolute loader simply loads the code produced by a linker or linkage editor at a predetermined starting address. In the case of an absolute assembler this address is specified in the source program by an ORG or equivalent directive. The XPC68000 assembler (Flight Electronics, 1990) produces object code, such as that below, in the form of so-called 'S' records, which contain (absolute) addresses telling the loader program where in store the program is to be loaded. Apart from the actual instruction op codes themselves, S records also give the number of bytes in a record and a checksum byte: all the bytes in S records are encoded as pairs of ASCII characters. In the example, the (start) address is A0000.

In computers, or EPROM programmers, for which code in this form is suitable, the loader would commonly be present in some nonvolatile memory, in which case it is available after power-up; it would not then need to be loaded prior to loading in the target program.

Source code

Address	Object code		Assembler code	
A0000	123C0044		MOVE.B	#$44,D1
A0004	207C00003000		MOVEA.L	#$3000,A0
A000A	227C00001000		MOVEA.L	#$1000,A1
A0010	1018	LOOP	MOVE.B	(A0)+,D0
A0012	D000		ADD.B	D0,D0
A0014	12C0		MOVE.B	D0,(A1)+
A0016	B001		CMP.B	D1,D0
A0018	6600FFF6		BNE	LOOP

Object code in the form of 'S' records

```
S00A00004B4A312E41534D20
S214FA0000123C0044207C00003000227C00001000E5
S210FA00101018D00012C0B0016600FFF60F
S804000000FC
```

The third line corresponds to code starting at MOVE.B (A0)+,D0

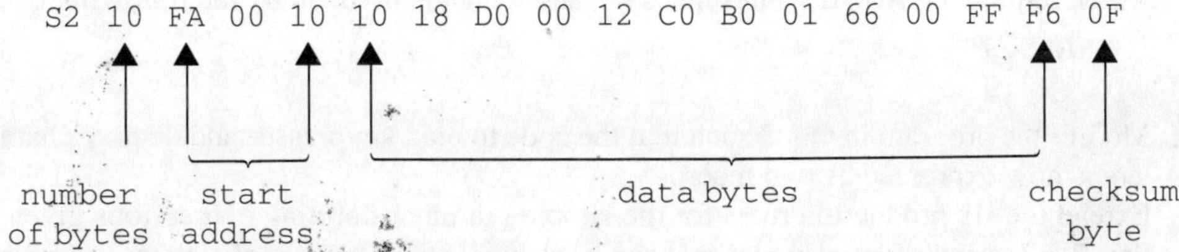

In loading code corresponding to the 'S' records above, the loader program would, after first reading past the S2 byte, read the following hexadecimal pair which gives the remaining number of bytes in the line into a counter, and then the bytes giving the address of the first data byte. The program would subsequently cycle in a loop, reading bytes corresponding to the actual instruction codes and storing each at the appropriate address, and decrementing a count of the remaining number of bytes. When the count of the remaining number of bytes to be read is zero, the checksum byte would by that time have been read and its value can be used to check the validity of the complete record. This process is repeated for all the remaining records in the file.

7.6.2 Relocating loaders

In many cases it is not convenient to specify precisely, at compile/assembly time, where in store a program is to be loaded. In many systems there would be a need to deal with several programs in store at any given time, and new programs would need to fit into unused parts of store, so the start address required by the program is not known until load time. The implication is that the loader will need to do some address calculation, perhaps adding an offset to an assumed initial start address of (say) zero, this commonly being the assumed start address in code generated by the linker.

In a system controlled by a multiprogrammed operating system there is another problem, albeit not one strictly concerned with the loader, which is that, once loaded, the code itself must be able to be moved around in store without having to be recompiled or relinked. In some cases the program may also need to be moved temporarily to and from backing store during execution. The instruction set of some simple types of computers, the 8085 microprocessor for example, does not allow this, and instructions such as the 8085's JMP 2000H refer to an absolute store address. 68000 code, on the other hand, can be relocated dynamically, by virtue of the provision of suitable address modes, including relative addressing, allowing **position independent code** (**PIC**) to be generated by translators.

7.7 EXERCISES AND REVIEW QUESTIONS

1. State the difference between an *absolute* and a *relocatable* assembler. In what circumstances would a relocatable assembler be preferred to an absolute assembler?
2. What is the difference between a one-pass and a two-pass assembler?
3. Write a 68000 assembler program which accepts ASCII characters read after a keypress, and then echoes the characters subsequently on the screen of a terminal. The keypresses are also to be stored in a buffer area of memory that can contain at most 80 characters, and an error message is to be displayed if the buffer is full. You should use the program given in 7.2.4 for guidance and assume the existence of a routine to read keypresses which returns the ASCII values in D0.B, and which is invoked by the following code:

```
TRAP   #11
DC.B   6
```

4. Modify the program in Qu. 3 such that the code to read keypresses and display characters is now expressed as two macros.
5. Extend the six production rules for the simple grammar defining expressions given in Sec. 7.3.3 to cater for: variables that can be strings rather than simply letters and bracketed expressions. In addition add a seventh rule which defines an assignment statement, employing the modified form of expression developed. Using this grammar give a syntax tree for the assignment x := y*(6+7) ;
6. What are the two main data structures generated from the lexical analysis phase of compilation?
7. Produce possible compiled 68000 code for the C function below, assuming the compiler makes full use of registers to store auto type variables where possible. Comment on any other optimizations that are possible.

```
int   comp_demo( char h )
{
      int   a,b,c,d,e,f ;
      a = h ;
      b = h ;
      c = h*2 ;
      d = h+56 ;
      e = h*45 ;
      second(a,b,c,d,e);
}
```

8. You are to install a PL/M compiler on a VAX computer, its source code is available on a reel of magnetic tape. The compiler is written in FORTRAN. State in general terms how you would carry out the installation.
9. Write a loader program in 68000 assembler capable of processing a file in S record format, for loading into the volatile memory of a microcomputer board.

REFERENCES AND FURTHER READING

Aho, A. V. and Ullman, J. D. (1988) *Principles of Compiler Design*, Addison-Wesley, Reading, Mass.

Bornat, R. (1984) *Understanding and Writing Compilers*, Macmillan, Basingstoke.

Calingaert, P. (1979) *Assemblers, Compilers, and Program Translation*, Pitman, Reading, Mass.

Flight Electronics Ltd (1986) *The Flight 68K Users Manual*.

Flight Electronics Ltd (1990) *68000XPC Macro Cross-Assembler User Manual*.

Kane, G., Hawkins, D., and Leventhal, L. (1987) *68000 Assembly Level Programming*, McGraw-Hill, Berkeley, Calif.

Knuth, D. E. *Art of Computer Programming*, vol. 1, Fundamental Algorithms.

Lawrence, P. D. and Mauch, K. (1987) *Real-Time Microcomputer System Design—An Introduction*, McGraw-Hill, New York.

8

INTRODUCTION TO OPERATING SYSTEMS

8.1 INTRODUCTION

A sizeable part of this book is devoted to operating systems, a reflection of the importance of this area to electronic engineers. The present chapter aims to give an overview of operating systems concepts, with a practical slant aimed at readers with limited knowledge in this area, who might typically be casual users of machines such as personal computers. Such users would need sufficient background in operating systems concepts to enable them to get the best out of their machines. Later chapters examine other, more advanced aspects of operating systems.

8.2 THE FUNCTION OF AN OPERATING SYSTEM

Perhaps the most immediate questions one should ask about operating systems are:

- What is their function?
- Who needs to know about them?
- Why find out about operating systems at all?

One short, albeit partial, answer to the first question could be:

- An operating system is a collection of programs that forms a computer's resource manager.

Computer hardware is complex even for personal computers, consisting as it does of processor, store and a wide range of peripherals such as disks, tapes, printers, modems, etc.

An operating system has to manage all these elements consistently and efficiently; an extremely taxing problem for any computer system large or small. It might be stressed that by *resources*, software elements are involved also. These include: files, shared data, programs and other elements to be referred to later.

As an example of the kinds of problem that have to be addressed, consider the act of a user entering a command at a terminal to print a file and the subsequent responses the computer would need to take under the control of its operating system. It may be that another user's file is already being printed, so the operating system first needs to check if this is the case. In many instances the operating system will arrange for the file to be temporarily stored on disk waiting for the printer to become available. Without describing the detailed procedures involved much further,[1] hardware in the form of the printer, a disk, and also the CPU come into play. The operating system must orchestrate the activities of all hardware elements to ensure that resources are used *correctly*, so that users do not get their printout interspersed for instance, and also *efficiently*, so that users do not have to wait until their file is printed before being allowed to enter another command.

Answers to the second and third questions posed depend on how particular users need to interact with a computer system:

- **Casual users** may—in some cases—need know little more than the log-on procedure, and how to invoke packages that they need to use such as compilers and editors. It is a common experience that many users are lazy in this respect, and try to make do with as little knowledge about a computer's operating system as they feel they can get away with. A relatively small investment in learning how to get the best out of an operating system, to use the resources of a computer to their full potential, would in the long term really repay the time taken.
- **Computer operators** are responsible for the day-to-day operation of a computer system involving activities such as getting a computer up and running (termed **booting**), loading tapes and other activities involved with running jobs.
- **Systems administrators** are involved in matters such as selecting an operating system and perhaps upgrading it from time to time; such decision making infers considerable competence in understanding the features and limitations of alternative, commercially available systems.
- **Systems programmers** include, amongst other activities, writing or maintaining the code for an operating system; and their level of understanding needs to be deeper than that of some other types of user. This involves considerable expertise in the real-time programming area. A typical task that such programmers have to deal with is writing software (called **device drivers**), needed when a new peripheral is installed. Users interact with a computer at the command level, or else in terms of I/O statements inside an (application) program. Systems programmers on the other hand deal more *directly* with the operating system. Operating system code contains **system calls** which are procedure calls requesting the operating system to carry out an action. (Examples of the use of system calls are given in the later chapters on operating systems.)

---

[1] The technique involved is called **spooling**, and is covered in more detail in the next chapter.

- **Embedded systems designers** are frequently responsible for the development of both the hardware and software elements in a system and this involves the use, enhancement or, in exceptional cases, perhaps even development of the real-time operating systems that will run them; considerable familiarity with systems programming is needed, and also involvement with other aspects of the real-time behaviour of the system. The needs of this class of user receives particular attention in various parts of this book, notably Chapters 12, 13 and 14.

8.2.1 Virtual machine concept

One of the more obvious tasks of an operating system is to present users with a manageable model of a computer, in which the inherently complex nature of the underlying hardware is hidden from view where necessary. This avoids the need for users to involve themselves with specifying low-level operations—such as movement of the arm of a disk drive to a particular track, recovering from conditions such as full disks, and so on. Users of first-generation computers had to operate at just this level in fact, since these early machines frequently had no operating systems; only people with some computer hardware knowledge were able to use these computers effectively. The simplified model of a computer presented to users is generally referred to as a **virtual machine**. One can get a simple idea of the virtual machine concept as the view that users have sitting at a terminal typing in commands, developing and running programs or packages, etc. Such a view is of an abstract logical machine in which key presses are turned into actions subsequently carried out by the hardware; and in which the (visible) response of the hardware is interpreted by the operating system before being displayed on a screen. In brief the operating system turns the bare hardware into a usable, hopefully user-friendly, system.

To get some comparisons, one could think of the rather primitive services provided by monitor programs (these can be regarded as a very simple form of operating system, and are described in more detail in Sec. 8.3.1). With monitor programs, users are responsible for operations at the level of physical store locations and registers. This is perfectly acceptable for the kind of dedicated applications for which some kinds of single-board computers are frequently used. In most other types of system, however, such low-level interaction on the part of users would be quite unacceptable; involving, as it frequently does, detailed knowledge of a computer at the level of its instruction set, register provision, etc., and also decisions about management of store, in terms of where code, data and stacks should be located.

In multiuser systems, it is essential to prevent errors in a particular program from affecting other programs, notably overwriting/corrupting their code or data. Attempts to access store outside the area allocated to a program must be capable of being flagged by the hardware and subsequently acted upon by the operating system. Another example would be prevention of the unrestricted use of certain instructions that could have a system-wide effect. What would *you* think of a computer, were its users allowed to run, without any restrictions, a program which included a `HALT` instruction? The 68000 has certain privileged instructions which programs can be prevented from using; any attempt to execute such instructions are dealt with by an exception-handling mechanism, a clean recovery then being made by the operating system.[2] Most computers have this type of provision.

---

[2] Appendix 2 gives more on the exception-handling mechanisms used in the case of the 68000.

8.3 OPERATING SYSTEM TYPES

8.3.1 Monitor programs

Monitor programs are primitive forms of operating system frequently found in some types of single-board computers, often in the form of dedicated embedded systems where code may need little if any changes for the lifetime of the computer, and often written to some nonvolatile form of memory. Monitors allow users to enter simple commands, to carry out actions such as memory dump or starting execution of a program from some given starting address. I/O provision is often relatively primitive, allowing interfacing to a small range of devices such as terminals and printers.

8.3.2 Single-user systems

These provide a virtual machine for a single user and are mainly found on personal computers, MS-DOS used on IBM-compatible computers currently being most common. This operating system is described in more detail in Sec. 8.6. Other examples of single-user systems are OS/2 and CP/M, the latter being an early operating system used on 8-bit microcomputers. Such systems allow the use of simple packages, which include compilers, assemblers and editors, although they are becoming increasingly more capable, to match the greater and greater complexity of applications which are used on ever more powerful personal computers. Emphasis needs to be on ease of use, in view of the wide range of people who work with personal computers.

8.3.3 Real-time operating systems

Real-time systems were previously given a brief mention in Chapter 2. Recall that some computers operate under time constraints which it is *essential* to meet. Operating systems used for such systems are called **real-time operating systems** (**RTOSs**), and have to be designed with reliability, and rapid response to external stimuli as prime requirements. There will often be a minimum of interaction on the part of operator or user, in contrast to many other types of system. The software that runs on a real-time system is much more specialized than is the case with many other types of computer, is dedicated to the application of the computer and will in many cases change infrequently, if ever, during its lifetime. Packages such as compilers or editors, etc., are generally not needed (it might be added that RTOSs are on some occasions used for development work, in which case some of these packages would be needed). An RTOS can in some ways be regarded as an enhanced form of monitor program, used when the features of the latter are too primitive for the intended application. Evidently electronic engineers will commonly be involved with systems controlled by an RTOS, either in the development or maintenance of the hardware, the software or more likely both. RTOSs are discussed in more detail in Chapter 12. Details of some specific RTOSs are outlined in Appendix 3.

8.3.4 General-purpose operating systems

General-purpose operating systems have the characteristic of providing a wide range of services to a correspondingly wide range of users. Packages that run under such systems include compilers, editors, debuggers, CAD packages, etc.—evidently a very long list, and as difficult to enumerate as the many applications for which a general-purpose machine can be used. Most users are familiar with the interactive mode of use with machines controlled by this type of operating system, but most general-purpose systems also provide a batch mode, where, once a job (possibly in the form of a reel of tape) is entered into the system, users have no contact with it until the processed data have been produced. This (noninteractive) mode of use was commonly the *only* provision in early systems, where users frequently had to prepare their programs on media such as decks of punched cards or paper tape, the output from the program being available for collection the next day in many cases. **RJE** (**Remote Job Entry**) systems allow for submission of data with I/O devices that may be remote from a central computer via a communications link of some kind.

8.3.5 Multitasking operating systems

This classification applies to any of the four types of operating system described above. With **multitasking** operating systems (sometimes called multiprogrammed operating systems), the appearance is given of running several programs concurrently. It should be mentioned that the word task is usually employed in this context. Think of a task as a program in execution for now, although the concept will be explored in more depth in the next chapter. Since most computers only have one CPU, evidently the effect must in fact be obtained by having a mechanism to enable the CPU to switch from program to program. Provided this is not done too slowly, programs appear to be running concurrently, albeit on a CPU that appears to be executing more slowly than in an equivalent situation in which no multitasking is employed. Multitasking is used on multiuser systems where several users can work on a computer simultaneously. Allowing several users simultaneous access to a computer is one advantage of a multitasking operating system, another is that if a computer makes a request to carry out some I/O—in many cases a slow process—the CPU can be switched to execute another program which *is* capable of running. The alternative would be to let the program idle until the I/O operation is complete before allowing it to recommence. The need to overlap I/O and processing has long been an issue with computers going back to the 1950s, when the problem of I/O activity forming a bottleneck was recognized as one of the main problems in obtaining acceptable performance from a computer.

Multitasking is employed in single-user systems as well as multiuser systems, and is making its presence increasingly felt in the personal computer area; notable examples are UNIX and OS/2, referred to in later chapters. The whole area of multitasking operating systems is discussed in more depth in the next chapter.

A distinction should be made between the term task and **job**. The latter refers to activities needed to carry out some goal set by users, such as processing data on a tape and producing the processed results on (say) a printer. Jobs may be broken into steps, and it is these steps that could constitute tasks in the sense described above. There is also confusion possible with the term **process,** a term unfortunately often used interchangeably with *task*. Another source of confusion is that some make a *process* into a special form of *task* or vice versa! **Task** is the term mainly used in this book, with one or two exceptions.

8.4 MORE ON RESOURCE MANAGEMENT

Spelt out in more detail, management of the resources of a computer must include some policy to determine which out of a number of programs can actually be allowed access to a particular resource. For instance it would evidently not be acceptable, in a multiuser system, for users to have unrestricted access rights to the disk holding the operating system. Additionally it is common to restrict how much of a resource is to be given; many types of system impose limits on the amount of disk space that can be allocated to individual users, for example. Having vetted a user's (or a program's) request for a resource, the operating system then needs to invoke some mechanism for its *allocation*. Finally, when the resource is eventually no longer required by the user/program, the operating system uses a mechanism to reclaim it, allowing other users/programs to then request access with possible hope of success.

The resources themselves may usefully be split into categories, to simplify the description of different operating system management functions:

- **CPU management**—An obvious example of the way the CPU needs to be managed in multitasking systems is the problem of choosing the next task to run. Suitable policies need to be devised, dependent on the type of service the computer has to provide. Tasks with an important real-time deadline to meet will likely have greater priority than (say) a task to deal with output to a printer.
- **Memory management**—This is especially complex in the case of multitasking systems, when several tasks may need to reside in store at the same time. Store management then involves dividing store into areas into which the various tasks can be placed. Measures are required to ensure that a task does not encroach on the memory space used by other tasks, otherwise corruption of code or data could result. Additionally, a new task should not be put in store unless there is enough room for it. Store management techniques are discussed in Chapter 10.
- **I/O device management**—Some I/O devices such as printers cannot be shared and so must be dedicated to a task until the task is ready to relinquish control over its use. However, other devices such as disks *can* be shared and then different I/O management techniques are appropriate. It is possible for several tasks to read a disk, for practical purposes, concurrently. Having said that devices like printers are not shareable, it is possible, however, in some cases to give the appearance—in a restricted sense—of making them shareable, by techniques like spooling. Recall that files to be printed are first stored in a buffer area on disk from where they are sent to the printer in the order presented. The only time users are made aware of the nonshareable nature of a printer is the point at which they have to go to the back of a queue of other users to wait for their printout!

Scheduling: keeping users happy—using resources efficiently The workload of a computer involves a varied mix of CPU and I/O activity; and one of the aims of an operating system should be to manage both these elements so as to avoid any bottlenecks they might give rise to. As far as proper utilization of the CPU is concerned, the situation is more complex than for example just aiming to minimize the time taken to execute some given program. Short execution time for one program may be at the expense of long execution time for another, and it is a function of an operating system to resolve such conflicts. **Scheduling** is the term given to the way an operating system allocates resources to tasks and jobs, using some coherent policy to resolve competing requirements. Scheduling should aim, with satisfactory compromises made, to achieve the following objectives:

- All tasks should be given a *reasonable* amount of CPU time; this is usually termed **fairness**. Firstly, some tasks may be deemed to be more important than others, and thus deserving of more CPU time; tasks with tight real-time deadlines, say. On the other hand, an operating system that gave little or perhaps even no CPU time to low-priority tasks, would clearly not be satisfactory either. Scheduling should aim to avoid being unfair to any task, no matter how humble. Secondly, many systems use some form of pre-emptive scheduling, where tasks are periodically interrupted and another task allowed to use the CPU, even though the original task may not have run to completion. Were this *not* done, some tasks that did no I/O could literally run for hours, or even days in extreme cases, without *ever* relinquishing use of the CPU! This aspect of scheduling is discussed in more depth in the next chapter.
- In multiuser systems, the response time for all users is minimized. Response times for interactive users sitting at their terminals should ideally be kept to well under 1 second, although under abnormal conditions, when the operating system cannot cope properly with its workload, this objective may not always be met. Unfortunately most users experience this from time to time! (How long would *you* be prepared to wait in such circumstances? 10 seconds? 1 second? 0.1 second?) The needs of interactive users can only be properly met if some form of pre-emptive scheduling is used.
- The utilization of all resources shall be maximized. Some tasks may spend most of their time doing I/O (tasks like this are said to be **I/O bound**). Such tasks should be allowed high-priority access to the CPU when needed, and thus allow the relatively short periods when they use the CPU to be interleaved with the I/O activities to be carried out. Other tasks, in contrast, make relatively heavy use of the CPU (termed **CPU bound**). If all tasks are substantially either I/O bound or CPU bound, resources will not be used effectively. Either the CPU will be working flat out with the I/O devices not being kept busy, or vice versa. A proper scheduling policy relating to priorities should aim to provide a system with a balanced load for its various hardware resources, and so avoid either of these two extreme situations.

8.5 FILING SYSTEMS

8.5.1 Files—what are they? why are they needed?[3]

The idea of a **file** has already been referred to briefly in earlier chapters. Essentially, a file is a named collection of logically related information, the name being suggestive of the folders in a filing cabinet, a quite helpful analogy in fact. The information so named could for instance be: the source code of a program, a letter prepared using a word processor or an executable program which can run after being loaded into store from disk. The file concept is a method used by computers to manage the information they store.

Computers will often store large numbers of files, it being not uncommon for some large

---

[3] This section is only intended as a practical introduction to filing systems, more detail is given in the chapters on storage management and UNIX.

computer systems to contain in excess of 100 000 files, and to better manage the complexity involved it may then be desirable to form files into groups according to some logical purpose or property that they share. Most systems have such provision, grouping files relating to individual users for example. As well as simplifying a user's view of his or her files, it also facilitates protection mechanisms—such as preventing other users from deleting one's files! Collections of files in whatever grouping has been selected are commonly called **directories**. A directory is essentially a table containing all relevant attributes of files with which they are associated, notably character strings representing the file names. With many operating systems, directories are themselves files.

Adopting the filing cabinet metaphor then: a filing cabinet would have a counterpart in the form of the storage device, such as a disk or tape. Filing cabinet drawers are comparable to directories containing folders corresponding to the files; and the files themselves can be broken down into more basic units, in the form of the information being stored. At the lowest (physical) level one would, for example, be thinking in terms of individual sectors of a disk, where the actual information is being stored, together with some means of linking together all the records that constitute a file.

Some naming mechanism is implicit in all this, and files in most computer systems are accessed by quoting a character string that represents them. Filenames must satisfy whatever conventions the operating system imposes. The rules for naming files for MS-DOS are given in Sec. 8.6. Section 11.6 does the same for the UNIX operating system. In many systems, filenames are divided into two parts—a filename **prefix**, which is indicative of the file's function, and a suffix, usually called an **extension**, which might (say) be used to distinguish Pascal source code; files produced by a word processor, executable files, etc. The two parts are often separated by a '.'.

8.5.2 Managing files

Operations on files are of interest to users right from the time they first start working with computers. At the very least, users must be able to create and, where appropriate, edit their own files, but additionally they would often want to share files created by others, and to have protection in the form of appropriate read, write and execute access to guard against unauthorized access, unwanted actions of other users, or their own mistakes. This, in essence, is the kind of provision an operating system must have in its filing system. Additionally, in keeping with the virtual machine idea, the filing system as presented to users should give them a logical view of their files, and should hide hardware detail such as the particular tracks and sectors of the disk on which the information is being stored. So, for example, to print a file, it should be possible simply to invoke a command such as `print XYZ.TXT`, leaving it to the operating system to access physically the areas on disk where the various records in the file are located, and then to arrange for the file to be printed out.

8.5.3 Hierarchical file organization

The need to group files in directories according to some logical function has already been mentioned. Grouping files according to users is virtually essential in any multiuser system, and this suggests a natural structure of one directory per user. Most single-user systems have provision for a directory structure too, MS-DOS is one example, and will be discussed later. The directories collectively can be pointed to by what is often referred to as a **root**

directory. Individual directories may have a sufficiently large number of files to justify carrying this structuring process further, and directories may point to subdirectories, which contain files of various classifications and perhaps further subdirectories. This is shown below where there are three layers in the directory structure; in some large systems, five or even more layers may be used. Finally, it is common to allow users the ability—with suitable restrictions—to create and delete directories, and also to give them appropriate names. The way UNIX organizes its hierarchical filing system is described in Chapter 11.

Identifying files by name becomes a problem in systems of any size, particularly in multiuser systems. Users should not be allowed to assign a name to a file on the assumption that a particular name is unique in the system. Other users can—and quite often will—use identical names for some of *their* files. The inference is that files need to be identified by *directory* as well as the individual filename. The usual approach is to quote a **pathname** which references all directories leading from the root directory to the directory containing the file, as well as the filename itself. The example below is from UNIX: the leading '\' is UNIX's way of referring to the root directory, `fred` is the directory containing Fred's files, `pscl` is one of Fred's subdirectories, while `xyz.pas` is a file in this subdirectory. Depending on circumstances, it may be necessary to quote the entire pathname in operations such as copy, delete, etc.

```
\fred\pscl\xyx.pas
```

8.6 MS-DOS AND THE IBM PC—BACKGROUND AND SOME HISTORY

Only an overview of MS-DOS is attempted here (the MS in MS-DOS stands for **Micro-Soft**, the company that introduced the operating system, while the DOS stands for **Disk Operating System**). There are a large number of books that discuss how to actually *use* MS-DOS, and enumerate all its commands and discuss practical matters such as file structure, etc.; Deitel (1990) is one example. MS-DOS will be discussed in terms of its role as an operating system for personal computers; comparisons will also be made with other operating systems used on personal computers.

MS-DOS, or simply DOS as some refer to it, was the principal operating system used for the IBM PC back in 1981 when this machine was first introduced. Since that time it has evolved considerably, to keep pace with developments in personal computer hardware, notably microprocessor technology, and also to satisfy the needs of the increasingly sophisticated applications for which personal computers are now being used. It is a single-tasking operating system, unlike some other operating systems used on PCs such as OS/2 and UNIX which are multitasking—this is increasingly seen as a restrictive feature of MS-DOS.

8.6.1 The influence of the IBM PC

The design and development of MS-DOS is inextricably bound to that of the IBM PC, and by implication to the 8086 microprocessor family which the machine uses. It is perhaps appropriate at this point, therefore, to give background to the architecture of the PC, particularly since developments in the microprocessor chips from the original 8088 to the latest chip in the family, the 80486, were matched by comparable developments in MS-DOS to exploit the performance of later generations of microprocessor.

The 8086 and 8088 In the early 1980s, the 64 kbyte memory limit with machines based on 8-bit microprocessors such as the 8085 and Z80 was seen as increasingly restrictive, and one of the reasons for the choice of a 16-bit microprocessor chip for the PC was to increase the amount of directly addressable memory. The 8088/8086 allowed 1 Mbyte of memory to be directly addressed; at the time these microprocessors were introduced this amount of addressable memory seemed perfectly adequate for the type of applications used by micro-processor-based computers. In addition, memory was, in comparative terms, much more expensive then than it is now; so, not unreasonably, it was thought uneconomic to have a larger address limit.

The 8086 uses a **segmented architecture** in which the 1 Mbyte memory space is divided up into a number of 64 kbyte **segments**. A **segment register** points to the **base address** of any given segment and, during each instruction fetch, the program counter contents are added to the value in the segment register to form the effective address. This scheme allowed for programs that could run in a 64 kbyte segment with a different program in each segment; alternatively, programs could occupy several segments—in fact these days many programs need more than 64 kbytes. Part of the reasoning behind the choice of the seg-mented address scheme was to retain compatibility at the binary code level with the 8080 8-bit microprocessor family; it being possible to run 8080 code in one of the segments of an 8086.

The 80286 The next stage in development from the 8086/8088 was the 80286, which appeared in 1984, to be used in the IBM PC/AT. Like the 8086, the 80286 is a 16-bit chip, but with considerable enhancements, including a higher clock frequency—12.5 MHz or greater—the 8086 on the other hand typically operates at a clock frequency of 5 MHz. Since most users needed to be able to run MS-DOS software, the 80286 was designed so that it could run 8086 code. A major enhancement was an increase in the address space from 1 Mbyte to 16 Mbytes, and this trend to larger and larger address spaces continues to the present day. The 80286 can operate in one of two modes. One, called **real address mode**, only uses 20 of the 24 address lines, and addressing is identical to that of the 8086. In real address mode the 80286 can run any program that can run on an 8086/8088-based machine. The other mode, called **protected mode**, supports multitasking. In protected mode it is possible to employ 16 Mbytes of store, with protection allowing several programs to run concurrently without being able to encroach on each other's memory space.

The disadvantage of this scheme is that the 80286 is not able to run in both modes at the same time. This meant that the 80286 could run standard MS-DOS software in real address mode, but could not then operate in protected mode. In protected mode, on the other hand, MS-DOS software would not run, although other operating systems, such as UNIX, that were multitasking could make use of protected mode. The practical result of this restriction was that most people who used an 80286 machine operated it in real address mode to be able to run MS-DOS software. The 80286 was thus for most users seen simply as a higher-speed version of the 8086. The remainder of the 16 Mbytes address space above the real address mode limit of 1 Mbyte is accessible only as extended memory, and special code needs to be included in programs to make use of it.

The 80386 and 80386SX 1986 saw the introduction of Intel's first 32-bit microprocessor, the 80386, which, together with the 80386SX, is found in many present-day PCs. The 80386 has 32-bit addresses and buses, while the 80386SX can be viewed as a cut-down, cheaper version of the 80386 with a 16-bit external data bus, although the data paths within the chip are 32-bits. Currently, machines based around the 80386SX chip are tending to eclipse those based around the 80286. One design aim with the 80386 was to avoid the limitations of the 80286 described above. As with the 80286, it is possible to run 8086 code and thus MS-DOS software—a mandatory requirement as far as the bulk of PC users were concerned. MS-DOS applications are run in so-called **virtual 8086 mode**; this allows memory to be partitioned into 1 Mbyte sections, each of which could run a separate program with each section protected from every other section.

The 80486 At the time of writing, the latest Intel chip in the 80xxx series is the 80486, introduced in 1989, and this chip continues the trend to both greater functionality and performance levels attained by microprocessors. The 80486 contains an integer unit fully compatible with the 80386, a floating point unit fully compatible with the 80387 coprocessor chip, a virtual memory unit and a cache memory. Currently, 50 MHz versions of this chip are available, with higher clock frequencies in prospect. The 80486 has between two and four times the performance of the 80386 depending on the application. The instruction set of the 80486 is fully compatible with that of the 80386, although there are a number of enhancements to the instruction set needed to support the cache function.

8.6.2 Developments in MS-DOS

The structure of MS-DOS and its development over the years reflects its initial application to the original version of the IBM PC, which had a comparatively primitive microprocessor in the form of the 8088, limited provision in terms of addressable memory and simple mass-storage devices. This has resulted in features not ideally suited to present-day versions of the PC, which commonly feature sophisticated 32-bit microprocessors such as the 80386 or 80486, large memory demanded by many application programs and peripherals such as fast, large-capacity hard disks.

MS-DOS Version 1.0, the first version, amounted to 4000 lines of 8086 assembler code requiring 12 kbytes of memory, and could only be used with floppy disk drives. Version 1.0 was small and very crude by current standards. One crucial restriction in MS-DOS Version 1.0, which persists with later versions, concerns memory management. MS-DOS was originally intended to allow for ten 1 Mbyte segments of RAM, and of the 1 Mbyte per segment, 384 kbytes was reserved by MS-DOS for ROM and RAM used for the screen display. Of the remaining memory, 640 kbytes was the maximum a program could use. The 640 kbyte limit has proved increasingly restrictive in recent years, with the need to run larger application programs, and with the needs also of more memory-hungry video systems such as Extended VGA and windowing systems such as Windows 3. The store map is shown in Fig. 8.1.

Different types of memory access under MS-DOS The limits on addressable memory

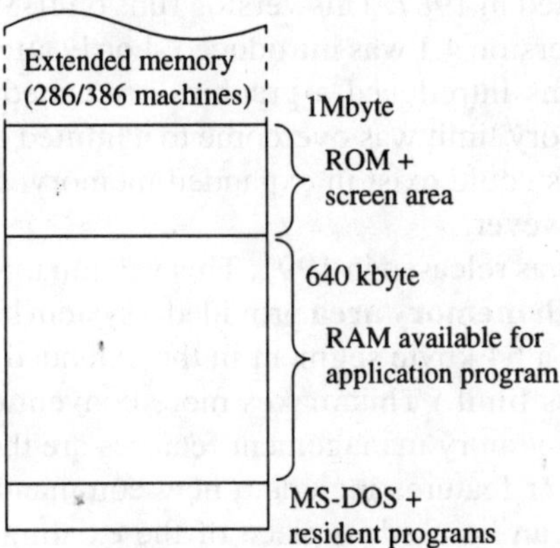

Figure 8.1 Store map for MS-DOS.

with the 80xxx chip family together with limitations imposed by DOS means that there is not—unlike systems based around the 68xxx family of microprocessors—a homogeneous flat address space extending over the entire addressable range. Memory is considered as belonging to one of three categories—conventional, extended and expanded:

- **Conventional memory** extends from address zero, has a maximum capacity of 640 kbytes and is the memory referred to above. MS-DOS programs can use conventional memory without the restrictions that apply to the other types of memory. Some device drivers use conventional memory, in which case only the remainder of the 640 kbytes is available for application programs.
- **Expanded memory** is only available on machines based around the 80286, 80386 and 80486 microprocessors. This type of memory is limited to 384 kbytes, extending from the top of the address space occupied by conventional memory. Special memory manager programs are required to access extended memory to prevent different programs using the same areas within it at the same time.
- **Extended memory** represents the remainder of the address space extending upwards from the 1 Mbyte address limit of expanded memory. This memory space is not generally available to normal MS-DOS programs, but can be used by disk cache, RAM disk, and other such programs.

MS-DOS Version 2 appeared in 1983, with the introduction of the PC/XT machine, and had features to support the needs of the hard disk used with the PC/XT, notably a hierarchical file system.

MS-DOS Version 3 appeared in 1983 to accompany the PC/AT, which was introduced at that time. By now the operating system had grown in size to 40 000 lines of code. A notable advance was the support for the 80286 microprocessor, this included a protection mechanism in which the address spaces of applications may be separated from each other. Version 3.1, introduced in 1984, had features that supported networking. Version 3.2 introduced in 1986 supported 3½ inch disks, this form of storage medium being much more robust and reliable than the earlier 5½ inch floppy disks and rapidly making the latter obsolete.

MS-DOS 4.0 was released in 1987. This version runs on IBM's PS/2 machine as well as on the original IBM PC. Version 4.1 was introduced shortly afterwards, since 4.0 had some major bugs! These versions introduced a graphics-orientated interface supporting windows. The 640 kbyte memory limit was overcome to a limited extent with Versions 4.x, in that data used by programs could exist in expanded memory. Code still had to be held in conventional memory, however.

MS-DOS Version 5.0 was released in 1991. This version makes it possible to locate the operating system in the **high memory area** provided a system has some extended memory. (The high memory area is a 64 kbyte segment in the extended address space starting just above the 1 Mbyte address limit.) This makes more conventional memory available for programs. This and other memory management features are the most important enhancements in Version 5.0. Other features include a new command shell, on-line help, and a full-screen editor, which can be used in place of the existing, rather limited line editor EDLIN.

MS-DOS Version 6.0 is the latest version of MS-DOS at the time of writing, released in 1993. Version 6.0 includes: a disk compression utility, provision for anti-Virus protection and a memory-optimization system called **MemMaker** which can make up to 200 kbytes more conventional memory available.

8.6.3 MS-DOS in use

An outline is given here of how MS-DOS is actually used, including a description of some of the more common commands. We noted earlier that no attempt is made to give an A-to-Z guide on the use of MS-DOS, but rather to provide information which may be quoted in later chapters dealing with operating systems, and generally give a general impression of the capabilities of MS-DOS.

The **command interpreter** is a program automatically loaded when MS-DOS is booted, i.e. after the machine is switched on, provided the operating system is in place on an appropriate disk. This program interprets users' commands and implements them. The command interpreter program is called COMMAND.COM. This program is responsible for displaying the DOS prompts such as C> or A> depending on the disk (other variants are possible depending on the way the operating system is set up to display prompts). Essentially you can think of a prompt as a request for users to type a command, this subsequently being passed on to other parts of the operating system for implementation.

COMMAND.COM allows users to communicate with the remainder of the operating system. The bulk of the operating system will in fact not be visible as far as users are concerned; the information presented by operating systems to users is on a strictly 'need-to-know' basis. The *hidden* parts of MS-DOS are loaded into memory when the computer is booted. System files interact with low-level software implemented in ROMs and which is called **BIOS** (= Basic Input/Output System). Communication to the hardware via its various ports is through the BIOS. Overall then, MS-DOS has an 'onion'-like structure, with the command interpreter as the outer layer and the hardware as the innermost layer. This is shown in Fig. 8.2.

Later versions of MS-DOS allow the COMMAND.COM program to be replaced by other command interpreters, some of which give a graphical user interface (GUI). This

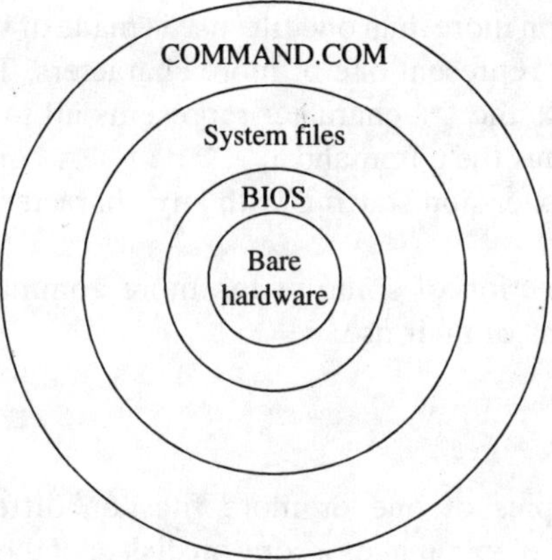

Figure 8.2 Structure of MS-DOS.

kind of interface required an extra layer of software communicating with the MS-DOS inner layers. Features such as windowing, mouse support, etc., are usual. Large amounts of memory and considerable CPU power are commonly required, and it is often the case that some GUIs only run effectively on machines based around the 80386 or 80486. Notable in the GUIs that include windowing features are: GEM (Graphics Environment Manager) from Digital Research, DESKQview from Quarterdeck and Windows from Microsoft.

Using MS-DOS commands As with most operating systems, MS-DOS commands are made up of groups of characters or character strings separated (delimited) by certain other characters. **Parameters** frequently form part of commands, and can change the way a command is executed. Commands can also contain **Switches**, which are parameters that turn on an optional function, and give a command more versatility. Switches are preceded by a '/' character. The prefix and extension in filenames are separated by a '.' character. The prefix should not have more than eight characters while the extension cannot have more than three. MS-DOS interprets files with the following extensions as being executable: .COM, .EXE and .BAT.

The example below illustrates the various possible elements in a command line. The command DIR (think of this as short for DIRectory), used as shown, lists attributes of a file in the directory d:. The line below the command is the response by MS-DOS: the third column is the size of the file in bytes, the fourth and fifth columns are the date and time, respectively, that the file was created or last modified.

```
C> DIR d:\ /OD

ABC      PAS   9052   02-24-91 10:00a
XYZ      FOR  10446   02-25-91 10:01p
```

In the original command, note the delimiter in the form of a space character after DIR. The parameter in this command is d:\, while the switch used is OD, which has the effect, in this example at least, of displaying *all* the files in the D:\ directory, in the date order in which they were last modified or created.

To have commands act on more than one file, use is made of **wildcard characters**. These are special characters that represent one or more characters. The '?' character represents any single character, while the '*' character represents all the remaining characters in a string in the command. Thus the command : `DIR d:*.?AS` gives a directory of files with any prefix and with any extension starting with any character, but followed by the string 'as'.

The list below is a selection of some of the more commonly encountered MS-DOS commands and illustrations of their use:

1. Operating on files
   (a) **COPY** makes copies of one or more files on different devices, e.g. `>COPY d:abc.txt e:xyz.txt` makes a copy on disk e: of the file abc.txt which is on disk `d:`, calling it `xyz.txt`.
   (b) **DELETE** deletes a file, e.g. `>DELETE abc.txt` deletes the file abc.txt in the current directory.
   (c) **RENAME** gives a file a new name, e.g. `RENAME old.txt nu.txt` changes the name of old.txt to nu.txt.
   (d) **BACKUP** constructs backup copies of one or more files complete with directories, e.g. `BACKUP D:\ A: /S` backs up disk D: to floppy disks in the A: drive. The /S switch backs up all subdirectories.
   (e) **ATTRIB** gives a file read-only access, e.g. `ATTRIB +R abc.txt` sets the access rights to abc.txt to read only. A '–' preceding the R parameter would set the read-only attribute to OFF.
2. Operating with Directories
   (a) **MKDIR** creates a new directory, e.g. `MKDIR \MYFILES` creates a subdirectory within the current directory.
   (b) **RMDIR** removes a directory (MS-DOS will not let you do this unless the directory is empty), e.g. `RMDIR MYFILES` removes the directory leading from the current directory.
   (c) **CHDIR** (or **CD**) makes a directory the current directory, e.g. `CD a:\FRED\WPR` moves to the subdirectory `WPR` leading from `FRED` and the root directory.
   (d) **DIR** displays information relating to files and directories (an example of this command was given above).
   (e) **TREE** shows the directory structure, e.g. `TREE` lists all directories and subdirectories in the current disk.
3. Operating with disks
   (a) **DISKCOPY** copies a complete floppy disk on a track for track basis, e.g. `DISKCOPY a: b:` copies disk `a:` on to disk `b:`.
   (b) **FORMAT** formats a disk for use by MS-DOS, and in the process deletes any previous information the disk may have stored. There are a large number of possible switches that can be used with FORMAT relating to different numbers of sectors, tracks, single-sided and double-sided, etc., e.g. `FORMAT B: /N:8 /T:80` formats a 720 kbyte disk in the `B:` drive, with 8 sectors/track and 80 tracks.
   (c) **VERIFY** verifies data as it is written to the disk, e.g. `VERIFY ON` turns verification on.

Tailoring the system environment The sections above outline what might be regarded as an essential core of commands needed to support activities such as program development. However, to give a more streamlined environment for users, allowing MS-DOS to be customized to particular uses, there are other provisions in MS-DOS. One important example is the provision for **batch processing**. This is a convenient way of implementing a sequence of commands for tasks that have to be realized without being involved with the detail of individual commands. The commands to be implemented are placed in a special file called a **batch** file and are then executed by running this file. Batch files can include constructs such as loops and conditional statements as well as simple commands, and form a rudimentary form of programming language. Using batch files can save considerable effort in keying complex command sequences. Batch files have an extension .BAT. A special batch file called AUTOEXEC.BAT is invoked by COMMAND.COM when the computer is booted, and this file is used to initialize the PC's environment. For example it can be used to display the current data, set the current directory, etc. Users would normally edit this file to customize their system to individual requirements. Another file called CONFIG.SYS is also read by MS-DOS when a machine is started (*before* AUTOEXEC.BAT is read). CONFIG.SYS contains commands to load device drivers, and also reserves space in memory for certain system functions. Again users can edit this file to customize their system.

8.7 SOME OTHER OPERATING SYSTEMS

MS-DOS is the operating system used in the present chapter as a convenient vehicle for illustrating some general concepts. To get a broader perspective, however, some other operating systems are described briefly below. The following are a (very) small selection in fact; there are literally hundreds of operating systems of various kinds, although many are somewhat specialized and would not be encountered by a wide circle of users. A more or less comprehensive list is given in Lane and Mooney (1988).

- **OS/2, a competitor to MS-DOS?**—OS/2 was the operating system introduced in 1987 to run on the IBM PS/2 family of computers, which was also introduced at that time. Originally OS/2 was aimed at machines that utilized the 80286 chip, this being used on the entry level PS/2 machine. Later Versions of OS/2 (2.x) are 32-bit implementations and can run on 80386/80486-based machines to exploit the full potential of these microprocessors. OS/2 removes some of the restrictions of MS-DOS, notably the limit on memory size, and is capable of addressing 16 Mbytes of memory in its original implementation (4 Gbytes on Versions 2.x). OS/2's other major feature is that, unlike MS-DOS, it is multitasking. Compatibility with MS-DOS software, at least with later versions than 3.0, is achieved by incorporating a mode of operation called **DOS 3.X BOX** in which it is possible to run one MS-DOS application within the OS/2 environment. With the advantages of these features, it was originally anticipated that MS-DOS would gradually fade away to be replaced by the much more capable OS/2; to date this has not happened, and in fact there are other operating systems that compete in the personal computer arena and make the future of OS/2 uncertain.

- **UNIX**—this operating system will be discussed in some depth in Chapter 11. It might be worth pointing out at this stage, however, that versions of UNIX such as ZENIX compete with MS-DOS and OS/2 in the personal computer market, although, unlike these two operating systems, UNIX is used on computers other than PCs. The primary advantage of UNIX *vis-à-vis* alternative operating systems is the ability of software written for it to run on a wide range of hardware platforms—literally from personal computers to supercomputers, this aspect is explored more fully in Chapter 11. The current position of UNIX, OS/2 and MS-DOS as far as personal computers are concerned is currently extremely volatile; one forecast (Schmidt, 1991) is that by the end of the decade the market will be divided into two types of software environment, MS-DOS and UNIX, with the position of OS/2 restricted to specialized areas. However, making long-term predictions like this is perhaps rather brave at the moment! Deitel (1990) gives an alternative view on this question.
- **VMS**—introduced by DEC in the late 1970s to run their VAX range of 32-bit superminicomputers. It is still in wide use, but currently only on VAX machines, although recently there have been reports that VMS has been ported to other hardware platforms.
- **MVS**—one of IBM's main operating systems used on the 360/370 range of mainframe computers. In view of the dominant position of this type of computer in the mainframe market place, MVS must be viewed as one of the most significant of the operating systems used for mainframe computers.
- **DR-DOS**—an MS-DOS compatible operating system produced by Digital Research. DR-DOS offers enhanced features over MS-DOS. Some versions of DR-DOS have a multitasking capability.
- **PDOS**—this is a real-time operating system used on systems that include those based around the 68xxx microprocessor family. This and other RTOSs are discussed in Appendix 3.

8.8 EXERCISES AND REVIEW QUESTIONS

1. Resource management is one reason for having an operating system: state the other reasons. For what types of system would an operating system *not* be necessary?
2. Quote an application to run on a PC where it would be an advantage to have the machine controlled by a true multitasking operating system.
3. Describe the purpose of the AUTOEXEC. BAT file on MS-DOS, and give examples of typical information it could contain.
4. List the relative merits of MS-DOS, OS/2 and UNIX as operating systems for personal computers. (The discussion in this chapter will provide some guidance here, but to give a full answer, reference could be made to other publications—this would be an especially valuable exercise in view of the rather volatile commercial situation with these three operating systems. Information in textbooks is always in danger of being superseded, especially so in a fast-changing area such as this.)
5. Distinguish between a *parameter* and a *switch* in an MS-DOS command.
6. State the advantages of a multitasking operating system. Can you visualize any *disadvantages*?

REFERENCES AND FURTHER READING

Bradley, D. J. (1990) 'The creation of the IBM PC', *Byte*, September 1990, vol. 15, pp. 414–422.

Deitel, H. (1990) *Operating Systems*, Addison-Wesley, Reading, Mass.

Lane, M. G. and Mooney, J. D. (1988) *A Practical Approach to Operating Systems*, Boyd and Fraser Pub. Co., London.

Lister, A. M (1988) *Fundamentals of Operating Systems*, Macmillan, Basingstoke.

Lister, A. M. (1989) *Fundamentals of Operating Systems*, 4th ed., Macmillan, Basingstoke.

Microsoft Corporation (1991) *Microsoft Users Guide and Reference to MS-DOS Version 5.0*, USA.

Schmidt, E. E. (1991) in A. Rosenblatt, 'PCs and workstations', *IEEE Spectrum*, January 1991.

Solomon, D. W. (1989) *Using DOS*, QUE Corporation, Carmel, Calif.

van de Goor, A. J. (1989) *Computer Architecture and Design*, Addison-Wesley, Wokingham.

9

MULTITASKING OPERATING SYSTEMS

9.1 INTRODUCTION

Chapter 8 gave a practical introduction to operating systems, multitasking operating systems being only briefly discussed. We now examine this type of operating system in more depth. It might be recalled that, with computers controlled by a multitasking operating system, several programs give the outward appearance of executing concurrently (pseudo-concurrently in the case of machines with a single processor). True concurrent execution is only possible in multiprocessor computers.

Designing a multitasking operating system involves the solution of several problems that do not occur with single-tasking operating systems. Some were mentioned briefly in the previous chapter. To summarize, the difficulties were:

- Getting acceptable response times for programs with tight real-time deadlines—interactive users need to get acceptable response times also.
- Giving programs fair and consistent access to a resource such as a tape drive, a printer, or the CPU itself.
- Scheduling programs to prevent potential bottlenecks caused by programs that are either CPU intensive or else I/O intensive, in other words preventing any program from 'hogging' a particular resource at the expense of other programs.

A problem *not* mentioned in the previous chapter is the situation when two programs may each need a resource currently controlled by the other, giving a condition where neither can run because it is waiting for the other program to relinquish control. This is called **deadlock**. A useful analogy to it is the way traffic can 'lock up' at a road junction, with two opposing cars needing the road space occupied by the other.

The present chapter aims to explain how a multitasking capability can be achieved, consistent with a solution of these problems and others to be discussed later.

9.2 TASKS—A MEANS OF STRUCTURING AN OPERATING SYSTEM

One of the more striking characteristics of operating systems in general, and multitasking operating systems especially, is the sheer complexity involved; *any* large-scale system needs some form of structure to handle its complexity, and operating systems are no exception. Early operating systems, it might be added, had hardly any structure at all; they were often referred to as **monolithic systems**. Such systems were difficult to write, invariably contained large numbers of bugs and maintaining them was a programmer's nightmare, especially so in view of the fact that systems programmers used to write code in assembler. Structuring an operating system is generally done in terms of the more obviously identifiable activities that it has to carry out: performing I/O to a disk, responding to input from a terminal keyboard, etc. Such activities can usefully be viewed as building blocks around which an operating system is constructed. The most tangible element of activities such as those described are the programs that need to run for them to be implemented. Additionally, there are data of various kinds, including register values; in total we are talking about all the resources needed by an activity as it executes. This in fact is the task concept very briefly introduced in the previous chapter. In summary, therefore, we are viewing a multitasking operating system as comprising a number of pseudoconcurrently executing tasks. Tasks are created by actions such as users entering a command, or in some situations by other tasks. Tasks interact when they need to exchange data or to synchronize with each other, for example to avoid the deadlock problem mentioned in Sec. 9.1.

9.3 MANAGEMENT OF TASKS

A variety of methods of managing tasks are possible, and techniques used vary, depending on the type of computer the operating system is controlling—not all approaches will be described here.[1] If we take the case of the interactive systems described in Sec. 8.4, tasks need to be **pre-empted** on a regular basis to give users a reasonable response time, with no task permitted to 'hog' the CPU excessively at the expense of other tasks. Clearly a mechanism is required by the operating system to stop a task executing after it has run for some 'reasonable' amount of time (termed a **time slice**), and then to pick another suitable task to run on the CPU. This it would proceed to do for its allotted time slice, or else until the task makes a request to carry out some I/O for example, when it would need to be put in a state of suspended animation until the I/O is complete. The duration chosen for a time slice is determined by a number of often conflicting factors. If too short, the time overhead involved in swapping from one task to another (called a **context switch**) will start to degrade response by absorbing an excessive proportion of CPU time. This problem becomes less important as the time slice is increased, but eventually a poorer response will result. For example, in a timeshared system, users may see an unacceptably long time between hitting a key on their terminal and subsequently observing the corresponding

---

[1] A more extensive treatment can be found in Tanenbaum (1987).

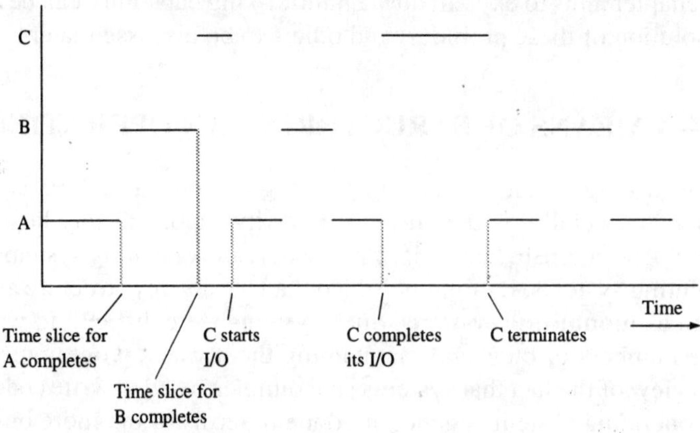

Figure 9.1 Multitasking three tasks.

character echoed on the screen. A typical value of time slice might be 100 ms, being a reasonable compromise between these two conflicting requirements, but there will be much variation depending on the type of operating system.

A possible sequence of events is depicted in Fig. 9.1, for three tasks A, B and C. It must be added that this is highly simplified compared with more realistic cases. The horizontal lines represent those times for which particular tasks are active.

9.4 TASK STATES

In the above—admittedly simple—illustrative example, it is clear that there are several possible states in which a task can be:

- either the CPU is executing it currently (termed **running state**), or
- it is unable to run because it is waiting for some event such as I/O to complete (termed **blocked state**), or finally
- it could be waiting its turn for the use of the CPU (termed **ready state**).

In visualizing the structure of an operating system, some formalization of these states, and possible transitions between them, is helpful. The **state transition diagram** given in Fig. 9.2 shows the various task states, and transitions between tasks.

At this point you should be starting to get a picture of an operating system as a collection of tasks, some of which are formed into a central core, often called a **kernel** (the interchangeable term **nucleus** is also used by some). The kernel's job is to manage tasks in a way best understood by reference to the state transition diagram. Specifically, one of the most important functions of the kernel is to select a ready task to be the next running task, i.e. to **schedule** the task. A common approach to scheduling is to form a queue of tasks in the ready state, making the task at the head of the queue the running task. This is termed

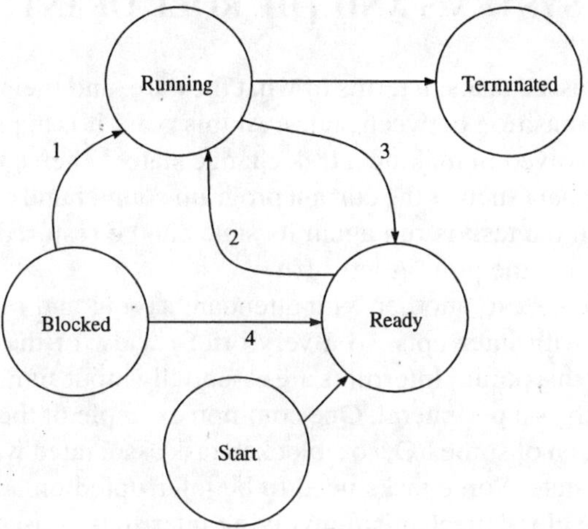

Figure 9.2 Task state transition diagram.

round robin scheduling, and the earlier example implicitly used this method. A running task ending its time slice (transition type 3) would go to the back of the ready queue, as would a task that had become unblocked (transition type 4).

More complex scheduling is possible. For instance it might be desirable in some systems to have tasks of several priority levels (Madnick and Donovan, 1974; Tanenbaum, 1987). Low-priority tasks would then only run if there were no high-priority tasks present. A high-priority task might be one for which there were tight real-time deadlines. Alternatively, tasks which have a heavy I/O requirement might be given a higher priority than tasks which make heavy demands on CPU time. This would enable I/O devices to be kept busy at the same time as allowing the CPU to be used effectively. Scheduling based on two or more queues is commonly used for many systems needing a priority scheme, and again it is helpful to conceptualize such a system using a state transition diagram as shown in Fig. 9.3, which depicts a system in which there are tasks of two priority levels. Scheduling tasks with several priority levels is the norm in a real-time operating system, and this aspect to scheduling will be discussed in more depth in Chapter 12.

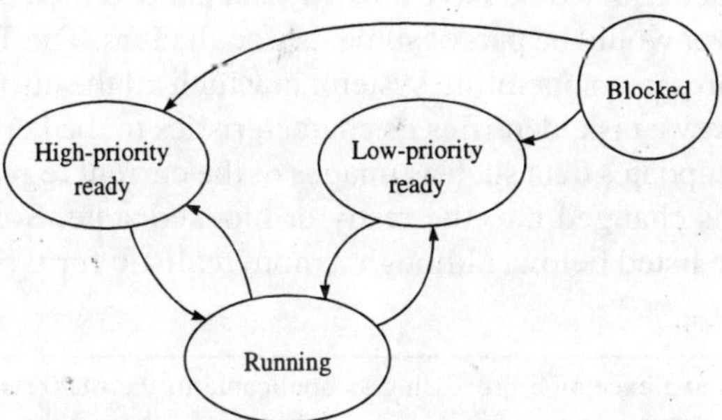

Figure 9.3 State transition diagram for two-level priority scheme.

9.5 OPERATING SYSTEMS AND THE ROLE OF INTERRUPTS

Thus far we have discussed tasks in terms of what they are, and their management, including possible types of transition between states. At this point it is appropriate to discuss the actual mechanisms involved in making a task change state. When a task is pre-empted, its current state including data such as the current program counter and other registers needs to be saved, so that when the task is run again its state can be restored, enabling the task to recommence execution at the point it left off.

Switching from one task to another with attendant saving and restoring of the state is usually implemented with interrupts, so a very brief review of the interrupt mechanism might prove helpful at this point.[2] Interrupts are essentially input signals to a computer from an external device such as a peripheral. One common example of their possible use would be to indicate completion of some I/O, so a blocked task associated with the I/O can then be changed to the ready state. Some tasks need to be interrupted on a regular time basis, in which case a real-time clock mechanism involving interrupts is used.

Following an interrupt, a computer carries out a sequence not unlike a subroutine call, except that, in place of the subroutine's call statement, the interrupt signal serves a roughly comparable function. Hardware can normally respond to interrupt signals in a time span of one instruction, so interrupts are a relatively fast way of responding to outside events in comparison to the alternative technique of polling. Following the initial interrupt signal, the program counter, and possibly other registers, and relevant memory contents are saved, after which an **interrupt handler**[3] is invoked after the interrupting source has been identified by the hardware. There will likely be a large number of interrupting sources, so implicitly a question of priorities needs to be resolved, with some interrupting sources temporarily being ignored if, at a prior stage, a more important handler has been invoked.

9.6 TASK DATA STRUCTURES

To recap: a task is essentially viewed as an executing program together with a set of register values plus other data of various kinds—we will call it collectively **state information**. In order to carry out operations such as pre-empting a running task, it is necessary to have available a table of state information defining a task, for the operating system to update or inspect as required. Each task would have its own such table termed the **task control block (TCB)**, which moreover would be part of some list for all tasks. The TCB is one of the most important data structures in an operating system, in which all the attributes of a task reside. The TCB is the only way a task identifies its characteristics to the kernel and to other tasks. The operating system updates data such as images of the current register values in the TCB when a running task is changed into the ready or blocked state. Some of the more basic elements in a TCB are listed below, although a more realistic form of TCB would contain many more elements:

---

[2] A review of interrupts and exception processing as applicable to the 68000 microprocessor is given in Appendix 2.

[3] **First-level interrupt handler (FLIH)** is another term used with IBM machines, and **interrupt service routine (ISR)** is yet another alternative term commonly used.

- Current state of the task (running, blocked, etc.)
- Unique identification of the task by name
- Pointer to code of task
- Pointer to the workspace of the task
- Elapsed CPU time consumed so far
- Register values
- Pointers to other tasks

There are a variety of ways of structuring a list of TCBs. One approach would be to form all tasks in a given state into a linked list. The running task could be at the head of a list of ready tasks, and the other tasks on the list could perhaps be ordered in terms of position in the queue of ready tasks.

9.7 TASK TYPES/HIERARCHIES

Recall that the task concept is used to give structure to an operating system: to make it simpler to design, use, or simply just to make it easier to comprehend. It is common to further structure operating systems by splitting tasks into different categories (Tanenbaum, 1987). Commonly this might involve classifying a task as belonging to one of a number of layers of tasks, with those on the higher layers calling on a service to be performed by a task on a lower layer.

At the lowest level we have the operating system kernel, this being the part that interacts with the 'bare' hardware, specifically the interfacing chips. Modules concerned with functions such as exception handling or task pre-emption are involved here. Some view the kernel as an **extended machine** (Deitel, 1989) in which the combination of kernel and (bare) hardware is a form of high-level machine complete with its own high-level machine code 'instructions'; the trap instructions of the 68000 can help in implementing this kind of model. The kernel is one of the few parts of an operating system that needs to be written partly or even entirely in assembler, involving as it does machine-specific details like interrupt enabling, and employing heavily used code such as interrupt handlers, and code concerned with task pre-emption. The kernel itself is often structured further by being split into layers, with modules such as interrupt handlers on the innermost layer, and memory management and other processor management in outer layers (Deitel, 1989). The **THE** operating system was split into a total of six layers (Dijkstra, 1968).

In a layered system, the kernel would communicate with what might be called **system tasks**, which provide services in support of **user tasks**; device driver tasks are one example of such a system task, these being responsible for managing the flow of data from I/O devices, making use of the computer's peripheral chips. The kernel hides details of the inner workings of the extended machine from modules in this layer. System tasks would commonly be written in a high-level language, and, in the case of UNIX, the language C is the language invariably used.

At the highest level are application programs of various kinds; packages such as compilers and editors would fall into this category, but also included would be command interpreters such as the COMMAND.EXE of MS-DOS, or the shell program of UNIX.

This layer, including command interpreters, sits outside the operating system proper and invokes services which the operating system can provide—opening a file would be one example.

9.8 TASK COMMUNICATION AND SYNCHRONIZATION

9.8.1 Race conditions and task synchronization

In many cases, tasks may be permitted to share common resources such as an area of store, or a file. However, certain hardware resources such as printers or tape drives are non-shareable in the sense that once one task has been granted access to them, other tasks must be prevented from using the resource until such time as it is no longer busy. These kinds of situation all raise problems of intertask synchronization and communication.

To illustrate the difficulties involved, consider the following problem. Two tasks—A and B say, require access to a counter in the form of a common memory location which they both increment from time to time. Assume each of the tasks include the following 68000 code for this purpose:

```
MOVE    COUNT,D0
ADD.B   #1,D0
MOVE    D0,COUNT
```

Consider the following scenario: task A starts to execute this code but only gets as far as the second instruction when task B starts to run, following an interrupt, say. The state of the first task will be saved in its TCB, including the value stored in register D0. Task B then runs, and increments the contents of location COUNT. Subsequently, task A runs again, restoring the previous register contents in COUNT. The count is now incorrect, being one less than it should be. An exactly similar situation to this exists where several tasks need to read or write to a common file.

Situations like that above, where several tasks are using a shared resource inconsistently, are called **race conditions**. It is absolutely essential that an operating system prevents any race condition from occurring. The key to doing so is to ensure that when a task is using a shared resource, other tasks are prevented from accessing that resource until the first task has finished with it. The concept is called **mutual exclusion**. Most tasks have sections doing computation that is not of relevance to other tasks in terms of shared resources, together with further sections doing computation that *is*. These latter parts of a task are called **critical sections**, and the operating system should ensure that no two tasks are in the same critical section simultaneously.

One superficially attractive approach to achieving mutual exclusion might be for a task to disable interrupts before entering a critical region, and subsequently to re-enable them after the end of the critical region. This approach can in fact prove hazardous, since, if for any reason the task inadvertently does not subsequently re-enable interrupts, the system would likely fail. Additionally it might often be difficult for the CPU to respond rapidly to critical events. Although disabling interrupts in this way are used in certain situations, it is not a practical approach in many cases, particularly in the case of user tasks. To use the method successfully, code for which interrupts are disabled must be very short.

9.8.2 Lock variables

A **lock variable** is a bit which is set or reset to indicate to other tasks if one particular task is in its critical region. When a task wants to enter its critical region it sets the value of the lock variable to 1. If a lock variable is 0, the task sets it to 1, and then enters its critical region. The idea is that, if subsequently another task wants to enter the critical region, it first tests the state of the lock variable, and will not in fact enter *its* critical region if the lock variable is 1. However this suffers from the same problem as the earlier example. If one task reads the state of the lock variable, which happens to be 0, and, before it sets the variable to 1, an interrupt occurs, another task then runs and also sets the lock variable to 1. Both tasks will then be in their critical regions together.

9.8.3 Protected lock variables

The **protected lock variable** is an extension of the lock variable concept, which overcomes the problem that the actions of setting and reading the lock variable are capable of being separated by an interrupt. The modification makes use of a special instruction provided on many computers, which carries out the action of both testing and setting a lock variable, but does so *indivisibly*. The 68000 microprocessor for example has an instruction called TAS (indivisible **T**est **a**nd **S**et), which is provided specifically for task synchronization purposes.[4] Briefly, the effect of a TAS is to read the contents of a specified store location into a register, and then to store a nonzero value at the same store address. The key point is that the two actions are made *indivisible* by the hardware. They cannot be separated by an interrupt, and thus not by another task.

The short assembler routine below illustrates how the TSL instruction can be used to test whether a task is in a critical region. The routine forms part of a task, and would be embedded in the task immediately prior to the code corresponding to the task entering its critical region, as shown. The TAS instruction copies the contents of location FLAG to register D0, and then sets the contents of FLAG to 1. The second instruction compares the *old* value of FLAG to 0. If it (i.e. the *old* value of FLAG) is 1, the lock must have been set previously by another task, so the program loops back to the beginning. Eventually the contents of FLAG will be set to 0 when another task leaves *its* critical region, so then the critical region can be safely entered, the lock variable being set. The task will eventually leave its critical region, and it can then reset the value of FLAG to 0.

```
              {other code}
ENTER$CRIT:   TAS   D0,FLAG ;loop before entering critical region
              CMP   D0,#0
              JNZ   ENTER$CRIT
              {critical region}
              MOVE  #0,FLAG ; reset lock variable
              {other code}
```

9.8.4 Semaphores

The previous approach, whereby a task continuously monitors the state of a variable in

---

[4] See Appendix 1 for more detail on the TAS instruction.

a tight loop, is called a **busy wait**, and has the disadvantage that it consumes CPU clock cycles while inside the loop. A more efficient method is for a task that cannot continue because it is waiting for another task to leave its critical region, to be put into a blocked state rather than being continuously active in a busy wait loop. Such a task would, after being added to the list of other blocked tasks, become dormant while in this state, and so would not consume any CPU time. Tasks in the ready state on the other hand will run from time to time, and thus use the CPU to the detriment of other tasks. The (blocked) task should only be put into the ready state when it is capable of continuing. This form of synchronization uses a variable called a **semaphore**, it is worth mentioning that some authors regard the protected lock variable concept discussed previously as a simple form of semaphore. There are two types of semaphore, a **binary** type which can have only two values 0 or 1, and a **counting** semaphore which can assume any integer values. An operating system will commonly employ a number of semaphores to deal with all of its synchronization problems.

Semaphores are implemented by two routines we will call `wait` and `signal` which respectively decrement or increment the associated semaphore variable. They can be compared to the protected lock variable routines to set and reset a lock variable. The `wait` routine decrements the semaphore if it is nonzero otherwise the task is blocked. The `signal` routine unblocks a task or increments the semaphore. A further consideration is that there may be `wait` and `signal` calls on a particular semaphore for *several* tasks. It is thus possible for more than one task to block following a `wait` on a semaphore found to be zero.

There evidently needs to be a policy for dealing with all tasks blocked in this way. Commonly they are formed into a queue of tasks blocked on a particular semaphore. On a subsequent `signal` involving the same semaphore, the task at the head of this queue is the first one to be put in the ready state.

We are now in a position to give the structure of the `wait` and `signal` primitives for counting semaphores. Note that `wait` and `signal` operations must be indivisible to properly enforce mutual exclusion. One way this could be achieved is by disabling interrupts until the relevant operation is finished. There is little hazard in disabling interrupts here since the code for `wait` and `signal` is relatively short and moreover will form part of the kernel, and so be less liable to problems than it would have been had the code formed part of a user task. In the pseudocode for the `wait` and `signal` routines below, the semaphore `sema` can be of type INTEGER for code written in Pascal, or any other appropriate type for other languages.

signal (sema)

```
IF semaphore queue for sema empty THEN
   sema := sema + 1
 ELSE
  BEGIN
  remove the task from front of semaphore queue ;
  make the task runnable
  END
```

wait (sema)

```
IF sema <> 0 THEN
  sema := sema - 1
 ELSE
  BEGIN
  add task to back of semaphore queue ;
  block the task
  END
```

9.8.5 Message passing

Semaphores are a way of solving the mutual exclusion problem that—in the case of a counting semaphore—also allows for a simple communication mechanism using the value of the semaphore as a counter that several tasks can use. Implicit in the use of semaphores, however, is that they must share a common store, something that would not necessarily be applicable in the case of a multiprocessor machine; naturally the same is true of a computer network with which some useful comparisons can be drawn. The other difficulty is that any errors in writing the code for a task, such as improper use of `wait` and `signal` primitives, can lead to incorrect behaviour allowing two tasks to be in critical regions simultaneously; such errors are all too easy to make. An improvement would be to provide a set of high-level constructs with the aim of simplifying the problems of synchronization and communication aspects of a task, by making sure that these functions are performed behind the scenes in a secure manner, not allowing programmers to have access to the relevant code. Possibly semaphores could be used by such primitives, but their use would be carefully regulated and not susceptible to misuse. A system such as this uses the **message passing** technique (Theaker and Brookes, 1982; Tanenbaum, 1987; Lister, 1989). To implement a message passing scheme, two basic primitives are typically used. They are often called `send (Task_Destination,Message)` and `receive(Task_Source,Message)`.

The first parameter in both cases is a string naming the relevant task, while the second parameter is the message, typically in the form of a record/struct of some kind. There is a wide range of possibilities as regards the form a message can take. For example, if a large amount of data is involved, it may be more efficient for a sending task to place the data in a file and then pass a message in the form of the filename as a character string or a pointer to the file to a receiving task, which would then know which file to access.

Synchronization methods There are various possibilities for synchronizing tasks with message passing. The `receive` primitive could block a task until a message is actually sent by another task; and the `send` operation unblocks the target task if it is waiting for a message. With systems that operate asynchronously, messages are buffered and a target task can continue immediately after receiving a message. For systems that are synchronous, there is no buffering, and a destination task will go into a blocked state until the message has been sent.

Mailboxes The form of message passing described above involves transmitting data directly between a sending task and a receiving task. A mailbox is an alternative form of message passing in which the data to be transferred is first stored in a FIFO buffer called a

mailbox. A sending task transmits data to a mailbox, in contrast to the earlier form of message passing where sending tasks address data directly to a receiving task. Mailboxes can be accessed by more than one sender and receiver; moreover receivers can be distributed across more than one processor. There are some similarities between mailboxes and the UNIX pipe concept to be described in Chapter 11.

9.9 DEADLOCK

Deadlock, otherwise called **deadly embrace**, was referred to briefly in the introduction, being a situation where two tasks wait for a resource currently under the control of the other to be released. Consider the example of two tasks which at various times need to access a printer and a tape drive. Devices like this are nonshareable, they must not be used by more than one task at any given time. If the two tasks request and obtain access to the tape drive and printer respectively, and subsequently the task with access to the tape requests access to the printer, it will be blocked waiting for the other task to release the printer. If, in addition, the task that has access to the printer requests access to the tape, it will also be blocked. This state of affairs could continue forever—the two tasks are deadlocked.

Deadlock prevention involves devising some strategy to prevent situations such as that just described from occurring. This has to be done in the context of certain resources requiring mutual exclusion. One possibility, however, is to override the mutual exclusion condition for restricted cases, for example tasks that only involve output to a printer, and to use spooling techniques for them. In effect we are making a nonshareable device look like a shareable device. With spooling, data generated by several tasks can be concurrently written to separate files. Each of these can be scheduled for eventual printing.

Other methods of deadlock prevention are possible (Deitel, 1989), but in some cases there may be no satisfactory method, and attempts at **deadlock detection** might then be made. This would need to happen dynamically. Achieving deadlock detection is easier said than done; although a system may have several tasks deadlocked, other tasks might be running normally, and the external indication of a deadlocked condition may not be obvious to an operator or user. In fact some *automatic* method of detecting this condition is desirable. The operating system needs to have available to it data pertaining to shared resources, which it can inspect to test for deadlock. This entails a list of the resources each task is using and, for each resource, a list of all the tasks that are waiting for its use. These lists can be inspected each time a request for a resource is made. Following successful detection of deadlock, some recovery procedure is needed. This could involve manually aborting the responsible tasks, although, if deadlocks occur on a regular basis, more fundamental action would be called for.

Unlike deadlock prevention, **deadlock avoidance** aims to deal with deadlock dynamically. As a technique it is less commonly used than the prevention approach, but has the advantage that no preconditions are imposed on allocation of resources, thus avoiding a potential cause of inefficient resource usage. However, with the need to respond to events *on the fly*, the algorithms used for deadlock avoidance are not always foolproof, unlike deadlock prevention, where the possibility of deadlock is, in effect, 'designed out' of the operating system before it ever runs.

As a final observation, it is interesting that the often thorny problem of deadlock is completely ignored by some operating systems, UNIX being a case in point. It is optimistic to expect that this is satisfactory in all cases, however. For example, in specialized custom-designed real-time operating systems, users need to involve themselves more closely in the inner workings of the operating system. Certainly, with real-time operating systems, deadlock has the potential of creating unacceptable problems; what may be little more than an inconvenience in some systems, could lead to catastrophic system failure in others, with the potential of causing damage or even creating life-threatening situations. With the increased use of multiprocessing systems, deadlock problems will get harder to tackle with more resources to deal with, and the added complexity that this entails in terms of deadlock prevention, avoidance and detection.

9.10 TASKS AND THREADS

The overhead time in swapping from task to task is considerable, so a form of task called a **thread** is used in some operating systems. This can be viewed as a **lightweight** form of task, a number of which could, unlike tasks, share a common address space and execute more efficiently as a result. A task might comprise a number of concurrently executing threads, and effectively we are exploiting such parallelism that might exist within a task. Sharing a common address space brings with it the problem of protection, and semaphores or something equivalent would be required to ensure proper mutual exclusion and synchronization of the set of threads that constitute a task. OS/2 is an example of an operating system using threads (Deitel, 1989). SunOS (Deitel, 1989) is another example, the term **lightweight process** is used there, however. In OS/2, the threads within a given task share all that task's resources such as files, memory, etc., but an individual thread does have some data that is private, such as register images. System calls are provided for thread management, such as `DosCreateThread` to create a thread, and `DosSleep` to make a task dormant for some specified period. A typical use of threads would be to have one as a foreground thread where user interaction is involved, and another as a background thread which performs the remainder of the work associated with the task.

9.11 EXERCISES AND REVIEW QUESTIONS

1. The value of a time slice in a multitasking operating system is a compromise between two conflicting factors, state what they are.
2. It was stated in Sec. 9.6 that a list of TCBs for ready tasks could be ordered in terms of position in the queue of ready tasks. Would it make sense to do this for blocked tasks?
3. Section 9.6 describes a set of data that a TCB could contain. Write a suitable C struct for this data, and illustrate how objects of this type could be formed into a linked list, as suggested in the text.
4. What is the difficulty the spooling technique is designed to solve?
5. Give a reason why a multitasking operating system would normally need to use pre-emption for management of tasks.
6. Compare critically the different techniques for dealing with the deadlock problem. (The material in this chapter gives a brief introduction to deadlock and it is suggested that other references be consulted to explore the question posed fully (Theaker and Brookes, 1982; Madnick and Donovan, 1974; Tanenbaum, 1987; Deitel, 1989).)

REFERENCES AND FURTHER READING

Deitel, H. M. (1989) *An Introduction to Operating Systems*, 2nd ed., Addison-Wesley, Reading, Mass.

Dijkstra, E. W. (1968) 'The structure of THE multiprogramming system', *Communications of the ACM*, vol. 11, pp. 341–346.

Lister, A. M. (1989) *Fundamentals of Operating Systems*, 4th ed., Macmillan, Basingstoke.

Madnick, S. E. and Donovan, J. J. (1974) *Operating Systems*, McGraw-Hill, New York.

Tanenbaum, A. S. (1987) *Operating Systems*, Prentice-Hall, Englewood Cliffs, New Jersey.

Theaker, C. J. and Brookes, G. R. (1982) *A Practical Course on Operating Systems*, Macmillan, Basingstoke.

10

STORE SYSTEMS—MANAGEMENT AND USE

10.1 INTRODUCTION

This chapter can be regarded as an extension of the earlier chapters dealing with operating systems; provision of a full chapter entirely devoted to storage management reflects the importance of this area. Principally we are concerned with store concepts in relation to their implications for the operating system, as befits a text dealing with software topics; however, there is common ground with 'hardware' in the form of devices such as disks, and also memory management chips, which system designers need to be aware of. The material discussed therefore serves to some degree in an integrative role between software and hardware. More specifically, we are going to look at the efficient management of storage devices; for example conditions that may lead to I/O bound systems, and techniques to overcome such problems.

A lot of emphasis here is placed on the needs of multitasking systems, one important example being ways of preventing tasks from getting unauthorized access to each other's memory space, and this issue—**protection**—is a vital consideration in any store management scheme. Multitasking systems usually allow several tasks to reside in main store at the same time, which brings with it the question of **allocation** of storage space. If some runnable task is to be loaded into main store, which happens to be already full, a choice of which task to remove to make room for it must be made. Tasks may also need to be moved around in store—this is called **relocation**.

10.2 MULTILAYER STORE ARCHITECTURE

Before looking specifically at management of store, it might be appropriate to give a very

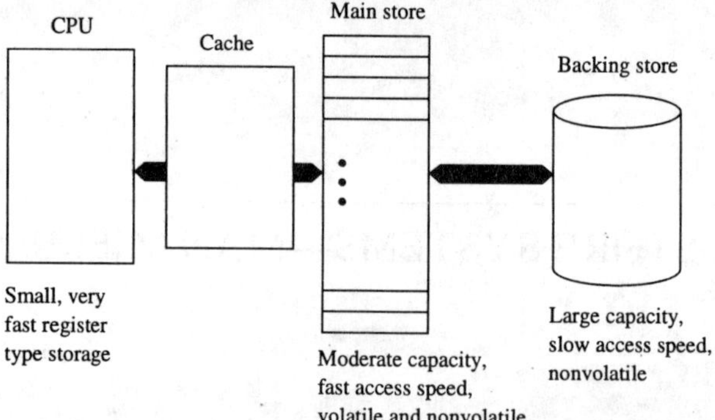

Figure 10.1 Layered store architecture.

brief review of some basic store architecture principles. A computer's store is usually split into two parts in the form of the main store and backing store, although many systems additionally have a fast buffer store called a **cache** interposed between the CPU and main store, with the idea of solving the potential bottleneck caused by the need for a relatively fast CPU to interact with a relatively slow store system. Put simply, a cache is a method of fooling the CPU into thinking that it is dealing with a fast store, not always successfully it must be added, but nevertheless in a way that is largely transparent as far as the CPU is concerned. A cache can be regarded as a hardware technique with little impact in terms of any effect on the operating system, and so will not be discussed further here.

The layered approach to store architecture (Fig. 10.1) is dictated by the conflict between speed and cost, combined with the need for a nonvolatile medium such as disk to act as a file store. Much of the later discussion will involve methods of managing a store system assumed to have this layered type of architecture.

10.3 STORE ALLOCATION, PROTECTION AND RELOCATION

It is possible to use store to hold simply one program, or perhaps several, and moreover, backing store may be used to hold parts of programs if it is inconvenient to hold complete programs in main memory (as opposed to its other use as a file store). In any event we need to discuss various methods of allocating store to programs, together with methods of protection and relocation.

10.3.1 Store allocation and protection—single-user systems

Early computers, and more recently personal computers, had their main store divided into an area used by the operating system, plus another used by the (single) user's program and workspace, with some store left over and unused. In this, and other types of store system, the operating system may be in volatile (RAM) memory, being loaded there from disk when the machine is booted. For this purpose a small bootstrap loader needs to exist in some nonvolatile store such as ROM. Other systems may have all, or part, of the code for the operating system in ROM, an approach adopted in some personal computers and in diskless embedded type systems where disks could raise problems. In some of these early systems, if programs were too large to fit into main store, they were executed by dividing them into parts called **overlays**; an overlay was loaded into store and executed, and then, after it had finished, the next overlay was loaded and executed.

Simple systems like this often have no protection at all. User programs are able to overwrite the part of store allocated to the operating system with no restriction, invariably with disastrous results! The operating system CP/M used in early personal computers is an example of such a system. Some single-user systems achieve a measure of protection by means of checks on addresses generated when a user program accesses store to do this (Fig. 10.2). A dedicated CPU register is provided which is initially loaded with the highest address used by the store; and the address generated by a program is compared to the contents of this register by the CPU hardware. If the address is less than that in the register, an exception is triggered, and the operating system then carries out some error-handling procedure. Subsequently, users may see some form of error message displayed on their terminal screens.

Finally it might be useful to talk about the IBM PC; its memory organization and protection were discussed briefly in Chapter 8. The memory map for the PC is different to that depicted in Fig 10.2, in that user programs are sandwiched between the operating system in RAM at a low area of memory, and the BIOS and other parts of the operating system in ROM, which end at the 1 Mbyte address boundary. Recall also from Chapter 8 that with a DOS-controlled PC, only one program can execute at any given time, although it is possible for several programs to be resident in the **transient program area** (**TPA**). Commonly, utility programs such as Borlands Sidekick operate like this, being activated by a specified key called a **hotkey**. This should not be confused with multitasking, users can only (manually) switch the machine from one program to another.

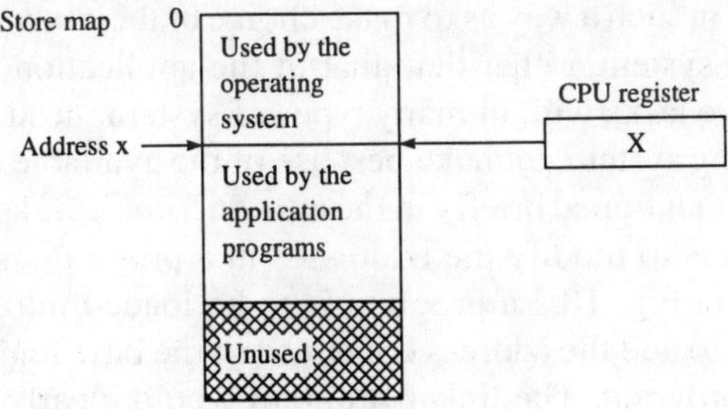

Figure 10.2 Simple single-user store protection system.

10.3.2 Store allocation—multitasking systems

In multitasking systems, it is usual to have a number of tasks residing in main store at any given time. One reason for this is the fact that frequently tasks will spend much of their time waiting for some I/O event to complete. For example, it is common for some types of task to wait in a loop while they read data from a file and then do some computation using the data that has been read. If, say, it takes 40 ms to read a block of data from disk, and 10 ms for the CPU to carry out computation on that data, the CPU would be idle for 30 ms waiting for the data to be read. With a number of tasks in main store at the same time, however, the operating system can swap from a task that is not runnable to another that is.

Multitasking systems *may* allow a task to remain in store until it finishes. However, in some systems, a task only stays in main store until it blocks for some reason, after which it is **swapped out** to disk by the operating system, on to a part of the disk often called a **swap area**. This allows another task to run earlier than would otherwise have been possible since there would then be room for it, in what would have been a full store. In systems that have a large number of tasks such as multiuser systems, it is common for there not to be enough memory to hold all tasks at once, and in this situation some swapping system would be mandatory.

We now consider some methods of allocating store to tasks for systems that involve swapping and those that do not.

Fixed partition systems Many types of store management systems have been developed for multitasking systems. One method divides store into a number of partitions of fixed size. Each partition is occupied by a single task, and the operating system switches from one task to another as described in the previous chapter. There would normally be a queue of suitably sized tasks waiting their turn to be loaded into a partition of adequate size. Fixed partition systems have the disadvantage that tasks will seldom fit *precisely* into a partition, so some store will be unused. Nevertheless, a fixed partition system was used by the operating system OS/360 on IBM mainframe machines.

Protection is more complex than for single-user systems—now, user tasks need to be protected against unauthorized access by *other user tasks*. To achieve this, some fixed partition systems employ two dedicated CPU registers holding the lowest and highest addresses of the current active partition, as shown in Fig. 10.3. Attempts at memory access outside these limits causes an exception and suitable action by the operating system.

In addition to providing a protection mechanism, it would be nice to have some means of loading tasks into store in such a way as to make choice of their *actual* location the responsibility of the operating system, rather than that of the application programmer. We have already noted that active tasks will, in many types of system, need to be moved around in memory by the operating system, to make best use of the available memory space—this is the relocation problem mentioned briefly in the introduction. One approach to relocation in fixed partition systems is to modify the addresses in a program as it is loaded into main store. For example, if in Fig. 10.3 a program is to be loaded into partition 2, the loader program could arrange to add the address contained in the base register to addresses in the linked version of the program. The linker program would clearly need to identify those parts of programs such as branch addresses that would need to be modified to make this

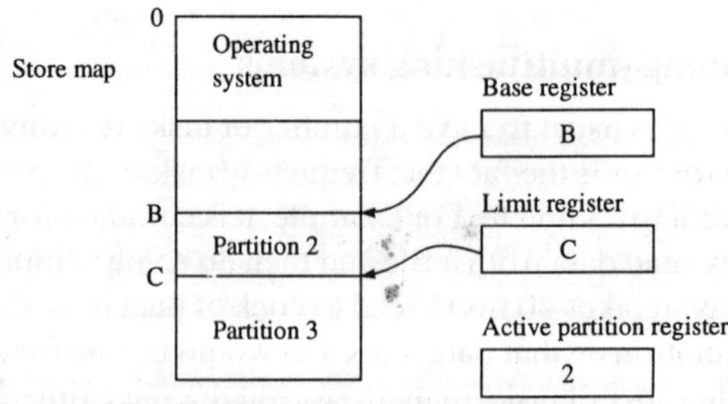

Figure 10.3 Simple base/limit register protection scheme.

possible. In this way the objectives of both protection and relocation are achieved with the hardware provision of base and limit registers. The Intel 8086 family of computers employs a scheme loosely similar to this in some respects.

Variable partition systems The most common way of allocating store in multitasking systems is to allow for the space for tasks to be allocated dynamically, rather than being constrained by having to fit into certain areas of store as in fixed partition schemes. A task is now to get as much store as it needs with no predefined constraints on the address boundaries at which it is be loaded. Additionally, if necessary, the operating system would remove another task to make room for an incoming task. Tasks enter and are removed from store in a way illustrated in the store maps shown in Fig. 10.4 which gives a short history of four tasks, showing the store occupancy of the tasks at five different times. The shaded regions represent unused store, more usually called **holes**.

It should be apparent from this example that, as a task finishes, it leaves a hole between the areas occupied by the remaining tasks. An example is (e) in Fig. 10.4 by which time `task2` has finished and `task4` has entered store. A hole between `task1` and `task4` will be created. This type of thing continues, and eventually it may be that tasks are separated from each other by a large number of holes, many of which might be quite small—possibly too small to hold any useful task. This characteristic is called **fragmentation**. The difficulty is that although the sum total of the space occupied by holes might be quite appreciable, *individually* the holes may be too small to be of any use.

Storage placement in variable partition schemes Store will have a number of holes following the kind of sequences described above, and it is the job of an operating system to

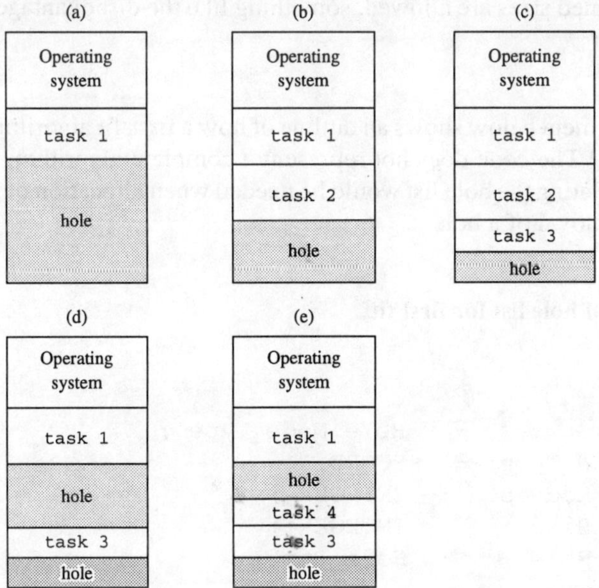

Figure 10.4 Store allocation with variable partitions.

select which hole to use for holding the next task to be loaded into store. All methods rely on the availability of a linked list of holes in which (typically) the start (base) address and size of each hole is specified. Several methods are possible:

- **Best fit**—the hole list is ordered in terms of holes of increasing size. A search is then made of the hole list, and the first hole large enough to hold the incoming task is selected. This evidently leaves the smallest amount of unused space.
- **Worst fit**—the hole list is ordered in terms of decreasing size. Again the hole list is searched as above, but this time the *largest* hole is selected instead of the smallest. This superficially seems to be an odd approach, in direct contradiction to best fit. The thinking behind it is that in choosing a large hole there is a greater probability of leaving a large hole for future use; with best fit, choosing a small hole might in the future leave a still smaller hole, leading to greater fragmentation.
- **First fit**—the hole list is ordered in terms of increasing hole base address. The hole list is searched, and the first hole large enough is used for the incoming task. This is a relatively fast method of allocating store, but it can lead to a situation in which a large number of small holes appear near the head of the hole list.
- **Buddy**—here a number of lists of holes are created, each comprising blocks of one particular size. The sizes of the lists are 1, 2, 4, 8, etc., words—i.e. powers of 2. A request for store of a particular size will result in the removal of a block in the relevant list. If the list happens to be empty, a block from the list with the next highest number of words is split in two, with one block being used to satisfy the original store request, and the other block being added to the list of the smaller of the two blocks. Allocation of store is fast with this method, but there is the problem of using store best, in that only blocks of certain predefined sizes are allowed, something like the disadvantage of fixed partition systems in fact.

The Pascal fragment below shows an outline of how a first-fit algorithm could be used to search a hole list.[1] The code does not represent a complete algorithm, for example, provision for manipulating the hole list would be needed when allocation of new storage space resulted in the removal of a hole.

Manipulation of hole list for first fit

```
TYPE
    Hole_Ptr            =   ^Hole_Node_Type ;
    Hole_Node_Type  =   RECORD
        Base_Address    :   INTEGER ;
        Hole_Size       :   INTEGER ;
        Next_Hole       :   Hole_Ptr
    end ;
```

---

[1] Chapter 5 gives details of linked lists in Pascal.

```
VAR
   Space_Found       :   BOOLEAN ;
   Ptr , Head        :   Hole_Ptr ;
   Size              :   INTEGER ;


...

...
Ptr := Head ;
Space_Found := FALSE ;
{ search for hole, using first fit }
while NOT (Ptr = NIL) AND (NOT Space_Found) do
   begin
   if Ptr^.Hole_Size >= Size then
      with Ptr^ do
         begin
         Base_Address := Base_Address + Size ;
         Hole_Size := Hole_size - Size;
         Space_Found := TRUE ;
         { code for case where hole COMPLETELY used, with the
           need to remove this hole from hole list, goes here }
         end ;
   Ptr := Ptr^.Next_Hole
   end
...
...
```

10.3.3 Sharing—allowing access to common programs

An essential objective in any memory management system is that of allowing tasks to have shared access to certain programs. For example, it is common in a multiuser system for several users to need to access packages like editors or compilers at the same time, and it makes little sense to provide multiple copies of the same program, one per user, with the resultant increased demand on storage space that this would entail. The operating system must make a decision on which programs can be shared; and nonmodifiable programs such as compilers would be an example of programs that could be shared, while modifiable data would not normally be shared.

10.3.4 Example of memory allocation in DOS

DOS organizes its free memory as a linked list of free blocks called a **memory arena**. Each entry in the list called a **memory control block** consists of 16 bytes, the function of each of which is listed in Table 10.1.

Table 10.1 Memory control block

Byte	Function
0	This byte is 0×90 if last block, $= 0 \times 77$ otherwise
1 to 2	0 if block not allocated, Process ID otherwise
3 to 4	Size of the block in units of 16 bytes
5 to 7	Not used
8 to 15	File name of owning program

Each memory control block represents a contiguous area of memory directly above it. Each control block acts as a pointer to the next control block in the list with bytes 3 and 4 giving an offset to the next block and serving as a pointer. When DOS receives a request from a program for some memory, it searches the free list to find a large enough block. Unsuccessful attempts to allocate memory to a program, for whatever reason, result in the message 'Memory allocation error' being displayed. The system then halts, and the unfortunate user has no alternative but to reboot the machine. Several possible strategies are open to users to allocate memory, although versions earlier than DOS 3.2 used first fit, this also being the default option used by later versions, which can, however, be overridden.

It might be worth noting at this point that .COM type programs will automatically be allocated *all* the available free memory, in the first free block large enough to hold the program. One would see this as sensible in the context of a single-user machine for many situations. In contrast, .EXE type programs are allocated the amount of memory specified in a field within the header which comprises part of an executable DOS program. Note, however, that when a program is linked, the linker will, by default, try to allocate *all* the available memory to the program, and if less memory is required, the appropriate command line switch should be used to specify the actual amount of memory required—this option is not open to programmers in the case of .COM type programs.

10.4 VIRTUAL MEMORY SYSTEMS

We have discussed methods of allocating store to tasks as and when they are needed, including methods of efficiently fitting tasks into available space in store. Some methods of achieving protection and relocation have also been covered. The discussion now broadens to cover memory management in situations where programs may be larger than the physical size of memory. We have noted one solution to this in terms of splitting programs into overlays. This is generally not an attractive method, involving as it does provision in code for dividing up programs into overlays. The more usual way of tackling this problem is by the use of **virtual memory** systems; we will see that the method is also closely bound up with protection and relocation, so these aspects of memory management will need to be revisited.

Essentially, the virtual memory concept involves making the combination of the relatively large-capacity secondary storage and the primary storage appear as a single flat monolithic address space as far as programs are concerned. This provides programmers with a simple model of the store system in which they need not be concerned with the *actual* location of data, leaving it to the memory management system to locate data either on backing store or in main memory. Programs use **virtual addresses** which are translated to physical addresses by the hardware. Moreover, the memory management system is designed automatically to carry out address translation in such a way as to prevent tasks from gaining unauthorized access to each other's address space. This means that, even if two tasks reference the same (virtual) address location, this would be translated to two different (physical) memory locations, or else to two different areas on disk. Finally, address translation is accompanied by relocation of tasks in a way transparent to programs, so the relocation problem is dealt with as well.

Conceptually, the situation is illustrated in Fig. 10.5 where the storage provision for two tasks is depicted (hatched regions represent unused memory). Blocks of code and data in the virtual address space of the two tasks can be distributed in blocks of real store, or else areas on disk, with tasks being split between main store and backing store in some types of system. Note that code or data that are contiguous in terms of virtual address space need no longer be so when the mapping to real store addresses has been carried out. A further inference is that virtual memory systems have solved the relocation problem, in view of the fact that the *hardware* has the task of carrying out the address mapping process. It might also be worth mentioning at this point that, although virtual memory systems frequently operate in the context of programs larger than the available memory space using backing store as described, this is not a *necessary* condition; the address mapping concept is equally applicable in a system without disks or where the virtual address space is no larger than the physical address space or even smaller for that matter! The advantages of the memory protection and relocation will still accrue.

10.4.1 Paged systems

In this type of virtual memory scheme the virtual address space is divided into blocks called

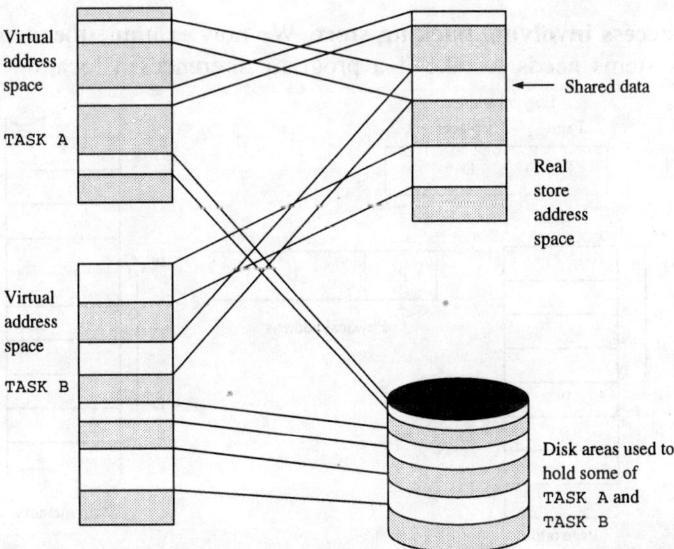

Figure 10.5 Relation between virtual and real store addresses.

pages—of 512 bytes in one type of computer—although there is wide variation in block size from one system to another. Those pages currently in main memory are called **page frames**, so that at any instant a task is partly loaded into page frames in main memory, with the remainder of the task being located in backing store. The paging mechanism must provide a way of identifying, for the current memory access, whether or not the data referenced is in main memory, and if not, transfer the page with the required data from backing store to main store. It follows that there must be a method of recording the presence of those parts of the task currently in pages in the main store. In order to determine to which page a program address is referring, virtual addresses are structured such that some high-order bits are used to reference a page, with the remaining low-order bits referencing a word within that page—called a **displacement**. Hardware must be provided that can process page numbers and displacements to carry out the functions described earlier.

A virtual address corresponds to a single contiguous address space, and, to convert from this to a physical address, the memory management hardware uses a store called a **page table**. Figure 10.6 shows how this system can turn an access to a location for which the virtual address corresponds to a page P and displacement D into a physical address. The physical address is computed by adding the displacement D to the value pointed to in the page table by the page number, namely P′.

It should be evident now that access to main memory in fact requires *two* store access cycles not one: the first to access the page table, and a subsequent cycle to access main memory itself. In the interests of speed, the page table might be implemented as a separate high-speed store rather than making the page table a part of the main memory, but this is really a hardware issue and will not be pursued further here (Deitel, 1990; Motorola, 1983). Division of a virtual address into a page number and displacement is entirely transparent as far as programs are concerned; it is up to the memory management hardware to make use of page number and displacement in mapping virtual addresses into physical addresses.

Page faults—access involving backing store We now examine the steps the memory management systems needs to take if a program references a location in a page not

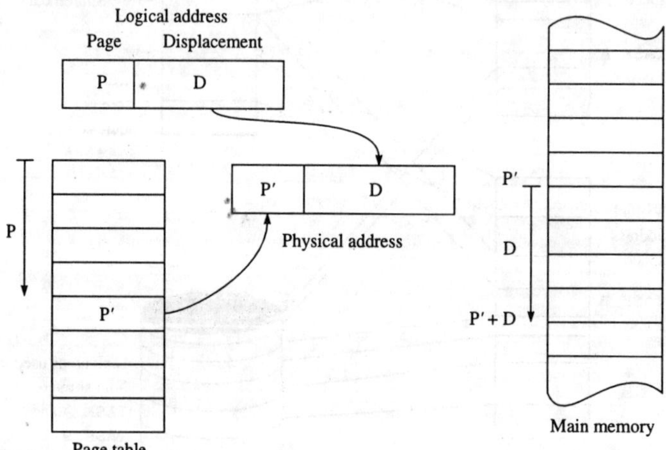

Figure 10.6 Paged memory system.

currently in main memory. This is signalled by a bit called a **page presence** bit forming part of the words in the page table. This bit is set to (say) 1 if the page is in main memory and 0 otherwise. If the location is *not* in fact in main memory, a condition called a **page fault** is said to exist. This triggers off an exception which, in the case of a read access, must load the relevant page from backing store into main store, and also update the page table to indicate the new page. Naturally the task involved must block until the I/O has completed.

The situation of handling page faults gets more complex if the main store is originally full, since a decision then needs to be taken on which page to remove to make room for the page about to be read from backing store. One strategy is to keep track of the page frames that are used infrequently, and to remove these when page faults occur. This is referred to as **LRU** (= Least Recently Used), and is one of a number of possible algorithms, a detailed description of which is given in most texts on operating systems (Deitel, 1990; Lister, 1988).

If the form of paging described above is considered from the time a task starts; then, as the CPU tries to fetch the first instruction, a page fault is generated to get the page containing the required instruction. Other page faults are generated when further instructions and data become needed by the program. In time, most of the pages needed by the task will be in main memory and the incidence of page faults will be less. This will be particularly true of programs for which the bulk of the time is spent executing a relatively small part of the total code, which would usually be true of programs with loops of various kinds—most programs in fact. Tasks need to have some minimum number of pages held in main memory if the CPU is to execute the program efficiently—referred to as the **working set** (Denning, 1968). This characteristic of programs is termed **locality of reference**. To sum up then: exploiting locality of reference properly involves having as much of the working set in main memory as possible, in which case page faults will occur relatively infrequently and the task will in consequence run efficiently. However, if this is *not* the case—memory may be too small for instance—there will be an increased amount of I/O to load required page frames in memory, and tasks then execute more slowly as a result. It is in fact possible to reach a situation where the computer is spending most of its time dealing with page faults, and relatively little time on useful computation. This condition is called **Thrashing**, and could well be a symptom of the fact that more memory needs to be installed to hold the working set.

How large should a page be? Choosing optimum page size involves tradeoffs between several conflicting factors. It is clear that, in general, typical portions of code, stack areas and other data will not precisely fill a page, so that some store is always going to be wasted. This argues against a large page size. On the other hand, a small page size implies a large number of pages and hence a large page table. Transfers to/from disk are often a page at a time with most of the time involved due to seek and rotational latency for the disk. This means that transferring a small page takes about the same time as a large page, militating against small page size. Optimum figures differ from one machine to another, although we quoted a page size of 512 bytes earlier, this being used on DEC's VAX range of minicomputers. Page sizes of 1, 2 and 4 kbyte are also used.

10.4.2 Segmentation—another memory management technique

If the structure of tasks is examined, it is possible to identify a number of different entities such as procedures, stacks, heaps and other types of data. It might be desirable to treat each differently, reflecting their different function: usually data would be divided into that which only the parent task is allowed to read, and that which other tasks are permitted to access; it would seldom be a good idea for tasks to have the ability to write to the memory space used by code for other tasks; we have already noted that if several users need to access a common package such as a compiler, it would be better to make the relevant code shareable between the tasks that need it. Considerations like these might suggest that memory management systems should treat individual parts of tasks differently, with the power to give different types of access rights (read, write, execute or combinations of these) as needed. Segmentation is a memory management method that has these objectives.

In most segmentation schemes, programs are allowed to reference memory by a **segment number** and a **displacement** relative to the segment referenced. The segment number is chosen by the programmer in accordance with the logical function of the code or its data. A virtual program address has a segment number appended to a displacement in something like the way a page number is appended to a displacement in paged systems. However, with paging, the virtual address from a programmer's standpoint is a single entity, and the use made of it by the memory management hardware is transparent to programs. The hardware to implement segmented systems is not unlike that of the paged system. An address mapping mechanism is required to translate a virtual address on to a real store address, and a **segment table** carries out a function comparable in some respects to that of the page table in paged systems. A segment number is used to index into the segment table; and one of the fields in the word referenced, called a **segment descriptor**, is used in combination with the displacement to form a real store address. Another field in the segment table gives the number of words in the segment. This allows protection against encroachment into other segments by comparison of the displacement in the program address with the segment length. Further bits in segment descriptors allow for whatever protection access rights are needed. Sharing segments is straightforward, just requiring a segment descriptor for the segment to be shared in the segment table of each task. The protection rights in each descriptor can be different.

10.5 MEMORY MANAGEMENT IN 68000 SYSTEMS

The 68000 chip was designed from its inception to give good support for memory management requirements, a far-sighted objective in view of the early date (1979) at which this microprocessor was introduced. Virtual memory systems can be developed, possibly using one of Motorola's memory management chips such as the MC68451, although some members of the 68000 family support the memory management function, on-chip. Memory management provision is seen as mandatory in many multitasking systems which use the 68000.

10.5.1 Protection mechanisms in the 68000

The 68000 chip has a protection mechanism applicable to multitasking applications, by the provision of two modes: **supervisor** and **user** modes. Supervisor mode allows full access to all the 68000's instructions, while, in user mode, certain privileged instructions cannot be used. An example of a privileged instruction is STOP, this loads a 16-bit immediate operand into the status register, and then waits for certain forms of exception to occur before fetching any more instructions. Attempts to employ this instruction in user mode would result in a **privilege violation** type exception. The other difference between the two modes is that there are separate stack pointers used respectively for user and supervisor modes. As the reader might have guessed at this point, supervisor mode is normally used when executing the kernel of an operating system, while user mode would be applicable for user tasks, which can be less well-behaved and thus more likely to cause a system to crash. Transition from user mode to supervisor mode is only possible by raising an exception—commonly initiated by a `trap` instruction. This allows entry into the operating-system kernel under carefully controlled conditions, permitting the operating system to 'vet' actions a user task might want to carry out, before allowing them to be executed.

The other protection mechanism built into the 68000, is a method of dividing the memory space of a system into separate code and data areas for both supervisor and user mode. This allows a total of 4×16 Mbyte separate memory spaces, all protected from each other. Access to each is signalled to memory by the microprocessor asserting a 3-bit code on output pins called FC0, FC1 and FC2. It is this code, in conjunction with the 24-bit address on the address bus, that determines which physical memory location will be currently accessed. The function code pins can be employed as part of the address decoding scheme of the memory system, but we will see later that they are also employed in connection with virtual memory systems made possible when a memory management chip is used. Table 10.2 enumerates the type of processor cycle applicable for each combination of the function code pins, although only four combinations are relevant to the present discussion of protection mechanisms.

Table 10.2 Processor cycle types for all combinations of function codes

Function code			
FC2	FC1	FC0	Processor cycle type
0	0	0	Undefined
0	0	1	User data space
0	1	0	User code space
0	1	1	Undefined
1	0	0	Undefined
1	0	1	Supervisor data space
1	1	0	Supervisor code space
1	1	1	Interrupt acknowledge

10.5.2 Example of Motorola's MC68451 memory management unit

Motorola's memory management unit (MMU) is a chip used in 68000 systems for store management in the context of a multitasking environment where a virtual memory capability is required (Motorola, 1983). The 68451 chip supports a paged[2] system of memory management. Segment size can be defined by users from 256 bytes up to 16 Mbytes

---

[2] The term *segment* is used in the manufacturer's data sheets rather than page, and the former term is used from this point on when referring to registers such as the segment status register.

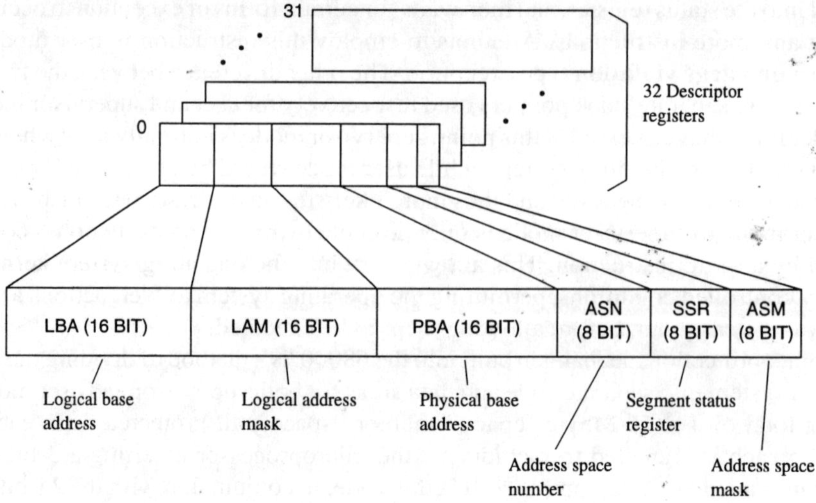

Figure 10.7 Descriptor registers in the 68451.

in steps of powers of two; this facilitates the implementation of the buddy replacement algorithm, should users prefer this to other algorithms. Additionally, the 68451 can be programmed to give segment sizes of different values; there is no hardware-imposed limit on page size as with some other systems. Up to 32 segments of virtual memory can be mapped into a system's physical memory space, and for this purpose 32 **descriptor registers** are provided. A field within these registers constitutes what we previously would have called an entry in a page table. It is possible to build virtual memory systems using anything from one up to eight 68451s, allowing up to 256 segments, although here only systems with a single MMU will be considered. Figure 10.7 shows the descriptor registers—some of its fields will be explained later.

Address mapping mechanism Figure 10.8 shows in outline the relation between the (virtual) address generated by the processor, and the physical address used by the memory system (the significance of the FC signals will be discussed a little later). The high-order

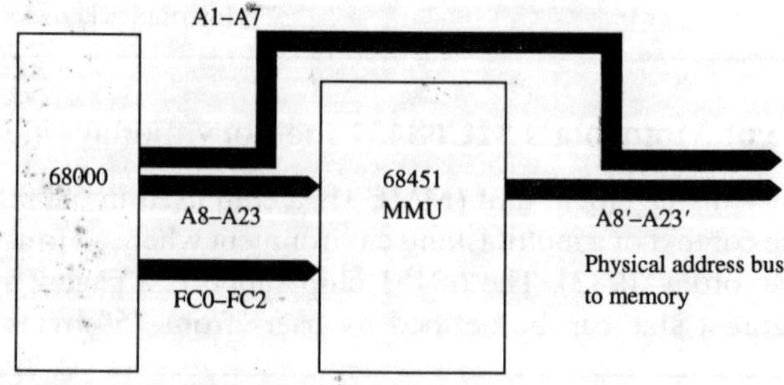

Figure 10.8 Simplified view of the MMU and its connections to a 68000.

address bits A8 to A23 are supplied to the MMU, while A1 to A7 are supplied direct to memory and form part of the displacement in the address (recall the role of the *displacement* in addresses from Sec. 10.4.1).

Remember that, in paged systems, certain bits in a virtual address are used in accessing the page table, which additionally has provision for the indication of page faults using a page presence bit. Here, the presence or otherwise of a page (or segment) resident in memory is tested using certain bits in the **logical base address (LBA) register**, which are compared with bits in the virtual address. The actual mechanism of generating physical addresses is somewhat complex, so we start off with a simple case where we assume for now that only 8-bit segments are required. The approach would then be to compare bits 8 to 23 in the virtual address with the entire contents of all the LBA registers (associatively—i.e. simultaneously—in the interests of speed) (see Fig. 10.9). A match with one of the LBA registers would indicate that the required segment is currently resident in store, and a calculation of the physical address would then proceed. Failure to find a match would result in the MMU asserting an output called `FAULT*`, which is normally connected to the 68000's `BERR*` pin. Assertion of `BERR*` results in a `bus error` exception, and the processor must then ensure that the data is accessed from backing store. The high-order bits of the physical address (bits 8 to 23 in this case), would be found in a **physical base address (PBA) register**, corresponding to the LBA register for which a match had been found. The complete physical address is found by incorporating the low-order displacement bits A1 to A7 which were routed directly from the 68000 to the store system.

An example of this is illustrated in Fig. 10.9, where a virtual address \$123456 is used by the MMU to generate a physical address \$800056, following the match between the (virtual) segment number \$1234 forming the high 16 bits of the virtual address, and the same number found in one of the LBA registers. The corresponding PBA register contained \$8000, thus forming the complete physical address \$800056 in conjunction with the displacement \$56. Note that \$8000 is the *lowest* address in the (256 word) segment.

To form segments of other sizes, a similar approach is adopted, but only a limited number of bits in the LBA participate in any address-matching exercise. In addition, only a

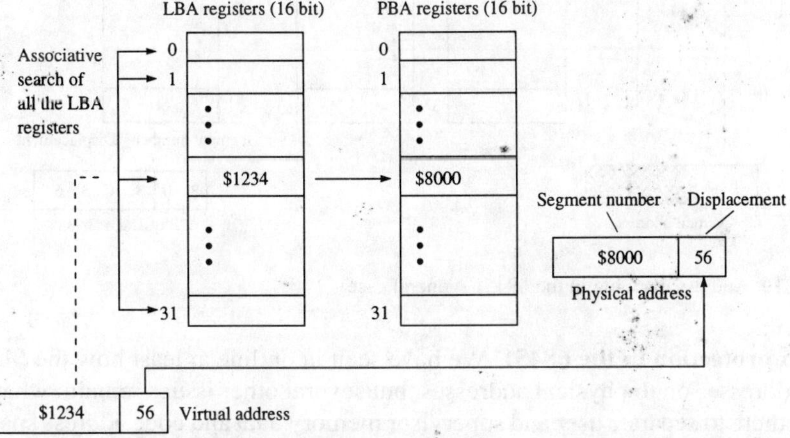

Figure 10.9 Address mapping in the 68451 (256-word segments).

limited number of bits in the PBA register are used to form a physical address. For example if we wanted (say) 16-bit segments, then it would be necessary to match 8 bits in an LBA register with the high 8 bits of the virtual address. The required bits are selected by a mask loaded into the **logical address mask (LAM) register**, which is a 16-bit register: this mask defines bit positions in the LBA that can participate in a matching comparison with the virtual address transmitted from the processor. Only bits in the LBA for which corresponding bits in the LAM are ones feature in any address comparison. Bits in the virtual address for which corresponding bits in the LAM are zeros are passed directly to the output of the chip where they form part of the physical address along with A1 to A7. Effectively we are controlling the number of bits in the LBA as far as any address comparison exercise is concerned. The end result should be that certain high-order bits in the virtual address are to be compared with the LBA masked with the LAM for all 32 descriptors. Calculation of a physical address to be sent to the memory system follows if there is a match. The mask in the LAM is similarly applied to the selected PBA register, thereby controlling the size of the segment.

Figure 10.10 illustrates a case where a 16-bit segment is involved. The cross-hatched parts of the LBA and PBA correspond to those bits for which corresponding bits in the LAM were zero. In the example, there is a match between relevant bits of the virtual address and the entry in the LBA for the 16-bit segment, i.e. $12. Thus the LBA for this descriptor is used to form the physical address. The other entry shown which corresponds to a 12-bit segment would *not* cause a match since the high 12 bits in the virtual address and the LBA entry differ. Indeed, it would be something of an embarrassment if more than one entry found a match in the LBA—the reader might care to think of a reason why!

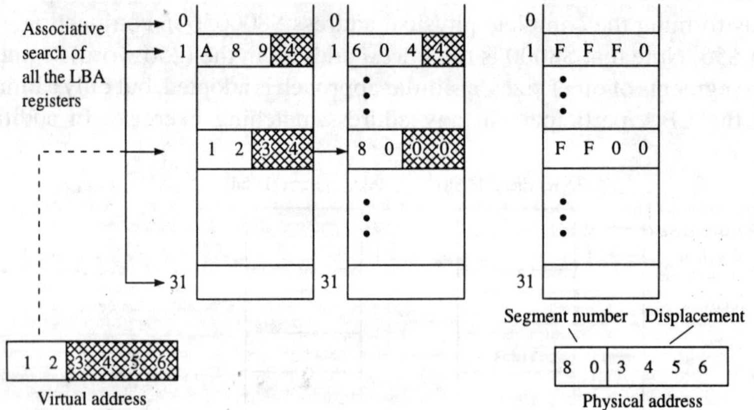

Figure 10.10 Address mapping in the 68451 (general case).

More on protection in the 68451 We have seen in outline at least how the MMU maps virtual addresses on to physical addresses, but several other issues remain: what mechanisms are there to separate user and supervisor memory data and code address spaces? How can we impose other types of access restrictions such as read only? Various fields within the descriptor registers determine these functions. We noted earlier, that the 68000 distinguishes four different types of memory access, as listed in Table 10.1, corresponding to memory areas for code and data in user and supervisor modes. The MMU is normally connected to the function code pins on the 68000, and uses the signals on these pins to distinguish between the four different types of memory access. The MMU can separate the address spaces of different tasks, identified by a code called an **address space number (ASN)**. The address spaces of tasks, as identified by their ASN values, are distinct from the address spaces identified by the function code signals; each function code defines a unique address space and, within each, several tasks can exist. Each task needs an ASN to distinguish it from another task.

The function code signals are used to index into a small 8-bit table called an **address space table (AST)**. The entry in this table pointed to by the function code signals is called a **cycle address space number (CASN)**, and is used in conjunction with elements in the descriptor registers to test for the presence of a segment that matches the current CASN. Each descriptor register contains two fields connected with address space matching. One of these, the **address space number (ASN)**, associates the corresponding segment with a particular task. A task is characterized in terms of those segments with the same ASN. The CASN read from the AST is matched with the ASN that corresponds to the descriptor register currently selected according to the address mapping procedure described previously. In the event of no match between the relevant ASN and the CASN, the MMU's FAULT* output is asserted—we have already noted that this is usually connected to the 68000's BERR pin, and would cause a bus error exception. Finally, an **address space mask (ASM) register**, is another register forming part of the descriptor register, used to determine which bits in the ASN should be masked with the value from the AST. This gives a mechanism for sharing address spaces between several tasks.

The example shown in Fig. 10.11 is for the situation where the FC pins have a state 001, so that entry of 1 in the AST is addressed. This value (80) is compared with that for the ASN corresponding to the segment currently addressed, and, since this is also 80, the match output signal is asserted.

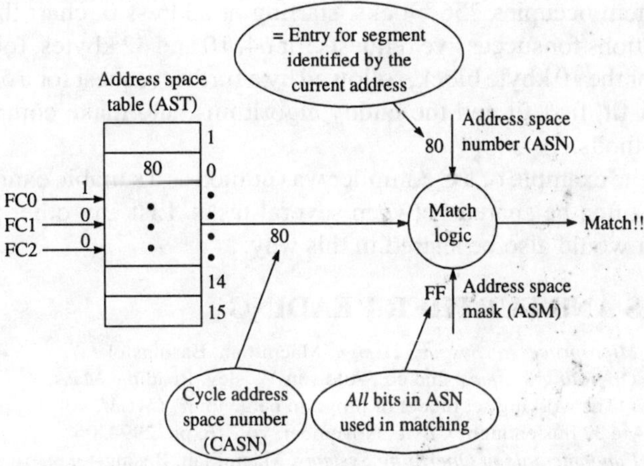

Figure 10.11 Illustration of protection in the 68451.

10.6 EXERCISES AND REVIEW QUESTIONS

1. The section describing storage placement in variable partition systems gave an outline algorithm for *first fit*; extend this by writing Pascal code for *best fit* and *worst fit* algorithms. Additionally you could make provision for changing the hole list in the case where a hole is completely used following allocation of some memory. The fragment shown on page 187 indicates where the code to do this would need to go.

2. A 68000 system uses a 68451 MMU. If a virtual address $611110 is generated by the processor and is to be used to access a 12-bit segment starting at the 'physical' address $E220036, show suitable entries in the PBA, LAM and LBA fields of one of the descriptor registers.

3. Figure 10.12 shows the virtual and physical address spaces for a 68000 system with a 68451 MMU. Give the appropriate entries for the following descriptor register fields: Logical Base Address, Logical Address Mask, Physical Base Address.

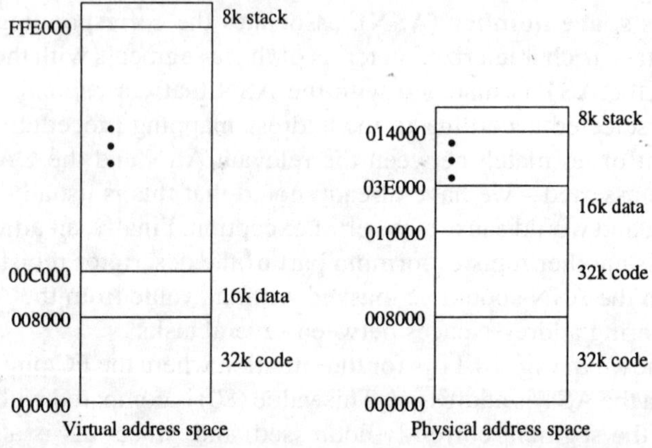

Figure 10.12 Address map.

4. A computer uses a memory management system with variable partitions. Store can be allocated in units in the form of blocks of 1 kbyte. Assuming for simplicity that the operating system occupies 256 blocks, starting at address 0, chart the change in hole sizes and locations for successive requests for 64, 10 and 32 kbytes, followed by release of the space for the 10 kbyte block, followed by a further request for a 5 kbyte block. Use best fit, worst fit, first fit and the buddy algorithms, and make comparisons between these four methods.

5. In Sec. 6.3.4, the example of a C compiler was quoted as a suitable example of a package which would often be shared between several tasks. List any other packages or data structures that would also be shared in this way.

REFERENCES AND FURTHER READING

Clements, A. (1990) *Microprocessor Systems Design,* Macmillan, Basingstoke.
Deitel, H. M. (1990) *Operating Systems,* 2nd ed., Addison-Wesley, Reading, Mass.
Denning, D. J. (1968) 'The working set model of program behaviour', *CACM*, vol. 11, pp. 323–333.
Krueger, S. (1989) 'Are 32 bits enough?', *Byte*, November, vol. 16, pp. 299–306.
Lister, A. M. (1988) *Fundamentals of Operating Systems*, Macmillan, Basingstoke.
Moote, R. (1989) 'Virtual memory—the next generation', *Byte*, November, vol. 16, pp. 341–350.
Motorola Inc. (1983) *Motorola MC68451 Memory Management Unit.*

11

THE OPERATING SYSTEM UNIX

11.1 INTRODUCTION

UNIX is a multitasking,[1] multiuser operating system—it is one of the major success stories in computing in recent years. UNIX is currently employed on a very wide variety of different hardware platforms ranging from supercomputers to personal computers, in which respect it is almost unique among operating systems, and this is one of the most important reasons for its popularity. A large number of software packages have been produced that can run under UNIX. The market for UNIX-based systems, software and services of various kinds is measured in billions of dollars annually. The commercial importance of UNIX is justification enough for a chapter describing it, particularly in view of its widespread use on workstations and the consequent interest in UNIX by electronic engineers engaged in areas such as ECAD. However, another aim in the chapter is to give a detailed example of the practical application of principles developed in previous chapters, in particular principles relating to multitasking operating systems. This should prove complementary to the discussion of MS-DOS in Chapter 8. UNIX offers performance and facilities considerably in advance of MS-DOS, with which incidentally it has some family resemblance as far as the user interface is concerned; however, in earlier versions at least, MS-DOS was not multitasking, and is generally quite primitive by comparison.

11.2 UNIX—PAST, PRESENT AND FUTURE

UNIX was developed by a small group at Bell Laboratories whose main objective was to design a computing environment aimed at software development, this specific objective

---

[1] As in previous chapters the term *task* will be used in preference to *process*.

often not being predominant in the design specification of other operating systems. One of the other design objectives of UNIX was to keep it simple compared to some other operating systems, leaving users to provide any additional features they needed. This gives users more flexibility, but naturally also more responsibility in any systems development work that they wish to carry out.

The history of UNIX began in the late 1960s, and it was designed in its initial form by a group which included a programmer named Ken Thompson. Thompson wanted to run some software he had developed on a DEC PDP-7 minicomputer, but found that the computer lacked the software support needed. UNIX was written to fill this gap. The first version of UNIX, which was entirely written in PDP-7 assembler, was a single-user system and was finished in 1969. Somewhat later than this, in the early 1970s, UNIX was mainly used on the DEC PDP-11, a 16-bit machine, and one of the most successful minicomputers ever manufactured. The use of UNIX as a commercial operating system did not gain momentum until the late 1970s, its application being mainly confined to academic institutions and within Bell Laboratories, until that time. Currently UNIX is making inroads into the mainframe and personal computer markets, although the mainframe field—for the present at least—continues to be dominated by IBM hardware and operating systems. The application of UNIX in the personal computer area, was referred to briefly in Chapter 9, with regard to the position of UNIX *vis-à-vis* MS-DOS and OS/2. Where UNIX seems to be strongest at present is in the workstation area, and it is estimated that by the mid-1990s over 80 per cent of workstations will be UNIX based. This prediction is of note to electronic engineers in view of their heavy involvement with workstations in areas such as ECAD.

One of the more confusing aspects of UNIX, and more importantly an obstacle to its use, is the proliferation of a number of frequently incompatible variants, often produced by companies with an eye to their own profitability at the expense of their competitors. Examples of companies who have produced proprietary versions of UNIX are: Sun with **SunOS**, Digital with **Ultrix** and IBM with **AIX**. Thankfully, early versions of UNIX are in the process of becoming obsolete, a notable example being Version 7 produced by AT&T in the late 1970s, which effectively has been replaced by their **System V**—(confusingly, Version 7 predates System V). At the time of writing, the latest version of System V is Release 4, generally referred to as **SVR4**). The future of UNIX seems more settled than in the past, and there looks like being two main versions: SVR4, just referred to, and **OSF/1**, a variant produced by a group of companies that includes IBM, Digital, and Hewlett-Packard. **OSF** stands for **Open Software Foundation**: the grouping was formed to come up with a version of UNIX which could act as a counterweight to AT&T's UNIX; the importance of the companies that comprise OSF seems likely to ensure the long-term success of OSF/1.

The IEEE have designed a standard called **POSIX** for the interface between an operating system and application programs, in the interests of portability (IEEE, 1986). Although the standard is mainly used in conjunction with UNIX systems, this is not mandatory, and for non-UNIX systems the POSIX standard allows for the translation of a target's system calls into system calls as recognized by the standard, these being 'UNIX like'. In addition, information returned from a target's system calls are translated into a form that also conforms to the standard. The real-time operating system PDOS to be discussed in Appendix 3 is an example of a system that conforms to the POSIX standard.

Another example of the tendency towards standardization in UNIX is the **System V interface definition** (SVID), introduced by AT&T (Deitel, 1984; AT&T, 1985). This standard, which appeared in 1985, is a definition of UNIX system V in terms of its system calls as viewed at the level of the actual source code call statements. Anyone wanting to supply UNIX system V must conform to this standard, and it represents a very powerful way of obtaining portability, irrespective of factors such as the particular hardware platform used. All UNIX system calls used in this book conform to the SVID standard.

In recent years, UNIX has been adapted for use in real-time applications, an area for which it had previously not been thought suitable (Keller, 1988). An example is the real-time version of UNIX marketed by the company Masscomp. Applications are typical of real-time systems, for example analysis of results generated by laboratory instruments, etc., i.e. areas where real-time performance is critical.

11.3 BACKGROUND TO UNIX'S PORTABILITY FEATURE

The development of UNIX and the language C were closely related, in fact UNIX was completely rewritten in C in 1973. This was done mainly to make it possible to port UNIX to a wide range of machines, something that would not be possible when UNIX was written in the machine language of one particular machine, namely that of the PDP-11. Currently it is possible to choose from well over 100 different hardware vendors without being locked into a single supplier. This is not true of a proprietary operating system such as VMS which—currently at least—can only be used on DEC's VAX range of computers. Portability was not a consideration when VMS, and other operating systems one could quote, were being designed.

UNIX has a central part called the **kernel** consisting of about 20 000 lines of code. 15 000 lines of this are machine independent and written in C. Of the remainder, 1000 lines are written in the assembler code of the target machine. A further 4000 lines of C code are also machine dependent. Porting UNIX to a new machine is a matter of writing the machine-dependent parts of the operating system. Although this is scarcely a trivial exercise and requires the services of skilled systems programmers, a port to a new machine is considerably simpler than would have been the case with most other operating systems, if indeed it were possible at all.

11.4 OVERVIEW OF THE UNIX FILE SYSTEM

UNIX is generally used with systems which involve large numbers of files, possibly running into the thousands. To manage the complexity involved, UNIX has a tree-structured hierarchical directory system (see also Chapter 8). The top level of the directory is called the **root**, and is indicated in commands such as `ls` (= list directory) by a '/' character. Subdirectories are signified by a name following the '/'. Subdirectories may themselves contain subdirectories, and these are indicated by a further '/' followed by the subdirectory name. For example binary files, such as the UNIX commands, are in the subdirectory /bin, so these files can be listed by the command `ld -l /bin`. If the /bin directory had a subdirectory called `mybin`, say, the files in the `mybin` subdirectory could be listed by the command `ld -l /bin/mybin`. Figure 11.1 depicts some of the most commonly used directories in a UNIX system.

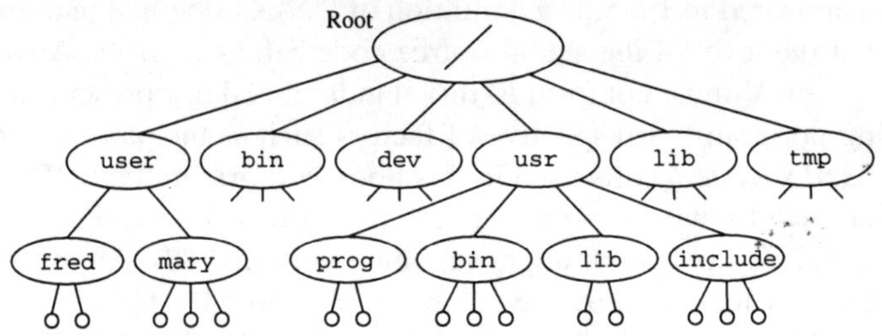

Figure 11.1 UNIX file organization.

Executable programs—notably commands—are stored in the `/bin` or `/usr/bin` directories. UNIX will search these directories for a program to execute a named command. Having two directories for binary programs arose for historical reasons, but nevertheless this has the advantage that more commonly used command programs can be put on the `/bin` directory which will be smaller and thus faster to search.

The `/tmp` and `/usr/tmp` directories are used for scratch (temporary) files. For example, many compilers generate work files only required for the duration of the compilation, and which may be subsequently discarded. Such files would normally be stored in the `/tmp` or `/usr/tmp` subdirectories.

The `/lib` and `/usr/lib` directories are normally store run-time libraries used during the linking process that follows compilation. However some systems use either or both these directories for oddments, such as the files that constitute the various passes of a compiler.

The `/usr/include` directory is searched by the preprocessor of a C compiler for files to be *included* in a C program, notably header files. Chapter 3 gave further details on this.

The `/dev` directory holds all the **special files** which represent the hardware devices in the system, such as disks and printers; devices are treated as files by UNIX, one of its characteristic features.

Users may get confused as to which subdirectory they are currently in, and the command `pwd` which displays the pathname to the current working directory could then be helpful. A pathname is the full name starting at the root and ending at the working directory, showing which subdirectories are traversed on the way between the two. In the previous example `/bin/mybin` was the pathname for the directory `mybin`.

11.4.1 Mountable volumes

UNIX based systems invariably have at least one disk—called the **system disk**—that holds important files required system-wide. The root filing system would normally be stored on this disk (or disks). Many systems also have removable media such as cartridge disks, floppy disks, or tape drives. Separate filing systems have to be created on each, and these have to be compatible with the UNIX filing system structure. Files on such media are integrated with the files on the system disk(s) by a process called **mounting**. Conversely, when a disk/tape is removed from the system, its file system is separated from the other files by a process called **dismounting**. UNIX commands for mounting and dismounting are called respectively `mount` and `umount`. The root of a mounted volume appears as one of the branches attached to the root of the system disk files. Thus, for the sake of argument, the files in the `/user` directory shown in Fig. 11.1 could reside on (say) a floppy disk. These could correspond to some user's private files, which are made accessible to the rest of the UNIX file structure following a mount command.

11.5 THE SHELL

The **shell** is the program that interprets users' commands; it has a comparable role to the COMMAND.COM of MS-DOS. In UNIX, unlike some other operating systems, the shell is a program separate from the kernel, it is just another application program as far as the kernel is concerned. This makes it possible for users to customize UNIX to particular requirements by using any convenient shell program they might care to use, or perhaps even write. For example, in a documentation environment it might be desirable only to offer users a restricted range of commands to edit, delete, display and print files. Shells that are in common use are the **Bourne** and **C shells**. The C shell, as the name infers, has a 'C-like' syntax.

The full command language interpreted by the shell is complex, but most of the individual commands are in fact quite simple to use, if somewhat cryptic, and seeming somewhat strange at first. A command line consists of a **command name** followed by none or more **arguments**, such as **options** and **filenames**, and separated by spaces as delimiters. The shell breaks up the command into its components; any files specified in the command are loaded, and the command arguments are made accessible to it. An example is the `ls` command for listing files in a directory. The example below lists the files in the directory `/bin/mybin`.

```
$ ls -l /bin/mybin
```

The `-l` (= long) is an example of a **flag**, which here gives a full listing of all attributes of the files such as size and date last modified. Most commands can be used with such flags to modify the precise action of a command. Flags are preceded by a '`-`'. In this particular example, without the `-l` flag, only the filenames would be listed, not their other attributes. There is more on this command in the discussion relating to directories in Sec. 11.6. A sample of a few other shell commands is given below. A standard book on UNIX would need to be consulted for a full list of shell commands (Bourne, 1987).

11.5.1 I/O redirection

One of the most powerful features of the shell is its ability to reassign the standard input and output files of a command. This is called **redirection**. By default, input is taken, in most systems, from the keyboard, but if instead it is required, say, to take input from a data file on disk, the *actual* input device can be specified on the command line. If as an example we want to use a program called `inquiry` to take data from a file called `mydata`, the command line would look like this:

```
$ inquiry < mydata
```

Output from a program is by default, in most systems, to the screen, but if output is to be directed to a data file this can be done as in the example below where the program `myprog` sends its output to the data file `outfile`:

```
$ myprog > outfile
```

Finally, both the standard input and output can be redirected at the same time, as in the following command where the program `myprog2` takes input data from the file `indat`, and sends its output to the file `outdat`.

```
myprog2 < indat > outdat
```

11.5.2 Filters

Many UNIX utilities can accept data from a file, process it in some way, and send the result to the standard output. Such a program is called a **filter**. For example, a UNIX utility called `dd` can be used to take a file with upper and lower case text and convert it to a file with only upper case as in the example in Fig. 11.2.

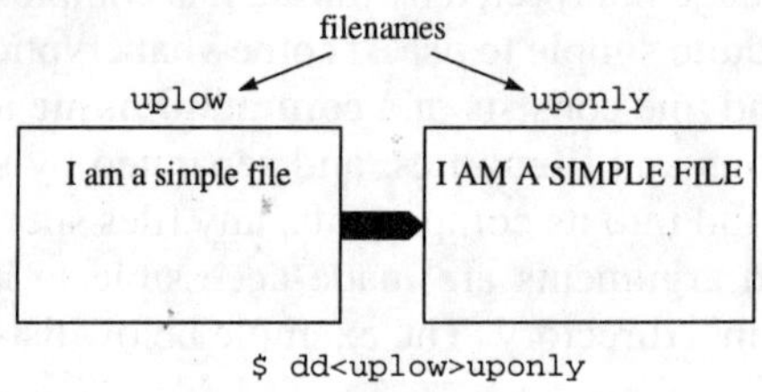

Figure 11.2 Example of a UNIX filter.

11.5.3 Pipes

Most UNIX utilities carry out comparatively simple functions; more complex functions are achieved by using a number of standard utility programs as building blocks. This approach, one of the original design objectives of UNIX, avoids the need for users to develop new programs for a particular application from scratch. Use of filters is further enhanced by the UNIX **pipe** facility. A pipe is used to pass the output generated by one command to another command, which then uses it as input. As an example suppose we had a list of names and addresses arranged in two fields in a file, i.e.:

```
name1 address1
name2 address2
name4 address4
etc.
```

Let us say that a list of names (only) in alphabetical order is produced. A utility called `cut` can eliminate particular fields in a file, while a utility called `sort` can put names in alphabetical order. The `pipe` facility can combine these two actions into one. Separately, the two commands would have been used as follows:

```
$ cut -f1 nameaddr > nameonly

$ sort nameonly > sortedname
```

However, using pipes the actions can be combined, as in:

```
$ cut -f1 nameaddr | sort > sortedname
```

The vertical bar '|' is the pipe character, instructing the shell to pipe the output of the `cut` command directly into the `sort` program as its input. If an operation such as that above is required frequently, a command file containing the appropriate command(s) can be written. (In UNIX, command files are called **shell scripts**, and will be discussed shortly.)

Superficially, it might seem that the command using the pipe is simply a concatenation of the earlier two commands, with the output from the first command being sent to a buffer (a temporary file), which is then used as input to the second command, the buffer then finally being discarded. In fact what happens is that *all* commands in a pipeline are run as concurrent tasks. Programs further towards the end of the pipeline get their input in stages, as the preceding stage produces output. A good analogy to this is a factory production line.

11.5.4 Pipe fitting

In many cases the intermediate data generated by a pipeline is not required, and, by default, UNIX does not save them. If, however, we need to hold on to the intermediate data, the **tee** utility can be used. In the previous example, if, for the sake of argument, a file with the list of unsorted names was required, the following command would save them in a file:

```
$ cut -f1 | tee unsortname | sort > sortname
```

11.5.5 Wildcard matching

Wildcards were discussed in Chapter 8, and are used in UNIX also. Recall that a wildcard is a special character that can stand for any character or string of characters. There are two wildcard characters in UNIX: the '`*`' character matches any sequence of zero or more characters in a filename. The '`?`' character matches exactly one character in a filename. A few examples should clarify the use of wildcard characters.

- `rm *` deletes all files in the current directory.
- `rm *.bin` deletes all files with a `.bin` extension in the current directory.
- `rm temp?` deletes all files in the current directory that start with the string `temp` and that also have *exactly* five characters in the filename.

11.5.6 Background processing

Background processing is a form of multitasking where users can run several programs at the same time after invoking a command in a particular way. The dialogue between a user and the shell is normally interactive. A user initiates one activity at a time, and then waits until it completes. The shell then prompts the user for the next command. However, some tasks are time consuming, and executing them interactively would tie up a terminal and the user unnecessarily. To eliminate this problem, UNIX allows users to request a **background task** execution. The shell starts execution of this task, and *immediately* prompts the user for another command. Effectively, the multitasking capability of UNIX is being *directly* invoked by users. Users request background execution by placing an '&' at the end of the command line. In the example below, the program `myprog2` takes input from the file `indat`, and sends its output to the file `outdat` as a background task:

```
$ myprog < indat > outdat&
```

This can be extended to deal with *several* background tasks invoked by a single command line. Individual commands are separated by a ';' as in the example below which compiles a C program and then runs it, the executable binary file, by default, being called `a.out` in UNIX:

```
$ (cc myfile ; a.out)&
```

It is also possible for UNIX—as opposed to users—to initiate background tasks. An example is the **print daemon** which is a task running continuously in the background, checking to see if any other task has made a request to print a file, and generally managing the spooling arrangements associated with printing files. Tasks requesting a file to be printed enter the name of the file in a special directory that the print daemon checks period-ically. After the file has been printed (on a FIFO basis), the print daemon removes the filename from the directory. It is common to have several daemons running in the back-ground in this way, another is a network daemon.

11.5.7 The shell as a programming language[2]

In cases where users find they wish to use identical sequences of commands repeatedly, UNIX allows the command sequences to be stored in a file called a **shell script**, and the entire sequence is invoked simply by typing the name of the shell script file. The example below does the following: firstly, changes to the root directory, secondly, deletes all files in that directory with extension `.tmp` and files with the name `scratch`, thirdly, reports on the amount of free space on all mounted disks (with the `df` command) and finally does a 'long' directory listing of the files in the root directory.

```
cd /
rm *.tmp scratch.*
df
ls -l
```

As well as allowing the unconditional execution of a number of sequential commands, a

---

[2]The information in this section relates to the Bourne shell, there are detailed differences in the case of the C shell.

shell script can employ conventional programming constructs such as loops or conditional operations. Additionally, it is possible to pass parameters to a shell script to make their use more general. For instance, a particular sequence of commands may be required for use with one of a number of different files, and it is clearly more convenient to make the shell script usable for all the files, rather than writing a separate shell script for each file. Thus the shell provides what is in effect a programming language making for more streamlined use of the operating system. It is also possible to employ facilities such as those described above at command level, they are not just restricted for use inside shell scripts.

Shell variables To facilitate writing shell scripts, **shell variables** are often used, these are names used to store character strings. Shell variables can be used to represent parameters passed to the shell script, or other variables internal to the script. In this latter role they may for example be used in assignment statements, or as loop counters. A variable is assigned a value by quoting the variable's name followed by '=' followed by the value it is to assume. Thus in the line:

```
fname = first.c
```

the value `first.c` is assigned to the shell variable `fname`. Such a line could appear either inside a shell script, or following UNIX's '$' prompt.

Another way of writing the earlier shell script would be to form a loop. The general syntax for this is as follows:

```
for name in arg1 arg2 arg3 ..
do
command list
done
```

So in the earlier example we could therefore write:

```
for fname in first second third
do
cc -o $fname $(fname).c
cat $(fname).c > /dev/lp0&
done
```

Note that '$' followed by a variable name instructs the shell to substitute the value of the variable in place of the variable name before executing the command.

11.6 FILE SYSTEM ORGANIZATION

Unlike many operating systems, UNIX does not impose any structure on the information contained in files, the data is simply a sequence of bytes as far as the operating system is concerned. Users may randomly address a byte from anywhere within a file. Most other operating systems structure files into blocks of words of some fixed size; however, this may lead to complications when the size of the block differs from that required by a program: an 80-column line for instance. Programs would in general then have to translate requests for blocks of data to the block size recognized by the particular operating system, a complication that does not exist with UNIX.

As with any operating system, UNIX allows users to identify files by name, in the form of a character string (up to 14 characters). Internally, however, UNIX keeps track of attributes of individual files using a number called the **i-node number**. The directory structure in UNIX is simple, each entry contains two fields: a filename, and the i-node number, which is a pointer to an **i-node**—see Fig 11.3. The i-node itself is a small table stored on disk until its associated file is to be used. The i-node holds all the information about the file, including the owner's UID, file creation data and file size. Some of the entries in the i-node table are explained in more depth below. When a file is opened, its i-node is loaded into main memory for fast access, and is subsequently restored to disk when a file is finally closed.

2 bytes   14 bytes

i-node number	Filename

Figure 11.3 UNIX directory entry.

All directories are treated as files by UNIX, and would contain one or more of the 16-byte entries shown in Fig. 11.3 depending on how many files there were in the directory. Attributes of directories may be examined using the `ls -l` command, an example, more fully explained later, is shown in Fig. 11.5. UNIX would need to access the i-node to extract the data displayed by the `ls` command.

Information relating to files on disk can be split into four parts:

- **A boot block**. When the computer is powered up, the hardware reads the boot block into memory and jumps to it. Once the system has booted, the boot block is not used further.
- **A super block** which has information describing the layout of the file system. This includes the size of the disk, and the boundaries of the other regions.
- **The i-list** which is a list of all the i-nodes.
- **Storage areas** for the data that files contain.

Each file is uniquely identified by a **major device number**, a **minor device number**, and an **i-number** (the index of the file's i-node in the i-node array). The major device number specifies the device class: floppy disk, tape, hard disk, etc. The minor device number identifies which of a particular class of device is being addressed. Device numbers can be examined in the `/dev` directory.

UNIX organizes a disk as a sequence of 512-byte blocks (1024 bytes in the case of UNIX System V). Blocks constituting a file may be scattered widely around the disk, and an important function of the i-node for a file is to keep track of where all its blocks are located. A list of the first ten blocks in a file is kept in the file's i-node itself. This list is thus adequate to access files of no more than 512×10 bytes. For files longer than ten blocks, the locations of the subsequent blocks are not stored in the i-node, but in another block called an **indirect block**. The i-node stores a pointer to the indirect block. The indirect block itself can only point to a limited number of blocks, so, for still longer files, the i-node points to a **double indirect** block, which points in turn to a group of indirect blocks, each of which finally points to the actual blocks in the file. UNIX also allows one further level of indirection (**triple indirection**) for very large files. The use of block numbers and indirect pointer blocks is shown in Fig. 11.4. It might be noted that to access files requiring the use of indirect blocks, the indirect blocks need to be loaded into store, involving an additional time overhead. This is not necessary with files of ten blocks or less, so accessing short files is more efficient.

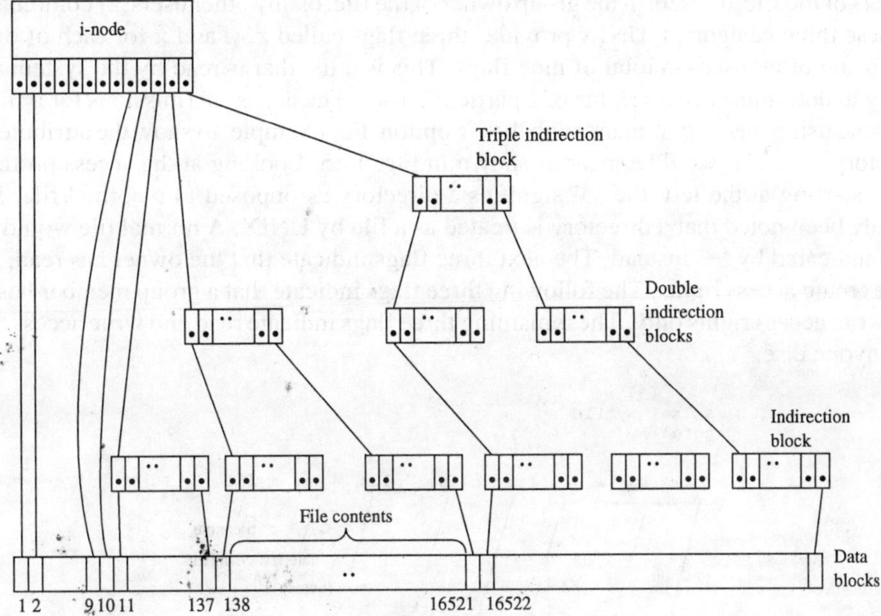

Figure 11.4 UNIX indirect file access mechanism.

Disk buffering In view of the relatively slow access speed of mass storage devices, such as disks, compared to main memory, it is possible that I/O activity can lead to a system bottleneck. To minimize this problem, UNIX allows a form of buffer in main memory (usually called a **disk cache**) to be used to store blocks of data read from disk, rather than transferring data directly from disk to the task that requires the data.[3] A succession of requests for data from a disk are, whenever possible, supplied from the cache, rather than from the disk, and are correspondingly faster. Data is only read from disk when the data required is not found to reside in the cache. Only the first request to read data from a disk block leads to disk I/O. Data to be written to the disk is, in the first instance, written to the cache. The disk is

---

[3]There is no suggestion intended that UNIX is special in this provision, in fact most operating systems can make use of disk caches, including MS-DOS.

only written to when it becomes necessary to replace the cache. The technique improves performance, since the bulk of programs tend to access the same blocks on a disk repeatedly. Data is only written back to the disk when the cache needs to be finally flushed.

File protection mechanisms UNIX has a file access permission scheme. Each file and directory has exactly one, and only one, user, who is the **owner** of the file. The owner is typically the individual who originally created the file. Each file also has a **group owner**, which is one of the user groups of the system. In addition to the file's ownership, a set of file access permissions is maintained for each file and directory. These permissions determine which users may access the file and what kinds of file access will be granted. Provision is made for read, write and execute access. The various types of access may be granted to owners of the file, or users in the group owner of the file, or any other users, or combinations of these three categories. UNIX provides three flags called r, w and x for each of owner, group and other user—a total of nine flags. This is a list that is read by the system when trying to determine access rights of a particular user. The access permissions for a file can be found using the `ls` command with the `-l` option. For example, to show the attributes of a directory, `myfile` would appear as shown in Fig. 11.5. Looking at the access protection flags, starting at the left, the 'd' signifies a directory as opposed to a normal file. It has already been noted that a directory is treated as a file by UNIX. A normal file would have been indicated by '-' instead. The next three flags indicate that the owner has read, write and execute access rights. The following three flags indicate that a group member has read and write access rights only. The remaining three flags indicate read and write access rights for anyone else.

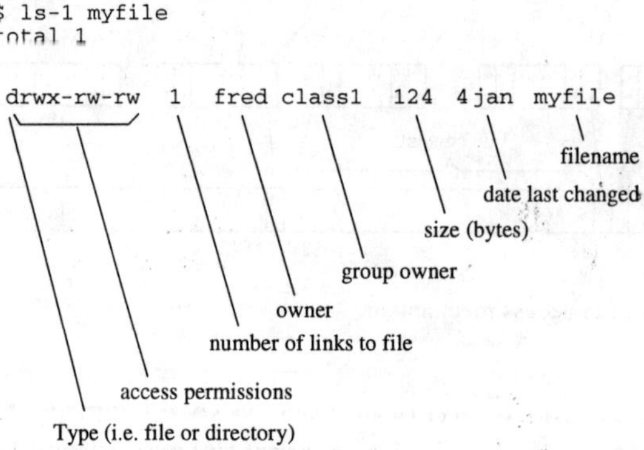

Figure 11.5 Typical directory information.

11.7 USERS AND SUPERUSERS—PRIVILEGES AND PRECAUTIONS

The checks on file access already described apply to the bulk of users on a UNIX system, with one exception, the **superuser**, who is allowed more freedom of action than other users. The superuser exercises system management functions and is allowed to carry out actions such as dismounting disks, deleting certain directories, and changing protection rights of other users. It is usual for the superuser to be given the user name **root**, and as the name implies the superuser owns the root directory; in addition the superuser owns the directories and files associated with the operating system. Naturally superuser privileges in a multiuser system should be very carefully guarded, and, in particular, access to the password associated with root needs to be restricted to those responsible for system management and no others: this is the most critical part of the security of any UNIX system. Passwords are stored in encrypted form in the system's password file, which also contains the following: username, user-ID, group-ID, user's home directory and finally the program that runs immediately after a user logs on, this normally being the UNIX shell program `/bin/sh`.

11.8 UNIX SYSTEM CALLS—INTERACTING WITH UNIX

When an application program wants to use the resources of the kernel, it does so by a **system call**, which is a call to a procedure to invoke the required action. System calls are effectively the software interface between user programs and UNIX. They encompass the areas of file access, task creation and termination, and intertask communication. The following discussion of UNIX system calls is illustrative rather than comprehensive, and a suitable text should be consulted for a complete list of UNIX system calls and their usage (Haviland and Salama, 1988; Deitel, 1984).

11.8.1 I/O system calls

I/O is dealt with by system calls that carry out the operations of opening, closing, reading and writing files. The example below is a system call to open a file. It returns an integer termed a **file descriptor** which identifies the file that has just been opened, and which here is assigned to the variable `fd`. If there is a problem, for example if the file does not exist, then `open` returns a value of -1. This is helpful in error handling.

```
fd = open( filename , mode ) ;
```

`mode` needs to have a value assigned depending on the access mode required: read only, write only, etc.

Once a file is open it can be read by the `read` system call:

```
nbytes = read( fd , buf , num ) ;
```

This will take `num` bytes from the file whose file descriptor is `fd` and put them in the character buffer `buf`, which needs to be an array of char. The call will return the number of characters actually read, which could be less than `num`, depending on the size of the file. If there are any problems a value of -1 is returned. Note that `read` implies sequential rather than direct access. An initial call to `read` will commence access from the start of the file. Subsequent `read` operations will be from the position just after the previous read. A system call `seek` is provided to allow for direct access (see Haviland and Salama, 1988, for more on this).

The system call `write` works in the reverse sense to the `read` call. This too involves sequential rather than direct access. A typical call statement is:

```
nbytes = write( fd , buf , num ) ;
```

The value returned by `write` is the number of bytes *actually* written to the file, or again
−1 if there is a problem.

Finally, to close a file the system call `close` can be used:

```
rval = close( fd ) ;
```

The `close` takes one parameter, the associated file descriptor. The call returns a value of
0 if there are no problems and −1 if there are: using an invalid file descriptor for instance.
An example of a C program that puts some of these system calls together is given below.
This somewhat primitive program simply copies a block of 1024 bytes from one file to
another. Code to properly handle error conditions has not been included in the interests of
clarity.

```
#define include <stdio.h>
#define BUF_SIZE 1024
main()
{
    char buf[BUF_SIZE] ;
    int fd_in, fd_out, rval, wval ;

    fd_in = open( "rd_file" , READ_ONLY ) ;
    fd_out = open("wr_file" , WRITE_ONLY ) ;
    r_bytes = read( fd_in , buf , BUF_SIZE ) ;
    w_bytes = write( fd_out , buf , BUF_SIZE ) ;
    rval = close( fd_in ) ;
    wval = close( fd_out ) ;
}
```

11.8.2 Task management system calls

The exec system call A single task may execute several programs during its lifetime by
using the **exec** system call. The kernel loads the new program into store, and resumes
execution of the task from the beginning. The name of the new program is specified in the
exec call. Unless there is some error condition preventing the creation of the new task, the
original task will cease to exist, being replaced by the task specified in the exec call, there is
no return from an exec call to the originating program. This is, of course, quite different
from the action of a procedure in many other types of program, where one gets accustomed
to procedures always returning to a calling program after they terminate. Files open for the
originating program remain open for the new program. The task also retains its user-ID and
group-ID.

The somewhat minimal C program below shows how to use the exec1 system call which
invokes the UNIX command ls to list files. The first parameter is the pathname associated

with the `ls` command. The last parameter is a form of null pointer to indicate the end of the parameter list, since there can be any number of parameters with this system call.

```
main()
{
    printf( "List of files in current directory\n" ) ;
    execl( "bin/ls", "ls", (char *)0 ) ;
}
```

The fork system call A new task in UNIX can only be created by a system call from an existing task called **fork**. The new task created is termed a **child** task. After the `fork`, both child and **parent** tasks are in execution. These two tasks are executing identical copies of the same program, with the same open files. Typically, after a `fork`, a child task will `exec` a new program, the overall result being to create a new task that is executing a new program. Both child and parent tasks are free to create further tasks in this manner, leading to a possible hierarchy of tasks. This approach is different to some other operating systems where a parent task creates a data structure for a new task which is effectively a skeleton, with no code and data associated with it until the new task starts to become active and creates some. The TCB (see Chapter 9) needed for the new task is formed and put into a list of tasks at this stage.

It may seem strange and over complex to have this two-stage method of creating a new task by successive use of the `fork` and `exec` system calls; why not just have a single system call which creates a new task? One reason is that shell programs can exploit the two stages to perform I/O redirection. Another reason is that there are many occasions when it is desirable to have several copies of the same program running on the computer, which would thus require the services of `fork`. For example, on a machine with more than one terminal, `init` will fork, and attach itself to every active terminal waiting for it to come alive. The modest, added complication in using the fork/exec mechanism is a small price to pay in terms of the increased flexibility of use achieved.

The short C program below demonstrates the use of `fork`. The system call `fork()` creates two copies of the original task, which then proceed to run pseudoconcurrently. The operating system returns a value of 0 from the `fork` call if the task is the child, and a positive integer if the task is the parent. This enables parent and child to identify themselves. The two copies of the program then simply display 'child' or 'parent' on the screen. When the program is run, alternate bursts of 'child' and 'parent' will be displayed indefinitely, in a way dependent on factors to do with scheduling, including the value of the time slice, and the number and character of other tasks that are active. It goes without saying that interspersing output in this way from several tasks is generally not a good idea in a more practical situation!

```
#define  TRUE  1
main()
{
    int pid ; /* -- task id -- */
```

```
pid = fork() ;
while( TRUE ){
  if( pid == 0 )
    printf( "child\n" ) ;
  if( pid > 0 )
    printf( "parent\n" ) ;
  }
}
```

Task synchronization and the `wait` **and** `exit` **system calls** The `wait` system call provides for task synchronization. It allows one task to wait while another related task finishes. It gets a mention here because it is useful when used in conjunction with `fork` and `exec` to manage a newly created task. The relevant points are illustrated in the program below which uses `fork` to create a child task that just executes the `ls` command. The parent task is put to sleep until the child task finishes by using `wait`, as shown. The `exit` system call— also used in the program below—terminates a task. It does things like closing file descriptors.

```
main()
{
    int pid ;

    pid = fork() ;
    if( pid > 0 ){
    /* -- must be parent so suspend until child finished -- */
        wait( (int *)0 ) ;
        print("OK files listed\n" ) ;
        exit( 0 ) ; /* -- exit returns 0 for normal exit -- */
    }
    /* -- must be child, so exec the ls command -- */
    if( pid == 0 )
    execl("bin/ls", "ls", (char *)0 ) ;
}
```

Intertask communication calls `fork` and `exec` allow us to create tasks, and, having done so, methods of allowing them to interchange data need to be considered. UNIX is richly endowed in terms of intertask communication methods. One obvious approach is to allow tasks to share files containing the data to be exchanged. This is one method much used in fact, but there are problems of mutual exclusion to solve—files or records within files may need to be locked while one task accesses it. By locking we mean that all or part of a file must be considered as a suitably protected, critical region, roughly as described in Sec. 9.8.

Signals Intertask communication using files can often be clumsy or inappropriate: say a user simply wants to stop a program running because it is in an infinite loop. This can be done by hitting the **interrupt** key (<DEL> or <CTRL-C> depending on the system). What is needed here is an intertask communication mechanism something like a hardware interrupt, in which the task within the kernel responsible for servicing the keyboard indicates to the program that is running that it must terminate. UNIX does this by a mechanism called a **signal** which can be thought of roughly as a form of high-level interrupt. In the example, the kernel would send a signal, referred to as SIGINT inside UNIX code, to those tasks belonging to the relevant user, which moreover are programmed to recognize it. The task(s) would then take appropriate action, which should be to terminate themselves. There are 16 different types of signal, and names such as SIGINT are defined inside a header file signal.h. Other signals deal with conditions such as illegal instructions, and bus errors. Particularly useful is a signal called SIGKILL. This signal, which can also be invoked at command level, is guaranteed to terminate the task that is its target. The same cannot be true of SIGINT which relies on the target task having code within it to respond to SIGINT (this process is called **catching**). The command that users can employ to send signals to tasks is dramatically called kill, an example is kill -9 6, which sends a signal 9—the value of SIGKILL—to a task of PID number 6. Naturally this command is very useful in terminating a rogue task which cannot be stopped in any other way! Obligingly, a background task, when invoked, will divulge its PID on the line below the command, so this can be quoted in a kill command if, *in extremis*, the task needs to be subsequently terminated. It might finally be noted that hitting the interrupt key will not terminate a background task.

The example below shows how to use and catch a signal within a program. The program just cycles in an infinite loop displaying a character '-' on the screen. The signal system call will enable the catching of the SIGINT signal, so that whenever the interrupt key is pressed the hitdel routine is invoked, displaying the string 'hit DEL key'. Some comparisons may be drawn between this routine and an interrupt handler—in fact, it can be regarded as its high-level counterpart.

```
/* -- defines values of signals -- */
#include <signal.h>
main()
{
    int hitdel() ;

    signal( SIGINT, hitdel ) ; /* signal enabled after this */
    while( 1 )
    printf( "-\n" ) ;
}

/* routine to catch SIGINT signal.
 note that the parameter is the signal number
 given in the signal system call - SIGINT here
*/
hit( int signal_number )
{
    printf( "hit DEL key\n" ) ;
}
```

Pipes Signals are used mainly to indicate error conditions or other abnormal events to tasks. However, to transmit data as such, some other approach is required. One method provided by UNIX is the pipe mechanism, a concept we have already met at command level. It is also possible to set up a pipe from within a program by suitable system calls. Recall that a pipe is a method of channelling the data resulting from one command to act as input data to be then acted on by another command. The underlying mechanism involves a FIFO buffer that is written to by the first command and read by the second. To use this method from inside programs, we evidently need system calls to write data to the pipe and also to read data from it. In addition pipes need to be created, so that UNIX can subsequently recognize a pipe when it is being accessed. The system call, not unsurprisingly termed `pipe`, is provided for this purpose. This takes a parameter in the form of a two-element array of ints, each of the elements is a file descriptor for the pipe, subsequently used in `read` and `write` calls. In the example below, a pipe is created, after which a fork is used to create a child task that reads a message from the pipe, the message being written to the pipe by the parent. No error checking is carried out in the interests of simplicity.

```
#define MAXMSG 9
main()
{
    int pipe_fd[2] ; /* The two pipe file descriptors */

    /* buffer to store data read from pipe */
    char mbuff[MAXMSG];

    pipe( pipe_fd ) ; /* create a pipe */
    pid = fork() ; /* now create a child task */

    /* task must be parent, so write message to pipe */
    if( pid > 0 ){
        write( pipe_fd[1], "message1\n", MAXMSG ) ;
        write( pipe_fd[1], "message2\n", MAXMSG ) ;
    }

    /* must be child to get here,
    so read message from pipe and display it
    */
    if( pid = 0 ){
        for( j=0; j < 2; j++ )
            read( pipe_fd[0], mbuff, MAXMSG ) ;
        printf( "%s\n", mbuff ) ;
    }
}
```

11.8.3 The shell and task management

The shell uses the UNIX task structure to implement user commands. When the shell receives a command, the following sequence of events occurs. The shell task forks a child task which will ultimately execute the command. The child task execs the utility program named in the command. Now two tasks are running for the user. If the command is to be run in the foreground, the shell task waits for its child task to finish. When the child task does finally finish, the shell task prompts the user for a new command. If the user requests the command to be run in the background, the shell task does not wait for the child, instead the shell task prints the task ID (or PID) for the child, and immediately prompts for the next command.

Pipes in the form described above can only be used when the associated tasks have a common origin—in the example, for instance, a child task created by its parent with the two communicating by means of a pipe. There are many situations where this kind of relationship will not be appropriate. With tasks such as print spoolers, for example, there will be no such common ancestry, and it is then necessary for server tasks such as print spoolers to be used by a succession of unconnected tasks, as and when they need to, retaining a permanent existence, independent of the other tasks using them. For this reason **named pipes** were introduced. These, unlike normal pipes, have a permanent existence, and so can act in a *server-like* role.

11.9 THE MAIL AND MESSAGE UTILITIES

In view of the fact that UNIX was devised at Bell Laboratories it is not surprising that the system makes it easy for users to communicate to each other across a telephone system. Two utilities to do this are `mail` and `write`. The `write` utility can be used to send short messages to users on another terminal on the same computer, or on a terminal connected to the system remotely over a telephone line. A typical command that sends a message from Bert to Fred is:

```
$ write fred
I want to see you NOW - bert <CTRL>-D
```

(Note that messages are terminated by the EOF (end-of-file) character `<CTRL>-D`.)

If users get tired of receiving unwanted messages, they can block them using the `mesg` utility: `$ mesg n`. Messages can subsequently be accepted again by the command `$ mesg y`. The mail utility enables users to send electronic mail (e-mail) to each other, either over a telephone network or to users on the same UNIX system. An example of this type of command is shown below (the '!' indicates a remote system name). A user called `fred` on the `woolwich` system is being sent a file called `memo`:

```
$ mail woolwich! fred < memo
```

11.10 EXERCISES AND REVIEW QUESTIONS

1. You are to choose between DOS and UNIX as an operating system to run on a personal computer. Discuss arguments for and against these two operating systems, citing instances of applications for which either would be a more suitable operating system.
2. Discuss the reasons for UNIX being a portable operating system.
3. Why is it necessary to write a part of the kernel of UNIX in assembler?

4. What is the advantage of loading the i-node associated with a file into main memory when the file has been opened?

5. Give a UNIX shell command to put the user names of people logged on to the computer, in alphabetical order, into a file. The shell utility called who will be required. This gives a directory in the following form:

```
fred console aug 6 08:40
bert tty00 aug 6 09:23
maggy tty01 aug 7 01:00
den tty06 aug 7 01:01
```

6. Give the UNIX command that would send a listing of all the files in the /bin directory to a file of name binlist.

7. A UNIX file has 30 000 bytes. State the number of indirect and, if needed, double-indirect blocks needed to access the data in the file. (Refer to Fig. 11.6.)

8. In Sec. 11.7.1 a crude program that uses system calls to copy a block of data from one file to another was described. Enhance the program by making it display appropriate messages if any of the system calls used fails. Further enhance the program by turning it into a file copy utility, with the two filenames to be quoted as parameters after the command.

9. A program produces a floating point exception by an attempt to divide by zero. Describe the recovery procedure that could be used in a UNIX system (a signal called SIGFPE would be sent to the kernel in this situation).

REFERENCES AND FURTHER READING

AT&T (1985) *System V Interface Definition*, issue 2, vols 1 and 2, AT&T.
Bourne, S. R. (1987) *The UNIX System V Environment*, Addison-Wesley, Reading, Mass.
Deitel, H. M. (1984) *An Introduction to Operating Systems*, Addison-Wesley, Reading, Mass.
Haviland, K. and Salama, B. (1988) *UNIX System Programming*, Addison-Wesley, Wokingham.
IEEE (1986) *IEEE Trial Use Standard: Portable System for Computer Environments*, IEEE Standard 1003.1.
Keller, L. S. (1988) *Operating Systems*, Prentice-Hall, Englewood Cliffs, New Jersey.
Tanenbaum, A. S. (1987) *Operating Systems*, Prentice-Hall, Englewood Cliffs, New Jersey.

In addition to the books listed here, there are magazines devoted to UNIX, and these are good sources for current developments in this rapidly developing field. Specific examples are *UNIXWorld* and the *UNIX Systems Magazine*; articles in other computer publications which refer to UNIX are very common. There was a useful series of articles on UNIX in the May 1989 edition of the magazine *Byte*.

Part IV

Systems design and development

12

REAL-TIME SYSTEMS—CHARACTERISTICS AND ALGORITHMS

12.1 INTRODUCTION

This chapter examines the characteristics of real-time systems, including a discussion of some of the more important algorithms needed in their design; in some ways it can be regarded as a continuation of the chapter on multitasking operating systems where we met related topics. Operating systems are, in fact, an important instance of a real-time system, questions such as concurrency and temporal deadlines being key issues. We now, however, adopt the viewpoint of someone *developing* a real-time application, and who would therefore be more concerned with choices involving real-time algorithms and design methods. Many of the concepts in Chapter 9 will naturally be relevant here—the concept of mutual exclusion, for example, is no different in principle whether applied to operating systems or any other type of real time system that has to deal with concurrent tasks.

The chapter is a bridge between material in previous chapters relating to real-time systems—including the material on real-time languages—and the topic of software engineering dealt with in the next chapter, where design of a software system is one of the principal topics covered.

12.2 ALTERNATIVE APPROACHES FOR REAL-TIME SOFTWARE DEVELOPMENT

It might be appropriate to start a discussion on 'algorithmic' aspects of real-time software by reviewing possible alternative approaches. The following simple example will serve as a vehicle on which to hang some of the relevant concepts: consider a system—a data acquisition system, say—which has to process input data from two different sources; both sets of data can arrive in a nondeterministic way, possibly simultaneously, and with different processing requirements.

12.2.1 Simple use of polling

A simple approach in designing a system like this is a polling technique, which gets the program periodically to test a status flag to determine whether any data is waiting to be processed, and then if so to enter one of the two different program modules depending on which type of data had been detected. This is essentially a sequential algorithm, and has the problem that further data that arrives while a previous set is being processed could be missed. Whether or not this is a problem in practice would depend on the time taken in processing the data and the attendant delay in reaching the part of the code where the status flag is being tested. The method certainly has the advantage of being conceptually simple and is popular for this very reason; however, in many systems a simple polling approach would not be acceptable because of the potentially unreliable way input data is treated.

12.2.2 Interrupt approach

A more satisfactory approach to the problem of making sure no data is missed could be an interrupt-based approach, with the arrival of one type of data causing an interrupt service routine to be invoked, and the relevant data processing then to be carried out. The other type of data could perhaps be dealt with by an infinite loop in the main program module. This would work best if the module with the relatively smaller amount of data processing needed were selected as the one to be dealt with by the interrupt service routine, because of the potentially faster response to input data which would accrue. This approach can create problems when data needs to be exchanged between the interrupt handlers, since the data will be held in a common memory, and the transfer of data needs to be made indivisible. One possibility is to bracket code that is responsible for transferring data between tasks, with instructions that inhibit and subsequently re-enable interrupts. However, if there is an appreciable amount of data involved, this could entail interrupts being disabled for a time long enough for real-time deadlines to be missed. Problems increase in more complex systems with more than one interrupt service routine and the likelihood of one being interrupted in midstream when an interrupt associated with a more important function occurs. In situations like these it may be that the kind of task management found in multitasking operating systems is needed, as considered in the following section.

12.2.3 Operating system approach

A third technique would be to make the application communicate with the hardware via an operating system, viewing the modules which process data as *tasks*, in the sense we have been reserving for that term in earlier chapters. Events such as input signals indicating the arrival of data to be processed would cause a transfer of control to the kernel of the operating system. Modules waiting for data to arrive would be in a blocked state, while the arrival of data would make the relevant task runnable, and eventually the relevant task would run, and the data processing be carried out. Tasks could communicate data between one another securely, using techniques such as message passing as described in Chapter 9, or methods discussed later in this chapter. Effectively, with this last approach we are using the facilities provided by a virtual machine that an operating system gives us—notably in terms of task management. Moreover, this is being done in the context of a real-time application, which would commonly exhibit concurrency in some form. As we have noted in a previous chapter, real-time operating systems (RTOSs) are tailored to this type of work.[1]

What *detailed* arguments are there for and against these last two alternative approaches? If the complexity of the system is sufficient to justify its use, the preference should be to use a commercially available RTOS for embedded-type systems in the interests of reduced development time, the likelihood of greater reliability which accrues from a widely used and thoroughly tested system, and the availability of support packages and development tools which frequently form part of the package offered by vendors of this type of software. Having said all this, however, one could quote many applications for which the use of an RTOS would either be overkill, or else, for various reasons, would not meet the performance needs of a system. As we have already noted, a simple polling approach is quite adequate in many simple applications, especially if fast response is of lesser importance.

12.3 TYPES OF REAL-TIME OPERATING SYSTEM

12.3.1 RTOSs and the alternatives

What we have been calling real-time operating systems are in fact usually divided by most people into two types, the simplest are termed **real-time executives**, and frequently distinguished from an RTOS proper. They are pared-down operating systems, often diskless, with code held in some nonvolatile storage medium such as ROM or EPROM, and which can execute very efficiently; a multitasking capability is usual. Real-time executives are encountered in many embedded systems, particularly where it is not practical to use disks, and where the system software needs instead to be stored in the form of firmware in ROM or other nonvolatile solid-state storage.

Real-time executives are relatively primitive, and programmers may need to put in considerable effort to make them usable; so, to ease programming problems, many developers go to the next highest level of sophistication in the form of a full RTOS. An RTOS has many of the features found in conventional operating systems, such as a disk filing system for example, but with special provision for real-time control. For instance it would usually be possible to configure such an operating system to give any required priority to the response to the various real-time inputs. Control over process pre-emption is also common. This kind of freedom would be hazardous for general purpose operating systems where users are less intimately involved with the operating system structure, and so are less likely to be knowledgeable about its operation. In fact one of the objectives in the design of a general purpose operating system is to prevent potentially inept users from tinkering with its characteristics.[2] The penalty in using an RTOS in a design rather than a real-time executive is the potentially poorer performance. For instance, in the case of the time to get a response to

---

[1] Some specific commercial RTOSs are discussed in Appendix 3.

[2] It might be noted that many operating systems allow privileged users additional powers to 'tune' the operating system or carry out other functions that need to be carried out system-wide, and that need to be carried out under strict control. We have met the *Superuser* of UNIX where this is the case; another example is the *System Manager* of VMS.

interrupts—that is the time from the interrupt to that at which the relevant task starts to execute—there is likely to be an order of magnitude difference for an RTOS compared to a real-time executive, although there are wide variations in figures like these from one system to another (Fritch and Cardoza, 1991). It might be noted in passing that performance figures such as interrupt response are continually improving with the availability of faster and faster hardware.

Finally, it is worth mentioning that operating systems may not always have the performance demanded by some applications. At the high end of the performance spectrum, some types of application require the hardware to execute as fast as it possibly can, with an absolute minimum of time overhead imposed by software. At this level, there can be no operating system at all, and the software, usually written in assembler, communicates directly to the hardware, without having to pass through any layers of operating system software. Such systems may, or may not, make use of the services of simple monitor programs, but will not be multitasking and so not impose any of the attendant performance limitations resulting from context switching for example.[3] It might be noted that irrespective of performance requirements, many systems are simple enough such that they do not require the sophisticated task management available in a multitasking system, and can simply be structured as a set of interrupt handlers in conjunction with a main program loop as described in the previous section.

12.3.2 Operating systems for multiprocessors

Systems which have multiple processors are becoming increasingly common in industrial, control, military and many other applications, mainly because of the improvements in system performance which they offer. As far as the operating system is concerned there is much common ground with operating systems which control uniprocessor type systems. Questions of task management are bound to be very similar in view of the very nature of tasks that conceptually are concurrently executing entities, irrespective of whether this is true concurrency as in a multiprocessor, or quasi concurrency as in the case of a uniprocessor. Multiprocessor operating systems do, however, have the additional responsibility of ensuring that the workload is distributed as evenly as possible amongst the processors, usually referred to as **load balancing**.

Implementation of real-time executives (or RTOSs) for multiprocessor environments could adopt one of three possible approaches:

- **Master–slave**—one processor, **the master**, together with its operating system is the main controlling element, and other processors, **the slaves**, will normally only execute one user program. Slaves will only implement low-level management functions—for instance low-level handling of interrupts. A failure of the master processor results in complete system failure—one of the main drawbacks of this approach for applications where fault-tolerant behaviour is important.
- **Separate executives**—each processor is responsible for managing its own resources and has its own copy of the operating system. This eliminates the reliability/fault-

---

[3] It might be added that, for some real-time applications, just about any computer is going to be too slow, in which case dedicated hardware would be required: signal processing is one example where this is so, although such applications fall outside the scope of this book.

tolerant problems of the master–slave approach. Conceptually the arrangement is not too dissimilar to a set of networked computers. The dedication of specific tasks to a particular processor may, however, result in one or more processors being under-utilized while other processors have several tasks that may be executing.

- **Symmetrical processors**—processors are viewed as having identical capabilities. The operating system may execute on any processor at any given time, thus providing better load balancing and also contributing to improved system reliability. This system is more complex than the other two because of the need for exclusive access to system resources. The approach has the merit that it does not lead to total system failure when one processor fails.

12.4 RTOS—A CLOSER LOOK

The main issues in the design of a real-time application are: performance, reliability and ease of use; and we now examine the role of an RTOS in achieving these goals. Remember, however, from what has been said before that the use of an RTOS will not always be appropriate.

12.4.1 Performance characteristics

One of the main distinguishing features of an RTOS compared to other types of operating system is its ability to respond more rapidly to external events, usually in the form of interrupts. Many operating system kernels may actually disable interrupts for brief periods for instance, in which case no response to interrupts would be possible until after they had been re-enabled again. In the context of a real-time application this could lead to an interrupt being missed. All RTOSs would normally quote some guaranteed maximum figure for a response to interrupt events.[4] Fritch and Cardoza (1991) quote a figure of 69 μs for the time from an interrupt to that at which a task starts execution in the case of the RTOS PDOS, although in other RTOSs this figure may be as high as 300 μs. Additionally, one would like the response to interrupts to be as deterministic as possible, that is there should not be much variation between the minimum and maximum interrupt response times. This simplifies life for the designer considerably. Response time could be jeopardized if a program had first to be swapped into main store from backing store as is done in many types of operating system, and in some RTOSs there is usually a facility called **memory locking**, which allows users to force some programs to be kept in main store.

Some operating systems, UNIX being an example, make a virtue of the fact that they are written in a high-level language, based on arguments such as portability as discussed in previous chapters. However, in an RTOS, there is much greater emphasis on performance, and for this reason many are written in assembler. RTOS kernels tend to be relatively small in size, so the programming problems involved in using assembler are less serious than otherwise would have been the case.

---

[4] A good discussion of performance measures such as this, including a comparison between different commercial RTOSs, can be found in Fritch and Cardoza (1991).

12.4.2 Scheduling—pre-emptive versus non-pre-emptive

RTOSs frequently make use of non-pre-emptive scheduling rather than the pre-emptive type found in many other types of operating system. This is done in the interests of giving tasks fairness in terms of CPU time. Pre-emptive scheduling involves a considerable overhead in saving the state of a task following pre-emption, and is commonly not acceptable in the kind of time-critical applications where an RTOS would be used. Having said this, however, all RTOSs must make provision for tasks of differing priority levels for reasons discussed in previous chapters and also below. By implication, the ability of a task to be pre-empted by one with a higher priority level must be built in. Thus it would be usual not to use pre-emption within a set of tasks, but only for those tasks with a given level of priority.

12.4.3 Choosing priorities for tasks

With many real-time applications it is possible to assign *fixed* priorities to tasks, i.e. priorities that do not change dynamically, and then to allow a task to run until it suspends itself while waiting for some real-time event such as I/O completion, or until it is pre-empted by a task of greater priority. In actually *selecting* a priority for a task, several alternatives suggest themselves. A simple method would give each task an individual numbered priority, the approach adopted in OS-9. Other systems have schemes in which tasks are grouped, and each group is assigned its own numbered priority level; some RTOSs have just two priority levels.

Some tasks are fairly obvious candidates as low-priority tasks. For instance, a printer task can afford to wait for an appreciable period while other tasks run, with little ill effect; all a task sending data to a printer is called on to do is fill the printer's buffer memory at what could be relatively infrequent intervals. A task associated with a terminal keyboard, however, would need greater priority than this, otherwise there might be a danger that users start to see unacceptable delays to keypresses. This kind of I/O would not require very frequent servicing, perhaps at intervals of minutes, or even hours, in the case of users typing a command, but a speedy response is required when this kind of input does occur. Some tasks will be CPU intensive, and there is a danger that they will tend to hog the CPU if given too high a priority, so such tasks could be given a lesser priority than (say) a task associated with some alarm function which might occur infrequently, but requires prompt action when it does occur.

Finally, it could be argued that considerations such as these apply—in general terms at least—to almost *any* operating system. Although there might be some truth in this, it is as well to remember that in the case of an RTOS we are faced with a situation where the developer is designing a system around an RTOS, and so is going to be greatly concerned with control over task priorities. In most other types of operating system, users often have to take what is on offer in the way of such things as task priorities—if indeed they can change priorities at all.

12.4.4 The role of an RTOS in the development cycle

RTOSs are sometimes used as part of a development system as well as in the primary role in controlling a real-time system. Most commercial RTOSs can be obtained with packages such as cross compilers and debuggers, on which the software for some target system—

possibly itself controlled by an RTOS—can be developed.[5] Software development can be carried out on any convenient host computer such as a PC and then downloaded into (typically) an EPROM programmer which would be used to program the chip containing the applications code. Naturally it is not essential to have the same operating system in the host machine and the target system; however, there is advantage to be gained in view of the compatibility between the two systems, for instance in primitives for real-time control.

12.5 ERROR DETECTION AND FAULT TOLERANCE

Looked at generally, almost all real-time systems need to be designed so as to allow some kind of failure mode to be activated following detection of a problem.[6] This should include a procedure to revert to the normal mode of operation after a successful recovery from the fault condition. Depending on the degree of fault tolerance required, some systems may use redundancy by employing duplicated hardware and/or software. One way this could work is to get the software to switch to backup hardware following detection of a fault.

12.5.1 Triple modular redundancy

Triple modular redundancy (TMR) is a very commonly used redundancy method, originally suggested by Von Neumann; it uses three identical hardware modules, typically processor boards, which are supplied with the same data and therefore normally generate the same output. The modules supply their output to a so-called **voting element** which accepts the inputs and delivers the majority 'vote' as its output (see Fig. 12.1). In other words, if two out of the three units generated the same data, this would be the output from the voting element, and the third unit would be ignored if its output differed from that of the other two.

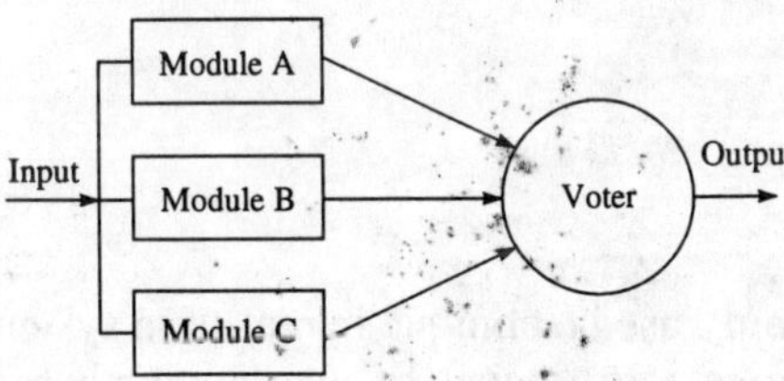

Figure 12.1 TMR voting scheme.

12.5.2 Dynamic redundancy

A system with dynamic redundancy comprises several modules, but with only one operating at any given time. If a fault is detected in the module currently operating, it is switched out and replaced by a spare. This infers that dynamic redundancy requires the consecutive actions of fault detection and fault recovery. The detection of a fault in

---

[5] Development systems are discussed in Chapter 14.

[6] The relevance of exception-handling techniques as discussed in Chapter 6 should be apparent here.

individual modules of a dynamic system can be achieved by one of the following techniques:

- Periodic tests
- Self-checking circuits
- Watchdog timers

In **periodic tests**, the normal operation of the functional module is temporarily suspended and a test routine is run to determine whether faults are present. A disadvantage of this approach is that it cannot detect temporary faults, unless they occur while the module is being tested. Self-checking circuits provide a very cost-effective method of fault detection. They are designed so that, for normal inputs, correct output is provided, or else an indication of a fault in a module is given.

Watchdog timers are an effective and commonly used method of fault detection. Times are set to certain values at pre-established points called *checkpoints* in the program executed by a module. A timer at a particular checkpoint counts down while the module performs its function, and is normally reset before the next checkpoint is reached. However, a software bug or a hardware fault will prevent the program from resetting the timer. The timer then issues an interrupt which causes automatic switchover to a spare module.

12.5.3 Software redundancy

This is used to detect software design faults and is a form of the TMR approach discussed earlier. The modules now, however, run three different programs written to the same specification by three different teams of programmers. This is evidently a very expensive technique, the cost of software being what it is, and moreover it does not detect any errors in the *specification* of the problem. Nevertheless the software redundancy idea has been used in practice, one example being the Airbus wing-flap control computer.

12.6 BUFFERING

12.6.1 Introduction

Buffering is a very commonly used technique in real-time systems. Essentially it refers to techniques aiming to improve performance by overlapping activities such as I/O and processing, which otherwise would be done in strict sequence, and thereby more slowly. Examples of buffering in the context of operating systems have already been discussed, one being the spooling technique used to deal with data to be sent to a printer. The use of a buffer memory *inside* a printer is another example. Here, data is loaded into the printer's buffer memory by the computer at a relatively fast rate, ahead of the time at which the printer mechanism can deal with it, and which the printer can subsequently access while allowing the computer to continue other processing—concurrently that is. Notice, however, that for large enough files the printer's buffer will fill up, with the possibility, in some simple systems at least, that the computer will remain idle waiting for the buffer to empty again. Admittedly this would not happen in most computers, where the operating system would allow the computer to run another task while it is waiting. Nevertheless, the example does illustrate a general characteristic of the buffering principle: this is that the idea of overlapping activities to improve throughput will not work effectively when a source sends data *continuously* at a fast rate via a buffer, to a destination which can only deal with it at a relatively slow rate. Throughput will be that of the slower element in the system. Buffering in this type of situation only smooths out variations in throughput from source to destination when the buffer is large enough to deal with all of the data involved.

The previous example considered a buffer external to the computer, but more commonly we are concerned with buffering implemented in the computer's memory, areas of memory being reserved by tasks for this purpose. It is very common to use buffering in conjunction with I/O devices, including the printer example just quoted. With unbuffered I/O, a task continuously sending data to an I/O device would need to block, frequently waiting for the transfers of data to complete. This would incur an overhead inside the operating system in connection with frequent blocking and unblocking of the task involved. A better method would be for the task to write all the data involved to a buffer ahead of the time at which the I/O device will deal with it. The task responsible for actually transferring data to the peripheral subsequently reads the data from the buffer and will only need to block if the buffer becomes empty.

12.6.2 Circular buffer

With the type of buffer discussed above we are using a FIFO, the size of which will vary, depending on the relative rates at which data is respectively being written and read. A **circular buffer**, in contrast, is of fixed size conceptually, as depicted in Fig. 12.2(a), where InPtr and OutPtr are pointers to the start and end of the buffer respectively. Adding data is accompanied by advancing InPtr in an 'anticlockwise' sense, while reading data from the end of the buffer advances OutPtr in an 'anticlockwise' sense. The cross-hatched region indicates the data that has been written to the buffer. More realistically the buffer is as depicted in Fig. 12.2(b), the end of the buffer being pointed to by EndPtr; when either InPtr or OutPtr have reached the end of the buffer, they are reset to point to the start of the buffer, i.e. StartPtr.

Problems can arise in circular buffers if the buffer is continually written to at a faster rate than that at which it is read, since InPtr can 'overtake' OutPtr, or conversely, if the

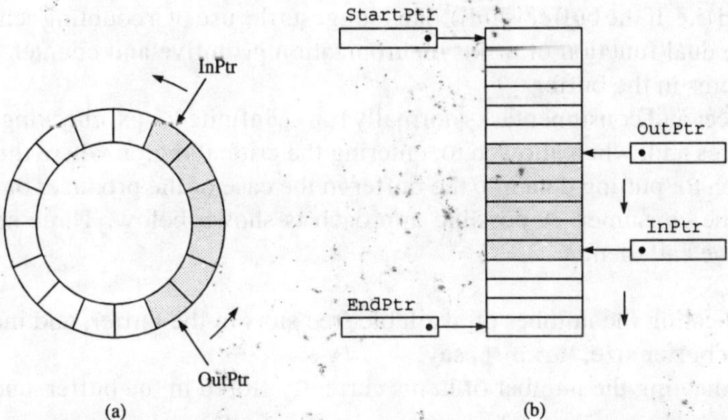

Figure 12.2 Circular buffer.

buffer is continually read at a faster rate than that at which it is written to, `OutPtr` can overtake `InPtr`.

12.6.3 Double buffer

Another variation on the buffering principle, **double buffering**, uses two buffers; a task wanting to (say) send some data to an I/O device, first sends the data to one of the buffers, and when full the data in the buffer is then transferred to the I/O device via a driver task. Concurrently, the originating task fills up the second buffer, until it is full. When the first buffer has been completely read, the I/O transfer then takes place using the data in the second buffer. The originating task now fills up the first buffer again. This process of using alternate buffers continues until all data has been transferred. With this scheme the originating task would only need to block if *both* buffers were full.

12.6.4 The producer consumer problem

Implementation of the buffering idea now needs to be considered, and the **producer consumer** problem is one name given to the approach used. Essentially, the situation is one in which two tasks have to exchange data via a buffer. One task *produces* data, sending it to a buffer. The other task reads the data from the buffer, or, put another way, it *consumes* the data. The producer could (say) be a task that generates data from the results of some computation within the CPU, while the consumer could be a task transferring the results of the computation to an I/O device. The two tasks run pseudoconcurrently, smoothing out variations in the processing rates of the consumer and producer. Several other problems need to be solved to make this scheme workable:

1. The buffer, being of finite size, must not be written to by the producer if already full.
2. The buffer should not be read by the consumer if empty.
3. Consumer and producer tasks must be prevented from accessing the buffer at the same time.

Evidently the tasks need some count of the number of items in the buffer, and they need to check if this count is zero (i.e. if the buffer is empty), or equal to the maximum number of items allowed (i.e. if the buffer is full). This suggests the use of a counting semaphore which can serve the dual function of task synchronization primitive and counter for the current number of items in the buffer.

The producer and consumer tasks normally run as infinite loops, checking and/or changing semaphores and, when allowed to, entering the critical region where the buffer can be read or written to: putting data into the buffer in the case of the producer or removing it in the case of the consumer. A possible approach is shown below. Three semaphores are employed—we call them:

- **free**—indicating the number of available free slots in the buffer, and initialized to the maximum buffer size, `MaxBuf`, say.
- **avail**—indicating the number of items currently stored in the buffer, and initialized to zero.
- **mutex**—ensuring proper mutual exclusion for the buffer.

Producer task

```
WHILE( TRUE )
   begin
   produce item to put in the buffer ;
   wait( free );
   wait( mutex ) ;
   write item to the buffer ;
   signal( mutex ) ;
   signal( avail )
   end ;
```

Consumer task

```
WHILE( TRUE )
   begin
   wait( avail ) ;
   wait( mutex ) ;
   read item from buffer
   signal( mutex ) ;
   signal( free ) ;
   process item further
   end ;
```

You should satisfy yourself that this scheme would never allow more than `MaxBuf`, or less than zero items in the buffer. You could try the effect of various sequences of read and write operations with the two tasks, examining the effect on the semaphores, and the associated changes in state of the two tasks controlled by the semaphore values. For example, if the consumer task ran first, following initialization, there would be a wait on `avail`, which, since it already had a value of zero, would cause the consumer task to block. It would require a signal on avail in the producer task to make the consumer runnable subsequently.

12.6.5 Mailboxes

Mailboxes were referred to in Chapter 11; recall that a mailbox is a FIFO buffer, to which a *producing* task transmits data, together with a signal to a *consuming* task that data is available in the mailbox. Mailboxes can be accessed by more than one producer and consumer. Mailboxes are a very commonly used method of task communication in much real-time software including RTOSs, in view of their relatively high efficiency.

12.7 THE READERS AND WRITERS PROBLEM

This problem, which is of considerable practical interest, refers to a situation where a number of tasks need to access data such as files, tables, etc., which need to be shareable between more than one task. Several tasks can be safely allowed read access at any given time, but if any task is writing to the data base, all other tasks must be inhibited from doing so. In other words there may be several readers but only one writer at any given time.

A possible solution to the readers and writers problem is given in Figs 12.3 and 12.4, which show two tasks for the `read` and `write` functions respectively. The algorithm must, in all `read` tasks, keep a count of the current number of tasks reading the data base, since the algorithm uses the occurrence of a count of zero in unblocking a task that wants to write to the data base. Semaphores are used to protect the variable holding the count, and to deal with mutual exclusion in relation to the data base which satisfies the previously stated requirements.

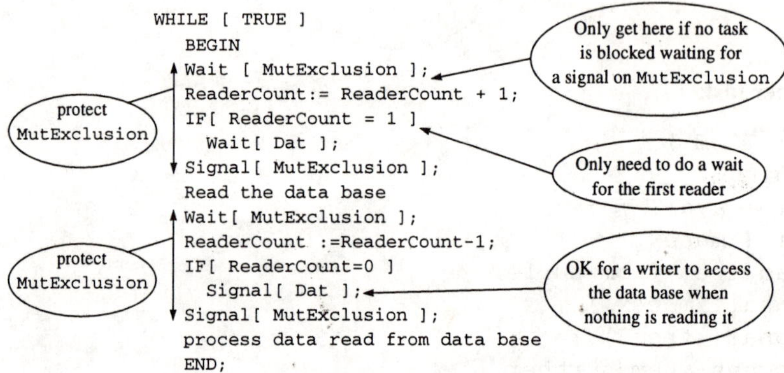

Figure 12.3 Reader task.

This algorithm has the feature that readers always get priority over writers, with the attendant possibility that, if readers access the data base heavily, the writer may not be able to update the data base at a fast enough rate, and in an extreme case it is possible that real-time deadlines will not be met as a result. This situation where a task is denied access for what is considered to be an excessive period is called **starvation**. Alternative algorithms should then be used that avoid this starvation problem giving greater fairness to writers (Axford, 1991).

```
WHILE[ TRUE ]
    BEGIN
    compute the data to be sent to data base
    Wait[ Dat ];
    Write to data base inside critical region
    Signal[ Dat ];
    END;
```

Figure 12.4 Writer task.

12.8 EXERCISES AND REVIEW QUESTIONS

1. Discuss the differences between an RTOS and other general-purpose operating systems such as UNIX.

2. In Chapter 9, message passing was discussed as one method of intertask communication, and it was stated that message-passing primitives could be implemented with semaphores. Show that the following are valid implementations of the `send` and `receive` primitives. Discuss the possibility of deadlock were the two wait semaphores to be interchanged.

(a) **send(Task_Destination, Message)**

```
begin
wait( Buffer_Space ) ;
wait( Mut_Exclusion ) ;
add Message to the tail of Task_Destination's Message FIFO ;
signal( Task_Destination.semaphore ) ;
signal( Mut_Exclusion ) ;
end ;
```

(b) **receive(Task_Source, Message)**

```
begin
wait( Task_Source.semaphore ) ;
wait( Mut_Exclusion ) ;
deal with Message from Task_Source ;
signal( Buffer_Space ) ;
signal( Mut_Exclusion ) ;
end ;
```

Note: `Buffer_Space` is initialized to the maximum number of items in a message buffer. `Task_Destination` and `Task_Source` name the tasks sending and receiving messages, while `Task_Destination.semaphore` and `Task_Source.semaphore` are semaphores related to their respective message queues.

3. For a system using a circular buffer as an intertask communication mechanism, give outline code for the producer consumer problem. *Hint:* You will need to make use of a variable giving the number of vacant slots, and also the number of used slots, in the buffer.

4. You are to design a system that can employ the services of an RTOS, and this has provision for tasks that can have three levels of priority. Your system needs to interact with the following devices: two dot-matrix printers, data to/from a modem via a serial port with a relatively small processing requirement, a terminal connected to another serial port, an 8-channel A/D converter which inputs data once per hour with a relatively small processing requirement, a single-channel A/D converter which inputs data once per 200 msec, and a single-channel D/A converter which is to output data every 1 msec. Assuming a separate task is to be associated with each of the devices, assign what you think might be appropriate priorities for the tasks.

REFERENCES AND FURTHER READING

Axford, T. (1991) *Concurrent Programming—Fundamental Techniques for Real-Time and Parallel Software Design*, John Wiley, Chichester.

Bennet, S. and Linkens, D. A. (eds) (1984) *Real-Time Computer Control*, Peter Peregrinus, London.

Fritch, D. G. and Cardoza, R. W. (1991) 'Benchmarking real-time systems', article published by Eyring Corporation, USA.

Houpis, C. H. and Lamant, H. R. L. (1986) *Digital Control Systems: The Hardware and Software*, McGraw-Hill, New York.

Jenkins, D. G. and Welland, R. C. (eds) (1990) *Software Engineering for Electronic Engineers*, Peter Peregrinus, London.

Lawrence, P. D. and Mauch, K. (1987) *Real-Time Microcomputer System Design—An Introduction*, McGraw-Hill, New York.

Levi, S. and Agrawala, A. K. (1990) *Real Time System Design*, McGraw-Hill, New York.

Thomas, H. W. (1982) 'Design and performance of a simple nucleus for real time control applications', *Software and Microsystems*, vol. 1, no. 6, October, pp. 161–166.

13

INTRODUCTION TO SOFTWARE ENGINEERING

13.1 BACKGROUND

Brief reference to software engineering was made in Chapter 5; we now examine this topic in more depth, using concepts developed in that chapter and elsewhere in the book.

Most people learn programming by writing relatively small programs, often no more than several hundred lines or so for many student assignments. This is fine to learn the syntax of a particular language to gain a basic understanding of some algorithms such as lists, stacks, etc., and to explore other programming techniques. In writing small programs, however, one fails to appreciate the nature of the difficulties in developing software of any size; commercial software products in excess of 100 000 lines of code are not uncommon. One of the more surprising facts of life for the beginner in programming is that, with large-scale software projects, only a small fraction of the time used to develop the software is devoted to *coding*; often less than 20% in fact. The bulk of the development time is directed towards aspects related to planning and testing. In this chapter we examine why this is so, and explore all the steps that are needed in the development of software, starting with ways of specifying its aims, and ending with its use by customers.

13.2 THE SOFTWARE CRISIS

In the 1960s, a radar station in Greenland reported a massive missile attack. Fortunately this proved to be mistaken and just as fortunately did not elicit a retaliatory attack! The radars and the computer systems were in fact being fooled by the fact that the moon was rising! More recently it has been reported that the software for a proposed computer system to handle the safety systems in a nuclear reactor is too complex for its correctness to be guaranteed. One could quote other horror stories where the sheer complexity of software is such as to cause concern about its dependability. Sometimes, as in the examples above, the consequences of failures in computer systems can be life threatening, but other spheres such as business applications for computers can suffer serious consequences of other kinds from the results of software problems, with customers getting unexpected £10 million gas bills, for example; and indeed there are numerous cases where failure of software controlling financial systems has given rise to extremely damaging financial loss. Software gets more complex to keep track with the ever-increasing size of computer-controlled systems, so the danger is that problems like these could get worse.

Bugs in programs have long been a common complaint with purchasers of software, but just as serious are the often massive cost overruns, delays in development and difficulties in program maintenance. All these problems, in particular their escalation in degree in recent years, have given rise to what has been described as the **software crisis** (Pressman, 1992; Ince, 1989). The consequences of not attacking this problem effectively are severe, both in terms of hazards—people have actually been killed by the consequences of software bugs in computer-controlled equipment—and, in financial terms, it is reported that $50 billion is spent annually on software in the USA alone, and a worryingly large proportion of this is not used effectively because of cost overruns.

The causes of the software crisis are complex, but what *does* seem clear is the need for the same kind of properly engineered approach to the development of software as that long used with any other product, be it a bridge, a car or a hairdryer. One definition of the term **software engineering** encapsulates the necessary principles involved as follows:

> The establishment of sound engineering principles in order to obtain, economically, software that is reliable and works efficiently on real machines (Dettmann and Johnson, 1992).

13.3 THE SOFTWARE DEVELOPMENT CYCLE

Software engineering involves a systematic approach to software development, and one possible scheme is illustrated in Fig. 13.1, which depicts a division into six sequential

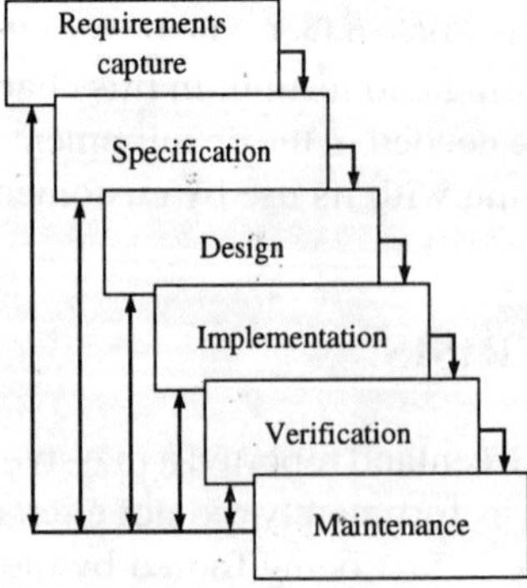

Figure 13.1 Waterfall software development model.

phases: capture of requirements, specification, system design, implementation (= coding), verification and finally maintenance; each phase provides an input to the next stage. Note, however, that, with this model of software development, the customer only gets to see a working product near the end of the development cycle, with the attendant problem that shortcomings will only be manifest at a late stage. In any event, customers have to be the ultimate arbiters of whether the software is meeting their needs, and inevitably several iterations through the various stages will be needed before a customer will be satisfied with the end product.

Some software engineers try to get a version of the software working at an early stage in the development cycle. This might be done to give customers a demonstration of perhaps a cut-down version of the final product in the interests of uncovering some problems as soon as possible. Following satisfactory performance of this (prototype) version of the program, it would then be possible to refine it and then go on to develop a final production version. This approach will prove particularly attractive in preference to the original waterfall type of development cycle in very complex software systems which require a lengthy development cycle, where the consequences of errors uncovered at a late stage could prove very expensive in terms of additional development time and the attendant cost. This model of software development is generally termed **prototyping**, and is illustrated in Fig. 13.2.

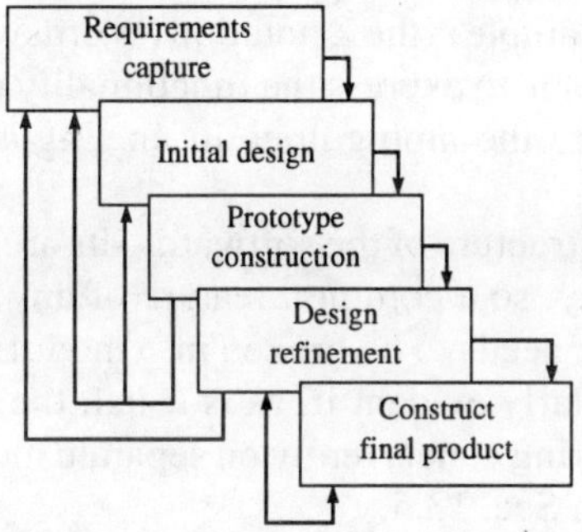

Figure 13.2 The prototyping development model.

13.3.1 The software development cycle—an overview

Requirements capture is a pivotal stage in the development of software, involving, as it does, input from customers who may have a very different background to that of the software developer. Customers vary enormously in their degree of knowledge of the needs of software engineering. Some will be able to give a very detailed specification of the function of their application backed up with extensive documentation describing requirements. Others may have little knowledge of the needs of software development, and it will therefore need much work to turn an informal, often inadequately defined, description of the application into a specification that a software engineer can use as a basis for design. Customers will express their needs in natural language terms, with all the potential ambiguity and vagueness that this involves. Customer and software developer will usually need to engage in a lengthy dialogue until both are aware of each other's problems and needs. As well as stating what the system is to do, customers may, and usually do, impose constraints on how the system is to be implemented. This might be in the form of some mandatory hardware platform—insisted on for reasons of cost, say; or perhaps a requirement to use a particular high-level language—we have noted elsewhere that Ada is the language demanded by the UK Ministry of Defence for writing much of the software used in systems delivered to them.

Specification follows from a statement of requirements, and represents some definitive statement of what the system is to do as agreed between customer and software developer in the requirements capture phase. This needs to be expressed in the form of a document describing the behaviour of the system in terms that the developer can use as a basis for a design. Such a document would frequently form the foundation of a contractual agreement between customer and developer, enforceable in law. In principle, the customer should not need to play a further role in the development of the software until it is ready for use, although, as we have already noted, software developers may prefer to demonstrate some features of the product in prototype form in the interests of shortening the development cycle. At the end of the specification stage it should be possible for a customer to say to the software developer: 'This is what I want, now go away and design it, and demonstrate the finished product to me in x months' time.'

As a final note, is has been increasingly recognized in recent years that more rigorous methods of design than those used in the past are increasingly important and formal methods of specification have been introduced, these being based on mathematical techniques used in proofs. One example is the Z notation (Wordsworth, 1992), a language based on set theory and predicate logic to express the functionality of a software-based system. Z aims to completely eliminate the ambiguities in an English language description of a problem.

Design involves us in the structure of the software—its architecture. Almost all software is of considerable complexity, so a common feature of any of the many different design strategies is that the software needs to be broken into modules in order to handle the complexity. This need is particularly evident if, as is usual, the project is a team effort, with different team members working concurrently on separate modules of the program. Design is discussed in more detail in Sec. 13.5.

Implementation (coding) should, if the design phase has been carried out properly, be the most straightforward step in the development cycle. The results of a design phase will typically, in spite of the fact that they will be expressed in language-independent terms, be a description in a form such as pseudocode, and not very far removed from the actual code. Well-structured programs should be a direct consequence of a well-structured design; and translation of design into code will be made simpler if a structured language such as Modula-2 or Ada is used.

Verification is essentially an examination of the behaviour of a program, tested against that expected from the specification. This will involve exercising the program with comprehensive sets of test data. There is a distinction between validation and **verification**; the latter involves making certain that the program, or program module being tested is *internally* correct. One author has drawn the distinction between validation and verification rather neatly as follows: 'Validation is building the right system, while verification is building the system right.' It should be stressed that verification and validation in various forms need to be carried out at *all* stages of the software development cycle.

Maintenance (Pressman, 1992; Ince, 1989; Jenkins and Welland, 1990), in many cases, is likely to represent the bulk of the cost of a software product over its working life—

frequently in excess of 50 per cent—so its importance can scarcely be underestimated. The maintenance phase is concerned with the following:

- Corrections to the software resulting from a customer's observations of any short-comings following the initial installation and use.
- Changes to the specification over the lifetime of a program.
- Adaptation to run on hardware different to that for which the software was originally intended.

Some changes in specification will be minor, but in general a review of *all* the stages in the development cycle indicated by Fig. 13.1 might be needed. Maintenance has been, and continues to be, a thorny problem for a number of reasons, most of which can be attributed to bad software engineering practice: programmers engaged in maintenance are often not the same individuals as those responsible for producing the original version of a program and it is a common experience that deciphering someone else's code is often far from easy. Badly documented projects including uncommented code, poor or absent manuals, and programs not designed to be readily maintainable are common facts of life making mainten-ance difficult.

13.4 DATA FLOW—ONE WAY OF EXPRESSING A SPECIFICATION

It has already been noted that there are problems in using natural language to express a system's specification. Implementors invariably use some kind of formalism that attempts to overcome the difficulties faced and bring to bear a greater degree of precision and rigour. One approach, called **data flow**, describes a system, partly in diagrammatic form, in terms of the nature of the data being processed, and the way data is transferred around a system. All computer systems can be regarded as data transformers in the sense that some input data has computation done on it, resulting in output data being produced; so the nature of the data in a system and the way it is processed is central to the understanding of any program. To carry this approach forward it is necessary to examine the nature of the data, and to analyse its structure; in addition, that is, to looking at the way data is input or output between various parts of the system.

13.4.1 An illustrative example—monitoring system

An example at this point might prove useful to clarify concepts. Consider a monitoring system to be used with an installation such as an unmanned water pumping station which sends data such as temperature, pressure, etc., at regular intervals (say once per hour) to a computer at a remote control centre via a telephone line and modem. The data, after various transformations, is to be displayed on the screen of a PC. The computer will have a file on disk to store an ongoing record of the data. Secondly, the remote station is to send data to the control centre immediately values outside certain limits are detected. This data is also to be stored in a file on disk. Certain data is to be displayed in suitable form on the computer's screen. Finally the operator is allowed to enter commands at the keyboard as to when graphs of the variation in time of particular data are required. This is an admittedly some-what sketchy description, in the interests of avoiding concepts being obscured by excessive detail. Nevertheless the example will serve to illustrate the concept of the flow of data in a system in general terms.

Figure 13.3 shows the essence of the problem described in a so-called **data flow diagram (DFD)**. This is a directed graph in which the boxes represent sources and destinations (sinks) of data, the circles (bubbles) represent transformations carried out on input data and the arrowed lines indicate the direction of flow of data. The parallel lines represent data stores, commonly in the form of a file or backing store although there are other possibilities. Two high-level functions can be identified at this point:

- Processing data input to the computer from its serial port
- Processing commands entered at the keyboard

These functions appear inside the relevant bubbles, while the lines are marked with details of the data with which they are associated. The numbers serve to give a convenient identification to the various processes,[1] and will prove of use at later stages in the development when the degree of detail is much greater, bringing with it the need to document carefully the various elements in the system.

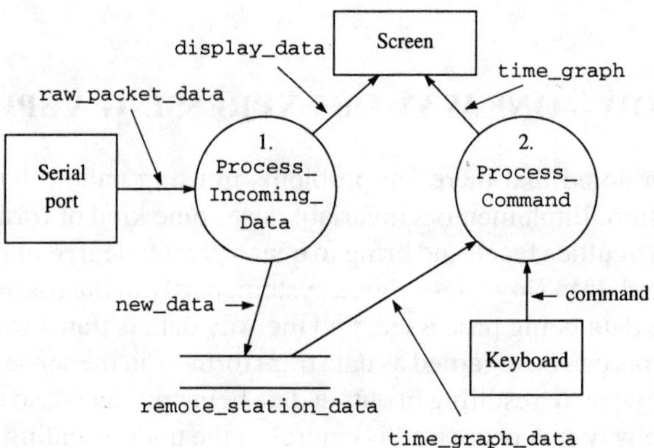

Figure 13.3 Data flow for monitoring system.

13.4.2 Refining the data flow diagram

The data flow diagram on its own only gives a partial picture: more needs to be said about the *nature* of the data, and also the internal structure of the processes described by the bubbles. The diagram shown in Fig. 13.3 only gives a top-level view of the system. At some stage though, customers or the developer will need a more detailed description of their system. This is achieved by 'expanding' each of the bubbles, so that the processes associated with all the resulting bubbles are identified with more detailed descriptions of the data flow, each bubble being described by its own data flow diagram. A description of the system at a lower level of abstraction then becomes possible. The process is termed **levelling**, and may need to be carried out several times, depending on the depth of detail

---

[1] The term *process* will be used in this chapter in preference to *task*, which is used elsewhere in the book, in view of the common preference for the former term in describing data flow concepts.

needed at any particular point. Customers and the developer are then able to engage in a dialogue about the software at one of a number of levels of abstraction.

In the case of our monitoring system the `Process_Incoming_Data` bubble could be decomposed into possible processes to:

- Deal with incoming calls from the modem;[2] the output from this would be a stream of data in the form of an array of characters generated from the bit serial data read from the serial port, and which will be called a *packet* from this point on. Packets are input to the system whenever a call has been established. This bubble will be called `Process_Incoming_Call`.

- Carry out the processing of raw data in a packet—remember that there are two types of packet: those associated with the regular transmission of 'normal' data, and those associated with out-of-limit alarmed type data. Packets need information to identify which of these two types is involved. Additionally some form of parsing would be necessary to make sure all the data in a packet is of the expected format.[3] The outputs from the process for alarm and regular type data would be to the data stores and also to the computer's screen. This bubble will be called `Process_Packet_Data`.

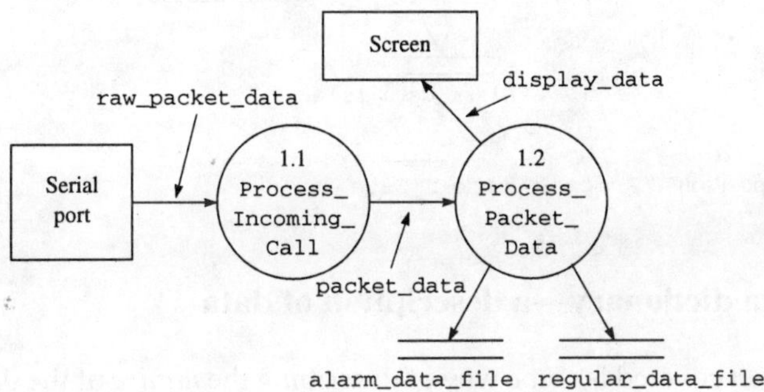

Figure 13.4 Structure of `Process_Incoming_Data`.

Note the numbering scheme to identify each bubble whenever a higher-level bubble is expanded. This is a way of documenting the relationship between a higher-level process and lower-level processes derived from it. It might seem a somewhat unnecessary embellishment here, but were the system developed further, much more detail would emerge, and then it really would be apparent that some explicit way of identifying the relation between the various processes is needed. Our system, in its current state of development at least, has been described by just two diagrams. In practice, a fully developed system would require a large number of diagrams, and a way of cross-referencing processes and other quantities between different diagrams is virtually essential.

---

[2] Handling data and control information from modems, and processing that information via computer serial ports is somewhat complex, and most of the relevant techniques are specific to the particular modem and computer used. It is not really appropriate to give too much detail here; readers who are interested in interfacing devices such as modems to an IBM-compatible PC can refer to books such as Mellor and Ward (1986).

[3] In a somewhat similar vein, recall that 68000S records are parsed by the use of the checksum byte which terminates each record—see Sec. 7.6.1.

Figure 13.5 shows the further stage of decomposition for `Process_Packet_Data` into a process that carries out the decoding and parsing functions on raw packet data in process 1.2.1, and then passes the two different types of data to processes that specifically deal with 'regular' data and 'alarmed' data, processes 1.2.2 and 1.2.3. These two processes output data to the computer's screen and also to data stores as shown in Fig. 13.5.

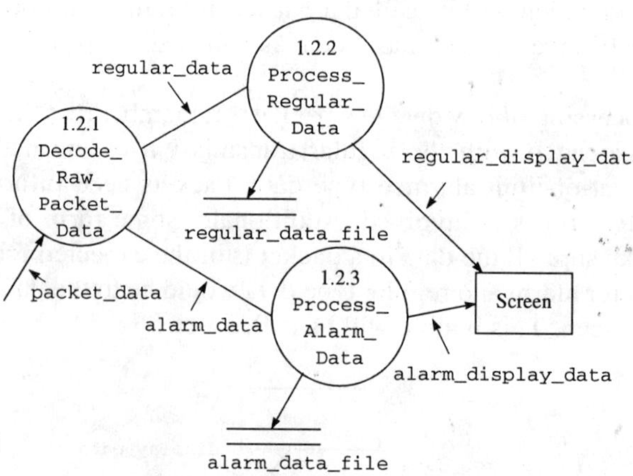

Figure 13.5 Decomposition of `Process_Packet_Data`.

13.4.3 The data dictionary—a description of data

The element not yet covered is some way of describing the *nature* of the data flowing from one process to another. To put this on a rigorous basis it is necessary to formalize the description of the nature of the data, just as we formalized the definitions of the inputs, outputs and data flow paths. Some writers use the idea of a **data dictionary** to do this (Ince, 1989). A data dictionary contains definitions of all data that feature in the data flow diagram. Some of the data will be primitive, in the form of characters, integers, etc., while other data will be structured, defined in terms of data of more elementary types—and doubtless eventually implemented as a Pascal record, C struct, etc., depending on the language. Data dictionaries are built up using a form of grammar that can loosely be compared to that used with compilers (see Sec. 7.3.5). A possible set of rules is shown in Table 13.1.

Table 13.1 Data dictionary rules

Operation type	Grammatical symbol	Comment
Sequence	'+'	= 'composed of'
Selection	'|'	'OR'
Repetition	$\{\}^n$	'n repetitions'
	( )	Optional data

Taking the earlier example of the monitoring system by way of illustration, the information entering the computer's serial port is bit serial data and would be transformed by the program into a string of characters, which we have been calling a packet. Assume now that packets are divided into the following elements: the packet type (i.e. indicating 'regular' or 'alarmed' data), characters representing the number of data items, characters representing the analog data itself and finally a NULL character signifying the end of a packet. Let us also say that all analog quantities and also the number of data items are 12-bit values, with each 4-bit nybble within these 12 bits encoded as an ASCII character. The binary quantities for analog values, as derived from three ASCII characters, in turn are eventually collected together as an array to be further processed by the program. Stopping at this point to avoid obscuring the argument with excessive detail, the types of data listed in Table 13.2 have been identified.

Table 13.2 Partial data dictionary for monitoring system

Data	Composition
Raw_Packet_Data	= Header+No_Raw_Data_Items_Raw+Data_List+NULL
Header	= Alarm_Data_Header | Regular_Data_Header
Alarm_Data_Header	= 'A'
Regular_Data_Header	= 'R'
No_Raw_Data_Items	= {character}^{3}
Raw_Data_List	= {Raw_Data_Item}^{60}
Raw_Data_Item	= {character}^{3}

13.4.4 Process description

Were the decomposition of the processes carried further, we would eventually have a fully expanded set of data flow diagrams and an associated data dictionary, defining the way that data is passed around a system and its nature; all the major components in a system would then have been identified. To complete the specification of the system, we now need to indicate the structure of the modules themselves or, to be more precise, the nature of the way that data is processed inside each of the bubbles in the fully expanded system; the text inside the bubbles of a data flow diagram only describes in outline the overall *purpose* of a process, not the details of the actions it has to carry out.

In describing the structure of processes, most developers use a form of language called 'structured English' or 'pseudocode', aiming to minimize the problems of using natural language: ambiguity, vagueness, etc., to which we have already referred. Structured English is built around the standard constructs of sequencing, looping and alternation, which form the basis for any algorithm that can be implemented on a computer and which in fact featured in earlier chapters. In practice a mixture of structured English augmented by natural language descriptions might be used to describe the internal operation of a process.

By way of illustration, the 1.2.1 process in Fig. 13.5 could be written as shown below using a Pascal-like syntax, although this is not essential; many other 'dialects' of structured English are used.

```
while forever do
begin
read another packet ;
if packet parses ok then
   begin
   case packet type of
      alarm type packet : process alarm packet ;
      regular type packet : process regular packet
   end
else
   Handle error condition
end ;
```

13.5 DESIGN—PRINCIPLES AND OBJECTIVES

Design follows from a specification of requirements as given in the documentation that describes the specification. In an ideal world, this document would give a definitive guide to the software designer; but realistically almost all designs will uncover ambiguities and errors not uncovered previously, so that it is necessary to iterate around the design and specification phases several times. The designer will, as we have already noted, need to live with any constraints customers might impose, such as language, hardware platform, etc. The end-product of a design should be information in a form suitable for programmers to use in actually implementing as code. This will involve the production of definitive, completely unambiguous documentation, which programmers can rely on absolutely. As well as this, documentation commonly also includes a **systems manual** for use by maintenance programmers—whose needs are different to those of the programmers carrying out the initial development—a **user manual** which the end user will require as a reference guide once the system is actually ready for use and thereafter, and finally a **tutorial guide** to help new users steer themselves through the system.

No attempt will be made here to give detailed descriptions of and comparisons between the many different design methodologies used in software engineering. Pressman (1992) and Jenkins and Welland (1990) give a reasonable account of most of the principle methods used. It is as well to point out though that the nature of the application will to some degree condition the choice of design methodology used, and in keeping with the emphasis on real-time systems in other parts of the book we will give special weighting to methods used in such systems (this is discussed later in Sec. 13.7).

13.5.1 Design—how can it be defined?

Design is no easy concept to define; however, we could say about it that:

1. It is a process of synthesis.
2. It is conditioned by the past experience and skills of its practitioners.
3. It *never* results in an error-free end-product at the first attempt.
4. It will always be influenced by the need to consider tradeoffs of various kinds.

The first point simply states the fact that there are always going to be a number of alternative design strategies possible, usually a very large number in fact; there is after all seldom one *unique* way to design a program for a given task, or to design *any* product for that matter. The range of alternative techniques is one of the factors that makes design such a creative activity; it is also one of the main reasons why it is usually difficult! Some choices might be easy: for instance at a late stage in the design a decision has to be made about the most suitable data structure to employ. From what has been said about data structures in previous chapters, it might be obvious that a linked list might be preferred to an array in a particular application—or vice versa—depending on circumstances. Other decisions could be much more difficult.

The second observation is a fact of life that is bound to influence choice of design strategy to some degree: a developer is not going to abandon a well-understood design method and all the expertise built up in using it for something else, without there being considerable advantage in so doing. Also, modern approaches in design methodologies attempt to automate some aspects of the design process, but will never completely eliminate the flair and intuition that the developer can bring to bear in a design.

The third assertion points to the reality that a design will initially contain errors, and thus will eventually need to be tested. Stated another way, a design needs to be set in the context of the other steps in the development cycle, being an iterative process that involves other steps in the software life cycle as we have already noted. A good design strategy will allow for changes to be made without throwing away large parts of an initial design.

Tradeoffs, referred to in the fourth point of the list of design attributes, can cover a wide range of characteristics; but perhaps most frequently refer to the need to make software products maintainable while at the same time still meeting the specification, including performance aspects such as real-time behaviour if needed. These are conflicting goals, and the importance placed on maintainability in recent years means that many designers lay stress on this at the expense of other aspects, although naturally there are quite a few exceptions one could quote.

13.5.2 Modules

As we noted earlier, a common feature of just about any design method is the need to divide program(s) into modules of some kind. Modules may correspond to entities such as Pascal procedures, C functions, etc., but in any event must be simple enough such that a designer can hold the structure in his or her mind without recourse to the aspirin bottle; this means that modules larger than about 50 or so lines of code are probably going to be too long. Conversely modules that are excessively *small*—say less than about 20 lines—will increase the complexity of the system's overall structure and thus also be the enemy of desirable attributes such as ease of maintenance. In summary, making modules too large or too small will make them difficult to develop, probably difficult to maintain, certainly difficult to test and is likely to lead to an increase in the number of bugs. Any design methodology must address this issue.

13.5.3 Coupling and cohesion

The composition of modules and the way they hang together and communicate with each other is naturally influenced by the design methodology adopted, and a wide number of possibilities exist. However, general objectives are to localize errors to as small a part of the program as possible and to minimize the amount of code that needs to be discarded following any required changes, particularly in the maintenance phase. Some general observations are possible in terms of attributes that aid these objectives, and are quite independent of any specific design methodology:

- **Coupling** describes the degree of interdependence of modules in terms of the data flowing between them, typically by the mechanism of parameters passed between procedures. An aim of the designer is to minimize coupling in the interests of maintainability and of localizing the effects of errors to as small a number of modules as possible. In other words, if changes are required for whatever reason, we would like those changes localized to as small a number of modules as possible. As one example, coupling would be very strong in cases where heavy use is made of global variables. We have the problem that if some faulty module changes the value of a global quantity to an incorrect value, and some time later another perhaps fault-free module subsequently accesses the same quantity, one might be misled into thinking that the fault-free module caused the problem. It is for reasons like this that global variables should be used sparingly, and data passed between the access points of modules in the form of procedure parameters.
- **Cohesion** refers to the logical unity, or lack of it, in the logical functions that a module is to carry out. For example if a program has to interact with an I/O device such as a disk, an aim in terms of cohesiveness would be to localize the interaction to a small number of modules, perhaps just one. Any calling module can then call on the services of some subordinate module to perform the I/O. Modules that carry out a multitude of disparate functions are going to be more difficult to test and maintain, with the danger that changes in part of a module may have unwanted effects in other parts of the module only loosely related or totally unrelated to the part being changed.

13.5.4 Top-down design

A top-down approach to design decomposes a problem into a hierarchy of modules, and as with any other approach the previous observations on coupling and cohesiveness need to be uppermost in the designer's mind. A central feature of the top-down approach is that we have a layered system for which internal details of lower-level modules should not be accessible to their 'parent' modules. This *information hiding* quality has already been discussed in Sec. 6.5 in relation to language support offered by languages such as Ada and Modula-2, and is inherent in a system that deals effectively with the problems of cohesiveness and coupling. In terms of *design*, information hiding has the aim of minimizing interactions between modules; and to achieve this goal we make careful choice of the interfaces between modules and the data structures that they should have access to.

As an example consider the design of a circuit analysis package. This commonly would have a 'front end', in the form of what is commonly termed a **schematic capture** module, in which a user's input circuit details are displayed in graphic form. The schematic capture module would supply relevant circuit data to a module that solved the circuit equations, which in turn would supply data such as the computed nodal voltages to a module—commonly called a **postprocessor** module—which typically would display frequency response information in graphical form. From a top-down view this would suggest the high-level structure depicted diagrammatically in Fig 13.6, where the position of the subordinate modules from left to right indicates the order in which they are invoked. Later stages in the design would add more and more layers in the module hierarchy.

A feature of top-down design is that specifying the fine detail of the design can be postponed until a late stage, which allows changes to be made before all the design details have been finalized. In the circuit analysis example, changes in (say) the postprocessor module should be possible with no knock-on effects for the other two modules. An argument against top-down design, however, is that errors in specification are not uncovered until late into the design, perhaps not until it is complete in some cases.

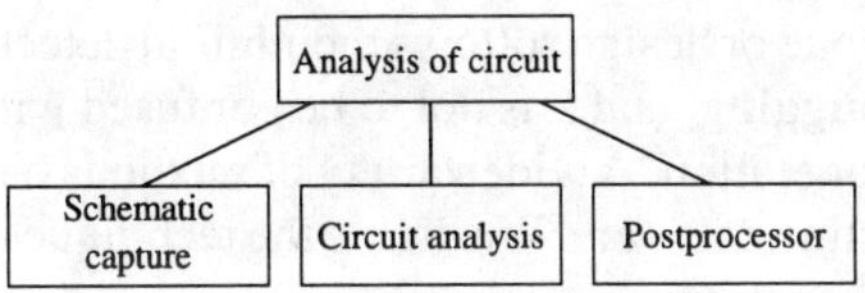

Figure 13.6 Top-level structure of circuit analysis package.

13.5.5 Bottom-up design

In many ways, bottom-up design is the converse of top-down design: we begin with the most primitive modules in a system and then assemble them in stages until a complete system is built. Coding for modules can start at an early stage in the design with this approach. Most design methods stress the top-down approach, but in some cases this may be overlaid with a bottom-up approach to the design of certain modules. For instance, the software responsible for producing the graphic display in our monitoring system resides at the lower levels in the hierarchy of modules in the program. It could usefully be designed, implemented, and tested bottom-up rather than waiting for the development of higher-level modules, provided the interfaces to the remainder of the program are well defined. This would allow concurrent development of this part of the software, possibly by individuals with special skills in developing this type of software. Pop-up menus, and graphs of various kinds are examples of the kind of function involved here.

13.6 TESTING

Testing has been mentioned briefly already—we discussed the need to form a set of tests as part of a system specification review. The whole question of testing now needs a detailed discussion, and it is necessary to define precisely what the objectives are, and to investigate possible testing methods. One might say that in view of all the care that should have been taken at the specification, design and implementation stages, testing should be almost unnecessary! Common experience indicates otherwise, however; it is not unknown for the effort put into testing a system to approach 50 per cent of the total budget. In any event, customers will not be satisfied with a system until its behaviour has been demonstrated to them in such a way as to exercise all visible features in an agreed set of acceptance tests.

With no apologies for stating the apparently obvious, it must be said that a prime objective in testing must be to devise tests that will have a high probability of uncovering errors; this is just one way of saying that effective testing methods involve one trying as hard as possible to see if a program will not work! It is also worth remembering that, although testing will uncover some errors, there is no absolute guarantee that it will uncover *all* errors—testing cannot show the absence of errors.

Specification, design, implementation and testing should be viewed as a circular process, and we need to examine the behaviour of a program in the light of its *expected* behaviour as set out in the specification, making changes in design and/or implementation until required and actual behaviour agree. Initially, concern is with methods that *detect* errors; rectification of any changes in code or design follows their initial detection. It is this latter activity that most people call **debugging**, and it is not to be confused with testing, which is a more general and higher-level operation. A wide variety of methods have been developed to give systematic methods of testing software. We divide the techniques into two categories: black box testing and white box testing.

13.6.1 Black box testing

In **black box testing**, the objectives are to test whether or not a module meets its specified goals, without regard to the internal structure of the module. The behaviour of a module being tested is observed solely in terms of output data it generates in response to suitable input test data. In the case of our monitoring system, for example, say we were testing the module that accepted data for a packet in the form of a character string (recall that this produced as output a list of analog quantities). We would only be interested in devising test data for packets that proved whether or not the processed analog quantities were being correctly output—say, as an array of floating point numbers.

13.6.2 White box testing

This can be viewed as the converse approach to black box testing, in the sense that it is the *internal* behaviour of program modules when exercised with test data that is observed. The two approaches are usually considered as complementary to each other, rather than alternatives to module testing. The objective is to exercise code through all its possible control paths, and also to fully exercise all aspects of the data structures. We are in effect in the position of someone 'looking' at the internal workings of some code as if through a transparent box. To exercise *all* paths in a program module involves knowing enough about the code to define them fully, it also involves generating—ideally at least—a comprehensive set of test data. For example there are problems such as *while* loops which are not properly tested if test data is such that the Boolean test used to decide if the loop executes is never TRUE. Defining test data that comprehensively tests programs in this way can be difficult, with programs of quite modest size having an extremely large number of alternative paths through the various looping and conditional parts that they comprise. The procedure in the example below illustrates this point. Test data that results in all possible combinations of *conditional1*, *conditional2*, etc., should wherever possible be devised if we are to exercise properly the procedure through all its possible paths.

```
procedure WhiteTestDemo( <parameter list> ) ;
{ local variable declarations }
begin
while { condition1 } do
   begin
   { statements }
   if { condition2 } then
      begin
      { statements }
      end
   else
      begin
      if { condition3 } then
         if { condition4 } then
            begin
            { statements }
            end
         else
            begin
            { statements }
            end
      else
         begin
         { statements }
         end
      end
   end
end { WhiteTestDemo } ;
```

We should also make sure to test the actions that a program takes in exceptional circumstances, not just in situations when 'normal' data is being supplied to it. (Recall that the
provision in some languages for dealing with exceptions has already been discussed in Sec.
6.7.) For instance, suppose a program is required to scan a file comprising a list of unsigned
integers, and from it we need to generate another file comprising those numbers less than
some threshold value. Test data should be created that includes values above, below, and
equal to the threshold value, but should also include any possible 'bad' data such as ASCII
characters, or negative integers, etc.; devising a *comprehensive* set of good and bad test data
can be far from easy.[4]

13.6.3 Unit testing

This is testing as applied to individual modules in a program, prior to the stage at which they
are tested as a unit. The method used is of the white box type. An inherent problem in testing
modules in isolation from other modules is that they must be used with special test software
that acts out the role of the modules with which the one under test will interact. The

---

[4] A worked example that illustrates these ideas appears in Chapter 14.

module will, in general, take data from certain modules and produce data for other modules. In the latter case so called **stub** modules are provided, these are vestigial procedures which accept data from the module under test, print relevant values, and also pass data back to the module if appropriate. A **driver** module plays the role of some main program that the module under test would normally interact with. Drivers and stubs are usually relatively simple pieces of code, in any event they only play a role in the testing phase and will eventually be thrown away.

13.6.4 Integration testing

Integration testing follows unit testing, and, as the term implies, the idea is to ensure that modules work correctly when they are interacting with each other. Modules are often produced by different members of a team, and, therefore, one of the effects of integration testing might be the uncovering of misunderstandings and bad communications between the individuals involved. Commonly, programs are split up into separate files with globally accessible variables and procedures, and some problems will only manifest themselves when modules are combined. Naturally some languages are better than others in the way they deal with quantities imported or exported between modules, as we have noted in the discussion on Modula-2 for instance. Integration is usually best done in a number of easy stages; attempts to integrate a large number of modules in one step frequently ends in tears, with the resulting complexity making it very difficult to track down problems.

Assuming we wisely adopt an incremental approach to integration testing, there are two possible ways to proceed: **top-down** or **bottom-up** methods of testing which in some ways mirror corresponding approaches to the *design* of programs. With top-down testing the first modules to be tested are the main calling program and modules that immediately interface with it. After the first phase of integration testing, these modules are treated similarly and some number of steps later all subordinate modules will have been tested. At any stage, untested modules need to be replaced by test stubs.

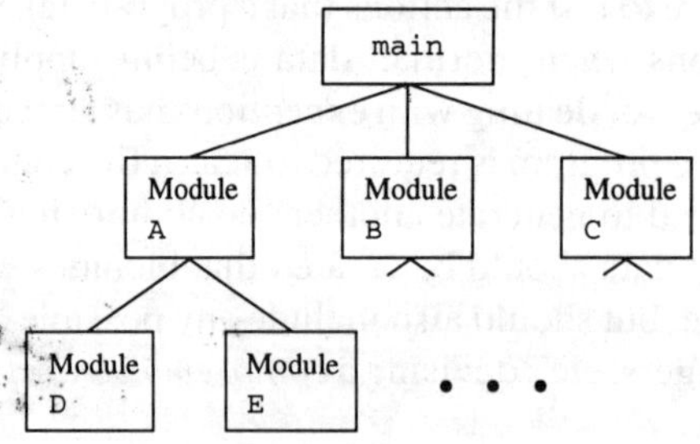

Figure 13.7 Top-down approach to module integration.

Several approaches to top-down integration are possible. In the illustrative system depicted in Fig. 13.7, we may opt to integrate the master module `main` with module A, leaving modules B, C, D and E to be replaced by test stubs. Alternatively, modules B and C may be integrated at this stage, depending on their relative complexity. A second stage could then replace the test stubs for modules D and E to complete the integration. With this sequence, there would not be a working program available until *all* modules had been processed. An alternative might be to integrate modules in the sequence A followed by D, in which case we would have some of the functionality of the final program available at an earlier stage. Customers may appreciate an early demonstration of certain features of the final product.

With bottom-up testing, as one might guess, the first modules to be integrated are those at the lowest level in the program structure, and these are combined into clusters, which themselves in later stages of testing form part of still-higher-level clusters; eventually clusters are integrated with the main 'master' program module, making a complete program. It might be said that bottom-up integration avoids the need for code used purely for test purposes such as the stubs needed in top-down testing; set against this, bottom-up testing has the considerable disadvantage that none of the functions carried out by the program can be demonstrated until the final module has been completely tested.

13.6.5 Alpha and beta testing

Good practice in developing software as described should in theory lead to a fault-free product, but in any event the acid test should be acceptance tests by a customer to prove to his or her satisfaction that this is so. In some cases there will be a large number of end users, however; one thinks of operating systems such as DOS, for example, where end users are numbered literally in millions. The different ways that a large and diverse number of users will apply a system are often extremely difficult to predict in advance, and many developers use a method called **alpha beta** testing to uncover problems in such cases. An alpha test is in many cases a form of acceptance test carried out at a developer's premises, and under controlled conditions. Subsequently a beta test is carried out by (ideally) a large number of selected end users in the environments in which the product would eventually be employed. The end users then record their experiences with the software and report them to the developer. The product is then finally released for use, following any changes resulting from feedback from the beta tests.

13.6.6 Debugging

Testing establishes that there is an error; debugging then tracks down the error. With many programmers, debugging tends to be a somewhat intuitive process guided by past experience in finding coding errors. However, there are several debug strategies that to some degree can put the whole process on a more systematic basis. In any event, if the other stages in the development cycle have been followed properly, correction of errors should only involve modest changes to the code, which moreover will be localized to one module.

Most compilers have facilities that help in the debug process: breakpoint insertion— stopping an executing program at a specified statement—and single stepping—stopping a program's execution after each statement—are two of the more notable facilities. (A more complete discussion of this appears in Chapter 14.)

One debug strategy called **backtracking** starts by noting the point in the code at which the error occurred, after which the code leading to it is handtraced, back to the point where the code is known to be sound. This might be in response to a program either crashing (terminating unexpectedly) or showing some other observable signs of distress. The 68000 assembler code example below illustrates the backtracking approach.

```
MOVE.W  $1234,D1
MOVE.W  $8000,D0
ADD.W   #$FF,D0
DIVU    D1,D0
```

The DIVU instruction divides D0 by D1 putting the result in D0. However, in this case an overflow is caused by the fact that the denominator (i.e. D1) is zero. For the sake of this argument we say the first instruction should have read: MOVE.W #$1234,D1, so in fact the instruction will get the data stored at location $1234 instead of loading the (immediate) data $1234 into D1. Location $1234 contained zero let us say, causing the overflow in the later DIVU instruction. Backtracking up to the DIVU instruction would have unearthed the bug, following examination of the contents of relevant registers.

The effectiveness of the backtracking approach obviously depends on the number of lines of code that have to be traced, and difficulties can be expected in backtracking over hundreds of lines of code.

Another strategy, **forward tracing**, is an approach used to detect bugs such as programs executing infinite loops. For example, in an assembler program the instruction BRA could have been used in error instead of a BNE instruction. With forward tracing the program would be advanced in steps, possibly one instruction at a time, examining key variables at each stage. In this example the loop counter should be decremented leading to an exit from a loop after a certain number of passes through it. Another common bug is shown in the program fragment below. The label LOOP should have been associated with the second instruction, not the first, so the loop counter will never get decremented:

```
LOOP:   MOVE.W  #$FF, D0
        SUBI.W  #1,D0
        BNE     LOOP
```

In the second example, which is given below, the fragment of C code has the kind of bug all C programmers seem to perpetrate at least once, which here is that the conditional statement (incorrectly) uses a '=' in place of a '==', so that x will be *assigned* a value 1, rather than being *tested* to see if it has a value of 1 as intended. It should be clear to you that the code will loop indefinitely.

```c
/* y greater than zero here */
while( y != 0 ){
    /* code which changes the value of x */
    if( x=1 ){
        y-- ;
        /* other code - which does not make y zero */
    }
    /* other code - which does not make y zero */
}
```

13.7 PROJECT MANAGEMENT—A BRIEF REVIEW

Project management in general is the subject of complete books (Wordsworth, 1992), so clearly it is not possible to do more than provide pointers to some of the more essential aspects of management techniques—naturally we emphasize aspects particularly applicable to the development of software projects. If we try to attempt to summarize project management, essentially the need is to establish the problems facing project managers and then examine approaches to solving those problems.

Project managers have the responsibility of:

1. Ensuring that assignments in a project are planned to meet deadlines and budgets.
2. Tracking progress to ensure that those deadlines are kept.
3. Identifying resource needs such as software packages which the development team might require.

Point 3 keeps track of quality control; often quality and factors such as cost conflict and compromises involving quality are areas that involve project management.

Many of the difficulties facing project managers arise from customers' requirements which all too often are stated in a vague or ambiguous way. Many projects require a considerable time before the requirement phase is complete and it is frequently the most difficult part of a project. Customers are also in the habit of making last-minute changes in their needs, and problems at this stage are difficult to plan for in terms of extra time and other resources needed. One of the hallmarks of a good manager is the ability to deal with the unexpected; to have the flexibility to make changes in project planning made necessary by unexpected events such as requirements changes from customers.

Also central to the role of a manager is the fact that they deal with people. This means that as well as planning and coordinating activities of team members he or she needs to recognize human qualities and in particular must motivate project members. Put more precisely, this involves providing a possibly disparate group of individuals with an environment where particular abilities are allowed to flourish, while getting the kind of cooperative effort essential with any team. Job satisfaction means different things to different people, and while for example some individuals work best under pressure, others do not. Some work best in isolation, others do better when working as part of a group. It is a function of a manager to recognize the personalities of individuals and organize their activities to get the best out of them. Most people appreciate being told the significance of the contribution of their work to the project as a whole, and need to be assured that what they are doing has an impact on the company's activities. Good managers will make sure that members of their project team are kept fully aware of individual contributions.

The human resources consumed by a project can be viewed in terms of the number of individuals involved and the time they require to complete the tasks assigned to them. However, the two factors are not interchangeable. It might be tempting to think that what one individual could achieve in (say) six months could be done by six individuals in one month. In realistic situations, team members will be communicating with each other to clarify aspects of a specification, to resolve problems and to integrate program modules, etc. Activities like these of themselves will add considerably to the time needed by projects; it is all too easy to underestimate the time involved and often not easy to plan for. The communication problem naturally gets worse as the number of members in a team increases.

13.8 SOFTWARE QUALITY—DEFINING IT AND GETTING IT

Software quality can seem a somewhat nebulous quantity, although by intuition one might start out by associating the kind of large-scale software used in—for example—air traffic control systems as needing high quality, while the kind of program written by a computer hacker for his or her own enjoyment and use as being of low quality. We need to be much more precise than this of course. Quality can in fact be viewed as a matrix of a number of quantities, although it is virtually impossible to get a single definitive list of just what these might be. A number of people have drawn up lists of factors that they regard as contributing to software quality (Pressman, 1992; Shooman, 1984). In any event quality attributes will be of varying importance for different types of project, and it is not realistic to expect a single definitive list to cover all possibilities.

Top of most people's lists of what they would view as quality attributes might be **correctness**—the degree to which a program meets its specification—and **reliability**—the degree to which the program executes its required function. Other quality attributes are discussed later, but some elaboration on correctness and reliability might be in order now, especially since confusion between them is possible. If we have a program that meets the customers' requirements, then that program can be said to be correct. However, it is still possible that the program may still be deemed unsatisfactory. Consider the situation where a program behaves *correctly* as we have defined it, but from time to time crashes. It is entirely conceivable that the specification is being met, but the behaviour of the program is evidently still unacceptable. It would be in this sense that the reliability concept would be used. Probably, in the kind of example cited, the (lack of) reliability would be the result of some deficiency in the specification. In other words it is possible that a program can be unreliable even if correct.

A program might be both correct and reliable to an acceptable degree: so what more need one then ask for, what have we missed out? *Cost* in fact is always going to be a consideration in any project, and the weighting given to cost as a quality attribute could commonly be as important as that given to correctness and reliability. Indeed if the price of getting a high level of reliability is excessive cost, compromises between the two may well be in order. It might well be possible to get a program that eventually works, i.e. is correct and reliable, but if the development methods used required many design changes leading to large cost over-run, the product would still be looked on as being of low quality. Similarly if the nature of the methods used in development make the maintenance phase difficult—attracting excessive cost penalties—one would also regard the product as being of low quality. Factors such as the standard of documentation would be a particularly important factor as far as cost of maintenance is concerned, notably in terms of providing the maintenance programmer with proper manuals.

The hardware and software resources needed in a product form part of any quality assessment, and these have both cost and performance ramifications. The cost of hardware resources has been getting less and less in recent years, and given less and less weighting as a contribution to quality as a result. It is a common experience that now, and increasingly in the future, the bulk of the cost in any modern computer project is related to the development and maintenance of the software. One would regard program performance in terms of the execution time and the amount of memory needed by a program, which are factors that need to be considered during the requirements phase of development, and contributory factors to quality.

Security is important to many types of user, and this can require much time and effort on the part of the developer in devising measures that prevent access to a system by unauthorized users. Customers such as banks would obviously place this factor high on their list of quality attributes, and *low quality* here could obviously lead to horrendous consequences. Equally well, though, in other types of systems security might not be an important issue at all.

13.8.1 Quality metrics

We have examined most of the ways of viewing quality, and although it is possible to come up with a set of attributes not quite as given here, it remains to devise some quantitative measures of quality. Ideally one would like to put a number to each relevant quality factor, and perhaps even to measure quality as some weighted sum of such numbers. For factors such as the size of code, or the number of lines of code produced per person per month for example, it is entirely possible to come up with a number to quantify a contribution made to quality. In other cases, quality is less easy to pin down, and some degree of subjective judgement is going to be inevitable. Factors such as ease of maintainability and reliability are examples of elements that are not at all easy to quantify. Some software metrics can be unreliable if not interpreted intelligently. For example, a metric such as 'the number of lines of code produced per person per week' could be said to be almost an encouragement to write long programs! More seriously, this—in some situations—would penalize code that was well written and efficient by virtue of being relatively short. In other words it is telling us more about the quantity of code rather than what one would more naturally regard as its quality.

13.8.2 The needs of quality control

As well as being able to *measure* quality, it is essential to monitor a project to ensure that quality standards are being adhered to. This is really no different in principle to the situation prevailing with the development of any other type of product, and quality control has long been a feature in organizations engaged in developing engineering products of all kinds. In cases where the size of the project is large enough, there might be a justification for a quality control department separate from the group actually undertaking the product development. Their role would be to monitor the development of the product to ensure that quality standards were being maintained.

A quality assurance department would normally report directly to management. They would give guidelines on quality methods to individual project teams, help prepare project plans—which state *how* quality is to be obtained—and subsequently monitor the compliance with quality plans throughout the course of a project. Naturally, being able to carry out all these functions implies a considerable degree of expertise in software development methods on the part of the quality assurance personnel.

13.9 STRUCTURED WALKTHROUGHS

One might be forgiven for thinking that the testing phase in the life cycle is the (only) point at which any problems in a design are uncovered and checks on quality made. The needs of software quality assurance are such, however, that this is seldom satisfactory. Any redesign following errors uncovered at the late stage in the development cycle at which testing is carried out will nearly always be expensive to rectify, and in some projects where this was in fact done the resulting cost overruns have resulted in major problems and even projects being cancelled. For software of any size, there is a need for monitoring the product for problems at *all* stages in the life cycle. If done rigorously, a large proportion of errors will be dealt with early, prior to the testing stage, and thus be eliminated in a cost-effective manner. One approach commonly used to achieve this objective is the **structured walk-through**. This essentially is a meeting in the form of a peer group review of some part of a software product. The form of structured walkthrough to be described below largely follows a set of rules suggested by Yourdon (1977) in connection with a particular design methodology—many other variants are possible. Used to best effect, a structured walk-through is a highly structured form of meeting where the person producing some part of the product—such as some code or a part of a design—presents his or her work to a group of interested parties for scrutiny and comment by them. Individuals attending the meeting should have well-defined roles; a typical set of responsibilities might be as follows:

- The presenter, who is responsible for actually producing the work to be discussed.
- A secretary responsible for documenting the findings of the meeting and any other aspect of paperwork involved.
- A chairman who steers the direction of the meeting, making sure that its terms of reference are adhered to.

Other interested parties may sit in on the meeting too, with perhaps less well-defined roles than those above, although questions such as documentation standards and quality control might justify the attendance of individuals devoted to these functions. As with just about any sort of meeting, the number of attenders needs to be kept to a reasonable size if all are to make a useful contribution. As well as defining roles for individuals, the structure and format of the meeting needs to be defined by a quite rigid set of rules. The producer should provide copies of all documentation for the product some reasonable time before the meeting is to occur. The chairman, secretary and others need to be selected by some fixed procedure acceptable to all. The duration of the meeting should be agreed in advance—say one hour—and the amount of material needs to reflect the time allocated. This might translate into some given number of lines of code, say, if a program is under discussion.

But what in *detail* are the objectives of the walkthrough precisely? Something that should *not* be an objective is the rectification of problems—this is the function of the individual producing the work. Rather the idea should be to *unearth* errors, following comments and suggestions by members of the meeting. At the conclusion of a structured walkthrough the chairman asks for recommendations as to whether or not changes are needed in the product, and if so to define what the changes should be. Recommendations should be recorded by the secretary, and circulated to attenders and other interested parties. This would then form an important part of the documentation of the project as a whole, and a means of checking for points at which changes have been suggested. It is common to use a form in which the subject of the walkthrough, its attenders, and actions to be carried out by the producer are recorded. A typical form is as follows:

Alpha Computer LTD Walkthrough report—Condition monitoring system (CS1A)

      Date   11/11/90
      Project—Southchurch Public Works Department, water pumping station remote
      condition monitoring system
      Report number 4
      Product under review—Digital alarm module

Attenders	role
F. Jones	chairman
A. R. Belmont	producer
W. E. Wardle	secretary
T. Smith	
I. S. Ash	

      Appraisal:

      Accepted with no change needed [   ]
      Minor revision, no further walkthrough [   ]
      Revision requiring further walkthrough [   ]

      Findings

As well as uncovering problems, a structured walkthrough has the beneficial effect of reinforcing communication links between contributors to a project. Another advantage is that high standards of documentation are encouraged. Inadequacies in documentation—such as departures from agreed standard methods of presentation—will be more publicly exposed, so the producer should be inclined to take more care than if he or she were just writing documentation solely for his or her own use.

13.10 SOFTWARE TOOLS

Advances in the techniques of software engineering have generally been accompanied by advances in the tools needed in software development. We have already met tools such as compilers, assemblers and linkers used in program development. But other higher-level tools have become available in recent years that directly support particular design methodologies. Typically such tools would operate at the level of data flow diagrams in the case of methods that use this approach. Many tools of this type allow users to develop designs using graphics in conjunction with the particular diagramming approach used in a particular method. Thus designers can draw data flow diagrams and then decompose them into lower-level data flow diagrams with checks possible between the data involved on diagrams that reference that data. High-level tools of this capability are, in effect, automating some aspects of program production. They are referred to as **computer-aided software tools**. It might be added in passing that some tools merely facilitate the diagramming of a design and do not handle the data they embody, in effect these are just specialized graphics packages.

The aims of some tools like compilers are reasonably obvious, but if we look at the aims of software tools in a general sense they have the function of streamlining and automating, where possible, the generation of program modules and, crucially, supporting software development *being carried out by a team*. These aims should extend to the maintenance phase of development and be accompanied by effective methods for the production of support documentation. The term **integrated project support environments (IPSEs)** has been coined to describe the complete environment needed in the context of software development by a team. In detail, an IPSE should facilitate effective communication links between team members, and control access to shared data such as program modules to prevent inconsistent use. They give support from the requirement phase through to the maintenance phase. A number of commercially available tool sets that implement an IPSE are available; many support one particular design methodology, although some IPSEs support more than one methodology. Support for the management aspects of software projects should be a feature too, so that, as well as encompassing tools such as editors or compilers, facilities supporting project planning and cost estimation should be a feature. A central concept common to support environments is the need to integrate the various activities in development, with a set of tools enabling this to be done in a consistent manner.

Tools for code generation Tools such as compilers have grown in sophistication over the years particularly in terms of the quality of code they can produce, but are of more relevance to the discussion here in terms of the development environment that commonly accompanies many compilers. For example, Turbo C integrates the edit, compilation, linking and debug phases of program development. Allied to a PC based around fast 32-bit microprocessors this makes for extremely rapid development cycles done entirely within a convenient menu-driven environment. Another example of a tool used in UNIX-based systems is **Make**. This automates the process of compilation and linking programs, although other functions can be carried out too. Make examines dates at which files were created in relation to object code, and recompiles or relinks accordingly, and ensures that the compilation/linking process is carried out consistently. Make might seem a somewhat over-complex tool, with only little benefit to productivity in the case of small programs, but when—as is common—we deal with projects for which the code might be divided into tens or even hundreds of files spread over a large number of directories it can seem invaluable. Turbo C includes a primitive form of Make.

Tools such as Make are fine for helping the development of program modules split up into a number of files—a usual situation for large projects—but need to be supported by tools that can deal with situations where code is produced by a number of individuals.

13.11 CONCLUSION

It must be stressed that only a somewhat brief outline of the principles and practice of software engineering is possible in a book of this nature. However, it should now be evident why software engineering is a major discipline of considerable complexity, the scale of which is evident from books such as Pressman (1992). References like this should be consulted if you are going to attempt the design of any large-scale software project. It would be a good idea if you could actually examine a sizeable software product, it being rather difficult to do this within the compass of the book. One example is an operating system called Minix, a UNIX variant. Tanenbaum (1987) describes the structure of this operating system and also gives its source code.

13.12 EXERCISES AND REVIEW QUESTIONS

1. In Sec. 13.2 examples were given illustrating the so-called *software crisis* caused by the complexity of software-based systems. Investigate other cases in which this kind of problem occurs—references such as Pressman (1992) and Easteal and Davies (1989) might be useful here.
2. Distinguish between the requirements capture and specification phases in the software life cycle. How would you see the differences between these two phases as viewed by a customer?
3. The following is a modification of the outlined specification of the data monitoring system given in Sec. 13.4.1:

   - An operator at the remote station is to now have the facility to control equipment such as pump motor circuit breakers and also to open and close valves.

   Give a modified version of the data flow diagram given in its original form in Fig. 13.3 which caters for this change.
4. Distinguish between verification, validation and debugging as applied to the software life cycle, and give examples of situations where these activities would occur in the case of the data monitoring system.

REFERENCES AND FURTHER READING

Dettmann, T. and Johnson, M. (1992) *DOS Programmer's Reference*, 3rd ed., QUE, Carmel, Indianapolis.
Easteal, C. and Davies, G. (1989) *Software Engineering—Analysis and Design*, McGraw-Hill, London.
Hetzel, W. (1984) *The Complete Guide to Software Testing*, Collins, London.
Ince, D. C. (1989) *Software Engineering*, Chapman and Hall, London.
Jenkins, D. G. and Welland, R. C. (eds) (1990) *Software Engineering for Electronic Systems Designers*, Peter Peregrinus, London.
Lawrence, P. D. and Mauch, K. (1987) *Real-Time Microcomputer System Design—An Introduction*, McGraw-Hill, New York.
Mellor, S. J. and Ward, P. T. (1986) *Structured Development for Real-Time Systems*, 3 vols, Prentice-Hall, Englewood Cliffs, New Jersey.
Naur, P. and Randell, B. (1969) *Software Engineering Conference Report by the NATO Science Committee*.
Pressman, R. S. (1988) *Software Engineering—A Beginners Guide*, McGraw-Hill, New York.
Pressman, R. S. (1992) *Software Engineering—A Practitioner's Approach*, 3rd ed., McGraw-Hill, New York.
Shooman, M. L. (1984) *Software Engineering*, McGraw-Hill, New York.
Tanenbaum, A. S. (1987) *Operating Systems*, Prentice-Hall, Englewood Cliffs, New Jersey.
Wordsworth, J. B. (1992) *Software Development with Z*, Addison-Wesley, Wokingham.
Yourdon, E. (1977) *Structured Walkthroughs*, Prentice-Hall, London.

14

DEVELOPMENT TOOLS AND TECHNIQUES

14.1 INTRODUCTION

This final chapter is, in some respects, a logical continuation of Chapter 13, and we consider in more detail the later stages in the development cycle—more specifically *testing*—a topic only briefly introduced in the earlier discussion on software engineering. Recall that the software life-cycle starts with requirements capture and, prior to the maintenance phase at least, ends with testing and validation. We concentrate here mainly on testing applied to embedded real-time systems; this involves both hardware and software development, so the problems involved in testing for both these facets need to be addressed. A variety of developments techniques and tools are used for hardware and software testing. **MDS** (= Microprocessor Development System) is the term given for the collection of software and hardware tools needed in the development of an embedded type system, and we will be devoting much space to describing what an MDS consists of, together with strategies for its use.

14.2 OVERVIEW OF THE DEVELOPMENT PROCESS

Software and hardware—these are the two parts into which we normally divide computer systems, and it would simplify life for the developer were it possible to develop and test each in isolation from the other; but to what extent will this be possible? How do we tackle the problem of integrating the two? Before addressing these questions in detail, we first review the nature and the attributes of the kind of software and hardware found in embedded systems.

- **Software**—although perhaps **firmware** would be a more appropriate term as far as many embedded type systems are concerned—is invariably tied to real-time interactions with the outside world, as we have been noting *ad nauseam* in various places throughout the book. Although certain code procedures can, for the purposes of testing, be treated purely as simple blocks of sequential code, in the final analysis, testing strategies need to deal with the often nondeterministic nature of real-time events like interrupts, and the effect they have on the software's behaviour. At a low level, the code will need to deal with the characteristics of peripheral and other chips with which it needs to interact, possibly involving debug at assembler code level, although we would like debug to happen at the high-level language level in which the bulk of the source code is likely to be written.

- **Hardware**[1]   We are dealing with a variety of different forms depending on the particular requirements of the system being developed. A common tendency is to use off-the-shelf boards such as **SBCs** (Single Board Computers), I/O boards, etc., and in so doing avoid the need for design at the circuit board level. Modern circuit boards based around 16- or 32-bit microprocessors are often of fearsome complexity, entail a great deal of specialist design skill, effort and time, and are probably best left to specialists to develop. It might be added that the trend to higher and higher clock frequencies, 50 MHz or more being used on some types of board at the time of writing, is going to make hardware design problems still more difficult in the future. However, custom-designed boards, in cases where there are specialized needs, are also in very common use, in which case the hardware element would form a significant part of the design of the system as a whole.

It is these two aspects, and in particular their interactive nature as elements in a real-time system, that underlies choices in debug strategies and tools that the developer has to make.

14.2.1 Hardware and software design-cycles

Design methodologies were discussed in Chapter 13, but, there, little was said about the specific role of the hardware element. Although the book makes little attempt to discuss hardware design in anything other than overview,[2] it is important to see how the hardware and software development elements need to mesh in with each other, and in particular to examine their respective roles in the overall design-cycle. Note, however, that the objective here is to explain better the ramifications for a design as far as the development phase is concerned, *not* to provide an exposition of any design method.

Figure 14.1 shows in outline an approach to system development. Starting from product specification, one of the earliest design decisions made involves the division into hardware and software components. Remember that one of the original aims of early microprocessor-based systems was the replacement of the relatively expensive hardware component by an equivalent but cheaper software counterpart, so the designer needs to think about the relative costs of the two. Having decided how to divide the system into its hardware and software parts, the two are then often developed in parallel, perhaps by different members of the design team. A certain amount of debug can take place on the hardware and software

---

[1] Recall that the different types of hardware were discussed in Chapter 2.

[2] Lawrence and Mauch (1987) give an excellent discussion of both hardware and software design of microprocessor-based systems.

elements separately. We would for example be using a software simulator to test as much of a program as possible, probably the *bulk* of the code in most cases in fact. Some of the more routine aspects of hardware testing and design can also be carried out without requiring the software to be available. Eventually, however, the software and hardware elements must be combined and then further debugging will be required. Often this can be difficult, involving as it does coping with the real-time behaviour of the system. Modifications following this phase need further iterations through the hardware and/or the software parts of the design.

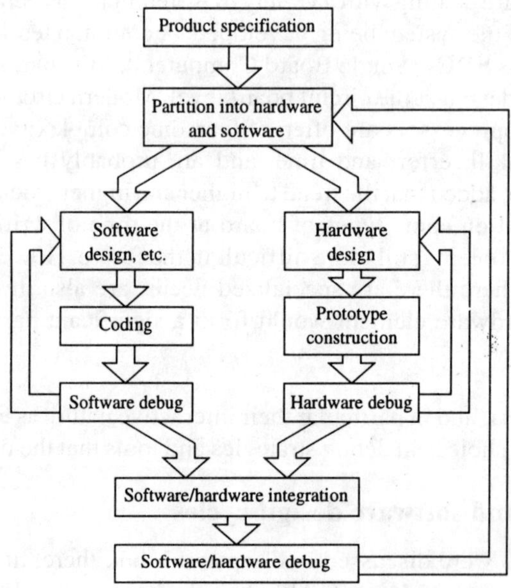

Figure 14.1 Design cycle for an embedded system.

14.3 DEVELOPMENT SYSTEMS—ELEMENTS AND TYPES

Many types of programmers become accustomed to a situation in which the code they develop actually executes on the same computer on which the original development was carried out. Debug facilities, commonly bundled with many compiler packages, can feature in this kind of development, an example being the debugger provided with Turbo Pascal which allows all the usual debug facilities: single stepping, breakpoint insertion, etc. With embedded systems, however, program development is done on a different machine (usually called the **host computer**) to that on which the code will finally execute (usually called the **target system**). The host will need to be capable of running the relevant specialized software needed, notably the cross compilers and cross assemblers described in Chapter 7, but also including specialized forms of debug packages to be described later.

14.3.1 Alternative approaches to debug—an overview

Debug in its various forms will be discussed in depth later in the chapter, but, summarizing, we have a choice in debugging a program: either we simulate its execution on a host computer, or debug it while it is running on the target system. Using the code produced by an assembler, so-called **simulator** packages aim to imitate the way the code would execute on the target system. However, in some situations this is bound to fail, since there are some (real-time) aspects of the program's behaviour that can only be tested when the program is actually running on the target system. Some simulators make a crude attempt to emulate the effect of an interrupt by allowing users to hit a certain key. Nevertheless, in spite of their limitations, most development systems have simulator packages, and large portions of programs can be tested on the host computer. This form of debug has the very considerable advantage of being able to use the facilities available on the host: large disks, access to other computers via a network, access to a laser printer, etc. Debug on the target system is an alternative to, or more usually complementary to, debug on the host, with a program's execution being controlled either by a target system's monitor program—assuming it has one—or else by specialized hardware debug tools.

14.3.2 Other elements in an MDS

The host computer is at the heart of any MDS, but there are other (hardware) elements, notably the in-circuit emulators and logic analysers used for hardware debug, which are described in Sec. 14.8. Another important component is the EPROM programmer used to program the chip(s) that will finally hold the target system's firmware. There would normally be a connection to the host allowing developed code to be downloaded into the memory of an EPROM programmer, which would then be used to 'burn' in the code in the memory chip(s). Figure 14.2 shows a typical setup for a development system, although there is much variation from one system to another. Various types of host computer are in use, and many systems use a single stand-alone host computer, often a PC, for relatively simple development work. Not depicted on the diagram, are the usual collection of tools needed for any hardware development, such as CROs, logic probes, multimeters, etc.

14.4 DEBUG FACILITIES—A SHORT LIST

We mentioned provision of breakpoints and single stepping in Sec. 13.6.6, but the discussion there was only intended as a very brief introduction to debug techniques: many more features are needed in practice, and some of the more common ones are as follows.

- **Breakpoints**—recall that this is a way of stopping the execution of a program at some specified address. Some 68000 debuggers implement breakpoints by replacing the original code with a software trap, a fact users might remember when they are testing programs. There is invariably provision for setting more than one breakpoint, although the maximum number allowed might be rather limited in some cases, perhaps as low as four. Provision for listing and removing breakpoints is usual.

  The more sophisticated debuggers allow breakpoints for events other than that of some specified address being reached. For instance an attempt to write or read within some specified range of addresses might be used to initiate a breakpoint. This might be

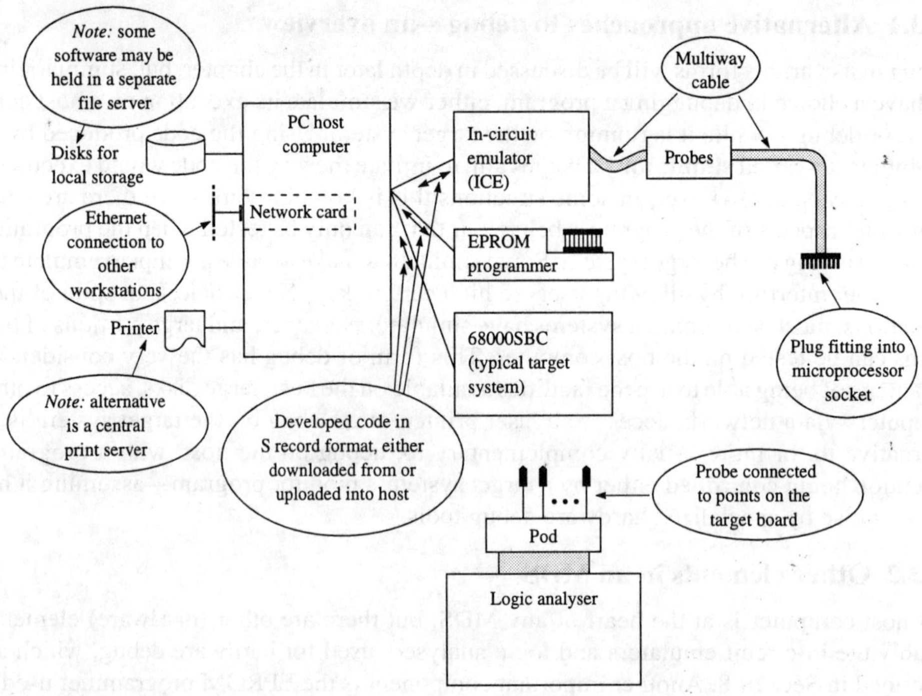

Figure 14.2 Hardware elements in a typical MDS.

worthwhile if the program made some attempt to read memory locations that had not yet had data written to them, which is not normally a fruitful exercise and nearly always indicates some kind of bug. Also, attempts to write to the memory area allocated to code are seldom a good idea, and usually indicate a bug.

- **Go**—a commonly used name for the command that starts execution of a program. Depending on the system involved, the start address can be specified either as part of the command, or else by previously loading the program counter with the start address. Some debuggers combine the actions of starting execution, and specifying a breakpoint at which the program is subsequently to stop.
- **Trace**—this is one of the names used to denote single stepping through a program. Sometimes there is provision for selecting the address at which tracing is to start, and the number of instructions to be executed before halting the program. Normally, the microprocessor's state would be displayed after each instruction.
- **Watchpoints**—similar to breakpoints, except that they are set to occur when the value of a variable is modified.
- **Display/modify memory**—if the required display is in the form of instruction mnemonics, this would in fact more usually be called **disassembly**, but provision for other forms of display for single memory locations or blocks of memory are usual, in hex, decimal, octal or ASCII character format. Many debuggers refer to this as **memory dump**. Also usual is a facility to load a specified value into one memory location, and to fill a complete block of memory with a required value.

- **Display/modify registers**—sometimes called a **register dump**, and in the case of the 68000 for example would give a display of the contents of all the address and data registers, and the status register (with individual bits identified by the flags they are associated with (overflow, carry, etc.)).
- **On-line help**—effectively a cut-down version of the information provided in the debugger manual, giving a somewhat minimal description of the various debug commands. This is surprisingly useful, especially on some of the more modern debuggers which can have a large number of commands, many of which might only be used occasionally and so possibly not be remembered.

This list represents no more than an essential core of debug facilities, and in addition it should be remembered that some systems allow debug at the source-code level, whereas the commands above are slanted toward use at assembler level; there will be more on source-level debug in a high-level language a little later. Also remember that in many modern embedded systems we are dealing with a multitasking environment and this creates further problems in the debug process, for which special packages have been designed, both at assembler and source-code level; there will also be more on this later.

14.5 DEVELOPMENT ON THE TARGET SYSTEM

Some target systems can act as their own development systems, provided they contain firmware and other facilities making debug possible. Many SBC boards for embedded systems have a monitor program with provision similar to that described in Sec. 14.4, usually including a one-line assembler. This gives a mechanism allowing code to be entered into the board's memory and its execution to be subsequently controlled by breakpoints, single stepping, etc. A rudimentary development system could typically comprise—in addition to the target system itself—a terminal connected to a serial port, plus any hardware (D/A converters, sensors, etc.) with which the board is to interface. A system like this would, however, be too primitive for most purposes; it is hardly a practical proposition to produce more than about 20 or so lines of code with simple one-line assemblers, the function of which is rather to prepare (very) short test programs, or else patch a few instructions or test data into RAM. A more practical approach would use a host computer—commonly a PC—to produce the code, which is then downloaded into the target's RAM memory. After this has been done, the on-board debug facilities can be used to test the code as it executes. This is distinct from the kind of debug carried out on the host, the latter being an exercise designed purely to test the software, in most cases at least.

A variation of development on a target system is development on a host computer for the situation where the host happens to be similar or identical to the target. A common example is the IBM PC, where typically the target might be some SBC based around the same microprocessor used in the PC. The developer is able to utilize the hardware in the PC: hard disks, printers and other peripherals which speed the development cycle. Some 68000 systems can also be used in this way, typically with a rack mounted 68000 SBC plus other (rack mounted) peripherals being used for development work, and the fully tested code being downloaded into a PROM programmer, and then blown into the EPROM chips to be used in the target. Figure 14.3 shows a host computer and target system that comes into this

Figure 14.3 Example of a host–target system.

category. The host (on the right) is based around three Force VME boards, a tape streamer for backups and importing new software into the system (centre), and a hard disk (on the right). The target system consists of three boards: one of these is a Force 68030 based SBC, another is an I/O board to extend the rather limited I/O capability of the SBC. It contains A/D and D/A converters as well as serial and parallel ports. The SBC has three serial ports, one of which is connected to the host system via one of *its* serial ports, with another serial port connected to a terminal.

An important potential problem with development on the target system arises out of the fact that the program executes in areas of the target's memory space different to that allocated to the firmware in which the code will finally reside. Put another way, we are using memory in a different way in the development phase to that in which the system will finally be used. This conflicts with the principle that any debug software, or hardware for that matter, should be entirely transparent as far as the behaviour of a target system is concerned.

Another kind of memory problem relates to the firmware to be eventually loaded into EPROM or PROM chips. A development cycle involves the following stages: production of code for downloading into an EPROM, plugging the EPROM into its allocated socket in the target system, and finally carrying out testing of some kind. This can be lengthy, and, if one considers that erasing an EPROM alone takes about 30 minutes, it is clear that it would only be able to try a small number of design changes to firmware code in a day. This would represent a costly bottleneck in terms of development time. The problem is often avoided by using an **EPROM emulator**, which is a volatile RAM memory module with a header plug designed to fit into the socket normally used by the EPROM chip. Code can be *directly* downloaded into the EPROM emulator from the host, and the EPROM would only make its appearance after the code to be loaded into it had first been fully tested using the emulator.

14.6 ILLUSTRATION OF DEBUG TECHNIQUES

14.6.1 Demonstration program

To illustrate the use of some debug techniques, we use a small 68000 assembler program, the listing file for which is given in Sec. 14.6.2. This is a somewhat manufactured example made brief to avoid excessive detail. The program was designed do the following:

- Read three ASCII characters and convert them to a 12-bit binary equivalent, which is then further processed by the program. The program loops indefinitely doing this. Prior to being converted to binary, the ASCII characters are tested to ensure they are valid characters for hex digits.

The program is not complete as it stands, and could be viewed as a fragment from a program in the process of development. The routines to read the ASCII characters and handle errors are replaced by test subroutines. In keeping with the real-time spirit of the other parts of the chapter, a dummy idle loop is included to allow for the time required by the parts of the code not yet written. This could prove useful—for testing I/O for example—if the partially complete code is to run on the final target hardware, and other parts of the software development is to be carried out in parallel with development of the hardware.

The code contains bugs, and the point of this exercise is to investigate possible strategies for tracking them down, and in the process demonstrate the use of some debug commands. Note that here the error-handling routine simply displays a message for invalid data; if the program were to be developed further, it would be necessary to write a proper exception-handling routine (exception handling was described in some detail in Sec. 6.7). The message 'ERR' is displayed if there are any problems, this being implemented on the particular monitor program used by a low-level system call in the form of a TRAP #11 instruction followed by a parameter that is a word of value 1; this has the effect of making the monitor program display the ASCII character assumed to be in D0.B on the terminal screen.

It is suggested that you carefully examine the program listing file before reading the text that follows, where debug methods for the program are discussed. Listing files can in fact be useful when actually testing programs, giving as they do information such as the physical address of instructions, instruction codes, and also a symbol table. Many monitor programs are somewhat primitive in terms of the information displayed when programs are being tested, and one is, in any event, with a standard terminal, limited to viewing only about 26 lines or so from the complete program. Programs can amount to literally hundreds or more lines of code so that viewing a relatively small window on the code as displayed on a terminal screen can create conceptual problems in attempting to relate the information displayed to the structure of the program as a whole.

14.6.2 Program listing file

```
1   000023                    ERRFLAG EQU     $23
2   400400  0400                      ORG     $400400
3
4                        *     CODE TO INITIALIZE PORTS ETC WOULD GO HERE
5   400400  6100001A        GETDAT  BSR     WAIT      SIMULATE DELAY DUE TO OTHER PROCESSING
6   400404  61000026                BSR     READ3     READ 3 ASCII CHARACTERS
```

```
 7   400408   61000036                 BSR      PARSE          CHECK IF VALID HEX DIGITS
 8   40040C   0C010023                 CMPI.B   #ERRFLAG,D1    ERROR FLAG RETURNED IN D1
 9   400410   67EE                     BEQ      GETDAT         ONLY PROCESS VALID DATA
10   400412   610000AA                 BSR      CONV           CONVERT TO 12 BIT BINARY
11   400416   610000CA                 BSR      USEDAT         PROCESS 12 BIT BINARY DATA (IN D1)
12   40041A   60E4                     BRA      GETDAT
13
14                    ********************************************************
15                    *                  DELAY ROUTINE                    *
16                    ********************************************************
17   40041C   263C000001FF     WAIT    MOVE.L   #$0001FF,D3
18   400422   048300000001     MORE    SUBI.L   #1,D3
19   400428   66F8                     BNE      MORE
20   40042A   4E75                     RTS
21
22                    ********************************************************
23                    *   ROUTINE TO SIMULATE I/O TO READ 3 CHARACTERS    *
24                    ********************************************************
25   40042C   207C004004E6     READ3   MOVEA.L  #TABLE,A0
26   400432   10FC0031                 MOVE.B   #'1',(A0)+
27   400436   10FC0032                 MOVE.B   #'2',(A0)+
28   40043A   10BC0033                 MOVE.B   #'3',(A0)
29   40043E   4E75                     RTS
30
31                    ********************************************************
32                    *    TEST IF ALL THREE CHARACTERS ARE VALID HEX     *
33                    ********************************************************
34   400440   207C004004E6     PARSE   MOVEA.L  #TABLE,A0
35
36                    *          HIGH CHARACTER
37   400446   61000028                 BSR      ISERR
38   40044A   0C010023                 CMPI.B   #ERRFLAG,D1
39   40044E   6700001A                 BEQ      ERROR1
40
41                    *          MIDDLE CHARACTER
42   400452   6100001C                 BSR      ISERR
43   400456   0C010023                 CMPI.B   #ERRFLAG,D1
44   40045A   6700000E                 BEQ      ERROR1
45
46                    *          LOW CHARACTER
47   40045E   61000010                 BSR      ISERR
48   400462   0C010023                 CMPI.B   #ERRFLAG,D1
49   400466   66000006                 BNE      OK
50
51   40046A   61000038         ERROR1  BSR      ERRMESS
52   40046E   4E75             OK      RTS
53
54                    ********************************************************
55                    * TEST IF A (SINGLE) CHARACTER IN D1.B IS VALID HEX. *
56                    * ERROR FLAGGED BY SPECIAL CHARACTER RETURNED IN D1.B *
57                    ********************************************************
58   400470   1218             ISERR   MOVE.B   (A0)+,D1
59   400472   0C010030                 CMPI.B   #'0',D1
60   400476   6C000006                 BGE      GTZERO         THEN OK SO FAR
61   40047A   60000022                 BRA      BADCHAR        OTHERWISE A BAD CHARACTER
62   40047E   0C010039         GTZERO  CMPI.B   #'9',D1
63   400482   6F000016                 BLE      CHAROK         CHARACTER MUST HAVE BEEN A DIGIT
64   400486   0C010041                 CMPI.B   #'A',D1        TEST LETTER CHARACTERS: 'A' TO 'F'
65   40048A   6C000006                 BGE      GTA            STILL OK
66   40048E   6000000E                 BRA      BADCHAR
67   400492   0C010046         GTA     CMPI.B   #'F',D1
68   400496   62000006                 BHI      BADCHAR
```

```
 69  40049A  4E71                      CHAROK  NOP
 70  40049C  4E75                              RTS
 71  40049E  123C0023                  BADCHAR MOVE.B  #ERRFLAG,D1  RETURN ERROR FLAG IN D1.B
 72  4004A2  4E75                              RTS
 73
 74          *********************************************************
 75          *              DUMMY ERROR HANDLING ROUTINE             *
 76          *********************************************************
 77  4004A4  103C0045                  ERRMESS MOVE.B  #'E',D0
 78  4004A8  4E4B                              TRAP    #11
 79  4004AA  0001                              DC.W    $1
 80
 81  4004AC  103C0052                          MOVE.B  #'R',D0
 82  4004B0  4E4B                              TRAP    #11
 83  4004B2  0001                              DC.W    $1
 84
 85  4004B4  103C0052                          MOVE.B  #'R',D0
 86  4004B8  4E4B                              TRAP    #11
 87  4004BA  0001                              DC.W    $1
 88  4004BC  4E75                              RTS
 89
 90          *********************************************************
 91          *   CONVERT 3 ASCII CHARACTERS TO 12 BIT BINARY IN D1   *
 92          *********************************************************
 93  4004BE  207C004004E6              CONV    MOVEA.L #TABLE,A0    POINT TO START OF AREA WITH CHARACTERS
 94  4004C4  1218                              MOVE.B  (A0)+,D1     GET MS CHARACTER, AND POINT TO NEXT
 95  4004C6  02810000000F                      ANDI.L  #$0F,D1
 96  4004CC  1018                              MOVE.B  (A0)+,D0     GET MIDDLE CHARACTER AND POINT TO NEXT
 97  4004CE  0200000F                          ANDI.B  #$0F,D0
 98  4004D2  E909                              LSL.B   #4,D1        SHIFT MS CHARACTER TO BIT POSITION 4-7
 99  4004D4  8200                              OR.B    D0,D1        COMBINE TWO NYBBLES
100  4004D6  E949                              LSL.W   #4,D1        MOVE HI TWO NYBBLES TO BIT POSITION 4-11
101  4004D8  1010                              MOVE.B  (A0),D0      GET LS NYBBLE
102  4004DA  0240000F                          ANDI.W  #$0F,D0
103  4004DE  8240                              OR.W    D1,D0        COMBINE WITH OTHER TWO NYBBLES
104  4004E0  4E75                              RTS
105
106          *********************************************************
107          *    DUMMY ROUTINE TO PROCESS THE DATA FROM CONV        *
108          *********************************************************
109  4004E2  4E71                      USEDAT  NOP
110  4004E4  4E75                              RTS
111
112          *     RESERVE SOME SPACE FOR THE THREE CHARACTERS
113  4004E6  0004              TABLE   DS.W    2
114  4004EA                            END

                -SYMBOL TABLE-
    PARSE   00400440              ISERR   00400470
    ERRFLAG 00000023                 OK   0040046E
    ERRMESS 004004A4               MORE   00400422
       WAIT 0040041C               CONV   004004BE
     CHAROK 0040049A             GETDAT   00400400
     ERROR1 0040046A             USEDAT   004004E2
      READ3 0040042C                GTA   00400492
     GTZERO 0040047E            BADCHAR   0040049E
      TABLE 004004E6
```

14.6.3 A possible debug strategy

Get some hands-on practice! Before reading the following sections on approaches to debug, it is strongly recommended that, if you have access to a 68000 assembler and in addition either a simulator[3] or a target system into which the assembled code can be downloaded, you actually prepare the program and run it in the way suggested below. Debug is very much a hands-on affair, and programs are too individual for anything more than general guidelines on possible debug methods to be given. You might explore your own test strategies after having worked through the following sections: no great claims are made for the particular approach adopted here, and other variations in technique are possible which could well be only marginally better or worse.

What sort of test data should be used? So what should be the approach to a strategy for testing the code? There is no unique answer, but, whatever approach is eventually adopted, the *first* task must always be to develop a set of test data; this should include all possible types of *bad* data as well as the ASCII data representing valid hex digits applicable in this problem. ASCII characters for decimal digits range from $30 to $39, and, for the (upper-case) letters A to F, the ASCII values are $41 to $46. Our test data should include *all* these values, and also values less than $30, between $39 and $41, and greater than $46. To be really thorough all three ASCII characters should be given the various data values described above. As a final note, it has to be said that for many programs there are just too many values of test data for it to be possible to test exhaustively the effect of each, and one is usually in the position of selecting those data that are most likely to unearth problems.

In what order should the modules be tested? Having decided on the various test data to be used, we need to identify the order in which the routines could be tested. For the sake of argument, let us say that the code simulated by the delay routine is not under investigation—in a practical situation it might represent code that has already been tested—and we first decide to test the routine PARSE. Having tested PARSE, we might then go on to test the routine CONV. The routine CONV is not yet being tested, and it can be temporarily bypassed by **commenting out** the call to CONV at address 400412 (by putting a '*' character in the first column, and so effectively skipping this routine).

An approach to testing PARSE To test PARSE, the requirement is to see if the error-handling code is invoked when appropriate, and so the program needs to be exercised with all possible 'good' and 'bad' data, as described above. Since at this stage the internal behaviour of the routines WAIT or READ3 is not of interest, a breakpoint can be placed after the call to READ3 at address 400408 (line 7). READ3 is written so that it supplies 'good' data in the form of characters that would be transformed eventually into the 12-bit number $123, in which case the error-handling routine should therefore not then be invoked. Putting a breakpoint after the call to READ3 gives us the ability to alter the data for the three characters, if needed, by patching values into the workspace area used to store the characters starting at address 4004E6 (line 113). A breakpoint could be inserted after the call to PARSE, say at address 40040C (line 8), to observe whether or not PARSE is working properly. Thus the

---

[3] There are a large number of simulator packages for the 68000 of varying complexity and sophistication. At the 'simpler' end of the spectrum, there are several reasonably good shareware packages which run on a PC.

program could be run up to the first breakpoint at 400408, and the monitor used to examine locations 4004E6 to 40044E8, which should contain '1', '2' and '3' respectively. If this is so, we can be sure that PARSE will be supplied with the expected data. Subsequently running the program up to the second breakpoint at 400410 would exercise PARSE.

Running the program to test PARSE If the program is actually run in the way just suggested, the program would be observed to behave correctly for the particular data used. At the first breakpoint, the expected data for the three characters is being stored at 4004E6–4004E8, and the ERR message is (correctly) *not* displayed. Alternatively, the latter can be verified by a register dump, which would reveal that the value of D0.B—being used as a flag for an error condition—does *not* have a value of $23 (= ERRFLAG) when the second breakpoint has been reached. The next phase involves using other data. Restarting the program from the second breakpoint would bring the program round to the first breakpoint once more, and the *modify memory* monitor command can then be used to put other data into memory for the three characters. Trying all the types of bad data suggested above will, as expected, produce the ERR message signifying bad data. However, if the character F is tried, which is a valid hex digit, the error message is incorrectly displayed. The bug is localized to the PARSE routine, and the approach at this point could be to uncover the bug by handtracing the instructions for this data; alternatively the program could be single stepped. Effectively the backtracking approach mentioned in Sec. 13.6.6 is being used, where we work backwards from a point at which a problem has been first noted—namely at the end of the PARSE routine in this case. The problem can be localized to about four or so lines of code within PARSE by inspection, and it is evident that the instruction at address 400476 should have been a BGT rather than the BGE used.

Summary We have shown how breakpoint insertion, in combination with single stepping, where closer control over a program's execution is required, can localize a problem and eventually isolate an offending instruction. This process is accompanied by judicious examination of key data, together with input of test data as needed. It should be stressed that the present example hardly demonstrates the full range of techniques that one might want to use in general, also the program and the nature of the bug here are rather simple compared to more realistic problems. If you have understood the principles used here it should be possible now to test the routine CONV; incidentally, this also contains a deliberate bug, albeit one that is quite easy to find!

14.7 SOFTWARE DEBUG—A CLOSER LOOK

We have examined debug facilities in Sec. 14.4, commented on the use of debug carried out on the target system in Sec. 14.5 and shown (software) debug in action in Sec. 14.6. Several other issues remain, however:

- Debug is best done at the source level of the language; we need to see why, and also how, it is done.

- The additional problems of debugging multitasking systems need to be examined.
- Finally, the question of combined hardware/software debug needs consideration.

14.7.1 High-level versus assembler-level debug

Section 14.6 dealt with debug, mainly from the viewpoint of the assembler programmer, although some important general principles emerged. This, however, avoids the reality that most programs are complex and thus need to be written in a high-level language. Naturally we would like to debug at source-code level too. Doing so would enable us to operate at the level of high-level quantities in symbolic form and to interact with data structures at the level of arrays, linked lists, etc. If you need any convincing about the relative advantages of debugging at source-code level as opposed to assembler-code level, you might care to revisit some of the examples given in Sec. 7.3; there, fragments of C code were shown, together with the 68000 assembler code that a compiler might generate from them.

One of the advantages of high-level languages *vis-à-vis* assemblers is that the compiler will eliminate the possibility of certain types of error that can *only* occur when writing in assembler; for example instructions concerned with loop management, where it is easy for the assembler programmer to introduce bugs, using (say) an incorrect BGE rather than a correct BGT, etc., but where a compiler would never produce errors. Because, in this sense, many kinds of error cannot occur at assembler level, there may be little interest in debugging a program at assembler level—i.e. at the level of register contents, etc.—and so little point in carrying out operations such as single stepping one assembler instruction at a time. Naturally there are situations where we *do* in fact need to debug at assembler level, particularly in real-time systems, but the need to do this is quite rare in most cases.

We briefly mentioned earlier the source level debugger associated with Turbo C; many other compilers also come equipped with debuggers. The Turbo C debugger is a menu-driven system, with several windows, one showing the source code, and another displaying variables selected for *watching* as the program executes. When single stepping through a program, it is possible to treat a procedure in effectively the same way as an instruction, as an alternative to single stepping through each statement in the function. Another useful facility is the ability to display the contents of the **call stack** which gives the order in which the various functions in a program are called, together with the values of the function parameters at the time of the function call.

14.7.2 Debug for multitasking systems

Debugging software which consists of a collection of interacting concurrent tasks raises some difficult issues, and it might be thought that the requirements of debug, which amongst other things can actually stop a program executing, will conflict with the need for a program to run as it would normally, meeting temporal deadlines, and exhibiting all the other normal real-time aspects of behaviour. So, is it always going to be possible to control a program with a debugger, *without* interfering with its real-time behaviour at all? The short answer has to be no: having breakpoints in programs will always disturb a program's normal execution, and it may be that a program will exhibit problems when running normally, while there may be no observable problem when it runs under the control of a debugger. If this is so, the kind of hardware debug to be described in Sec. 14.8 may help, since some hardware debug tools allow close control over the hardware not possible with software debuggers, no matter how sophisticated.

Debuggers for multitasking systems are called—by some software vendors at least—**dynamic debuggers**. One example, the iRMX Dynamic Debugger, used with systems controlled by Intel's iRMX operating system, allows one or more tasks to be debugged, while other tasks continue to run normally in real time. Observation of tasks or other objects queued at mailboxes and semaphores is possible. This type of debug allows developers to examine the execution of an individual task, as well as interactions between tasks, while at the same time minimizing the disturbance to the software as a whole.

Source-level dynamic debuggers are available that have most of the features described above, and, as the name infers, allow debug at source-code level. They allow single stepping through the code for a task, one high-level statement at a time. Provision for the display of source code is usual, and program variables can be referenced by their high-level symbolic names, enabling values to be displayed or altered. Most source-level debuggers have a mechanism for interacting with a program at assembler level, should closer control or inspection be required. One thus has the option of single stepping one assembler instruction at a time, if required. Examples of source-level dynamic debuggers are PXRAY used with the RTOS PDOS, TRscope used on VRTX32 systems and Soft-Scope used on iRMX systems.

14.8 HARDWARE DEVELOPMENT AND DEBUG

14.8.1 The logic analyser—an introduction

Some tests on digital hardware involve an examination of the waveforms of signals such as those on the system's buses. However, instruments such as CROs have limitations in making such measurements, both in terms of the limited number of channels available, and in problems relating to triggering. A further limitation of CROs is that the kind of time-domain information given is not always useful in the type of tests appropriate to problems specific to a computer. For example a typical test would involve determining whether or not a particular memory address was being read correctly by a microprocessor. Errors like this are not 'electrical' in the accepted sense, and information such as rise time, ringing, etc., is irrelevant. The kind of errors being dealt with are in fact errors in *data* at some instant. What we really require is an instrument that will determine both if the data is correct, and also if it occurs in the expected sequence. The phrase **data domain** (in contrast to time domain) has been coined for measurements of this kind.

Logic analysers were devised to make the kind of data domain measurements not possible with instruments such as CROs. Simple, early forms of logic analyser were little more than CROs with a large number of channels displaying 'cleaned up' traces of the signals being measured. Most modern logic analysers, however, allow for a wide variety of methods of displaying the data being captured, including disassembled instruction mnemonics in the case of signals derived from a data bus. The other important feature of logic analysers, and one of the more significant ways which distinguish them from a CRO, is the sophisticated triggering arrangements possible.

Figure 14.4 shows a typical setup where a logic analyser is being used to test a single-board computer. An HP1650A logic analyser (on the left) is hooked up to a 68000-based SBC (in the centre). Probes from the analyser are connected to a module (in front of the SBC), one of whose functions is to customize the analyser to the characteristics of the particular microprocessor used. A probe from the module has a header plug which connects to the microprocessor's socket. The header plug is designed such that the microprocessor—which is first removed from its socket—can be replaced on top of the header plug and functions as it would when it is connected normally. The SBC is controlled by the terminal shown on the right. Not shown is the host computer, which would be used to prepare code for downloading into the SBC's memory, or else into an EPROM programmer, or PROM emulator.

Figure 14.4 Logic analyser used to test an SBC.

14.8.2 Example of data domain measurement

Consider the case where a 68000 microprocessor is executing a `MOVE.W $8000,D0` instruction, the first word of the instruction byte being stored at the location starting at $400400. One approach to verifying whether or not the hardware is functioning properly at this point would be to display relevant bus signals. Some of these signals are listed below, in the sequence in which they should appear. (Note that the 68000 has a 24-bit address bus and a 16-bit data bus.) Note also that the information displayed relates to what Motorola term **bus cycles** (three read type cycles here), each of which would comprise several clock cycles.

Address bus	Data bus	Comment
400400	3038	Read (16-bit) op Code
400402	8000	Read Address (word) of operand
008000	20F0	Read contents of word starting at $8000 to D0
400404	...	Next instruction

A typical fault might be that the value finally loaded into register D0 (i.e. the number $20F0) is incorrect, perhaps because of a faulty memory chip. In other situations, particularly programs involving branches of some kind, we would be interested in the *sequence* of logic states, not simply their values *per se*. For example, a program might be cycling in a tight loop, continually polling the state of an I/O line, and only proceeding further when the I/O line is at some specified state. If there is some problem, this could be picked up by carrying out an analysis of the states in the neighbourhood of the relevant fragment of code. To summarize: the need is for an instrument capable of capturing and displaying a set of logic states and their sequence, as well as being able to do this at a fast enough rate to keep up with the clock rate of the target system and finally being able to store data for a sufficiently large number of samples to make proper diagnosis possible. This kind of measurement is quite different from simply capturing a set of raw voltage–time waveforms, and it helps focus more directly on problems involving data.

In most cases with the 68000, display of signals on a bus cycle basis will prove adequate, but sometimes it might be useful to examine signals for each clock cycle instead. The previous example comprised a grand total of twelve clock cycles minimum,[4] but the logic analyser was not shown to be capturing this fine detail of the instruction's execution. Figure 14.5 gives details relating to the MOVE.W $8000,D0 instruction, showing a selection of bus signals for all twelve clock cycles. Depending on the information required, a display something like that shown in Fig. 14.5 might be preferred to the textual form of listing given in the previous type of listing, for example to examine timings between signals, in which case the logic timing analyser to be described later would be used instead of the state analyser.

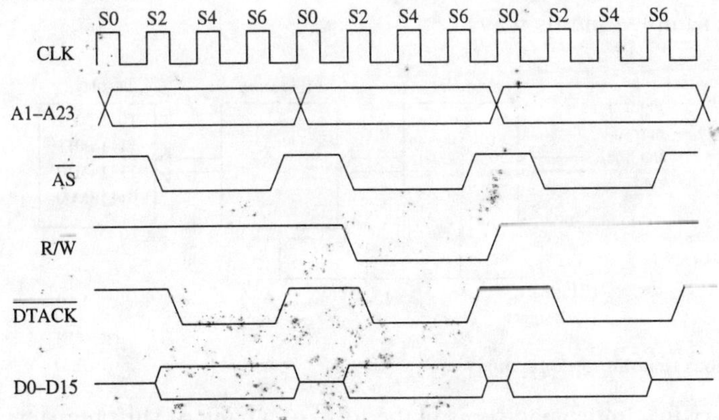

Figure 14.5 MOVE.W $8000,D0 instruction cycle.

---

[4] Wait states may be required in read or write cycles with the 68000 depending on the device supplying data to the microprocessor, in which case the total number of cycles will be more than the twelve shown in the three bus cycles depicted in Fig. 14.5.

Bus cycle	Clock cycle†	A1–A23	AS*	RW*	DTACK*	D8–D15	D0–D7
1	1	X	1	1	1	X	X
	2	400400	0	1	1	X	X
	3	400400	0	0	0	38	30
	4	400400	1	1	0	38	30
2	5	X	1	1	1	X	X
	6	400402	0	1	1	X	X
	7	400402	0	0	0	80	00
	8	400402	1	1	0	80	00
3	9	X	1	1	1	X	X
	10	008000	0	1	1	X	X
	11	008000	0	0	0	20	F0
	12	008000	1	1	0	20	F0

Notes:

† The display assumes there are no wait states.

The '*' in signals such as AS* signifies an active low state (this is a convention commonly used in Motorola literature).

'X' signifies tri-state.

14.8.3 Logic state analyser principles

A logic state analyser comprises elements that successively sample, store and finally display state information (Fig. 14.6). The intervals for which measured data is sampled are governed by an external clock signal, usually the system clock of the system under test. In the previous example, it was in fact implicit that the state information displayed in the diagram was sampled in this way.

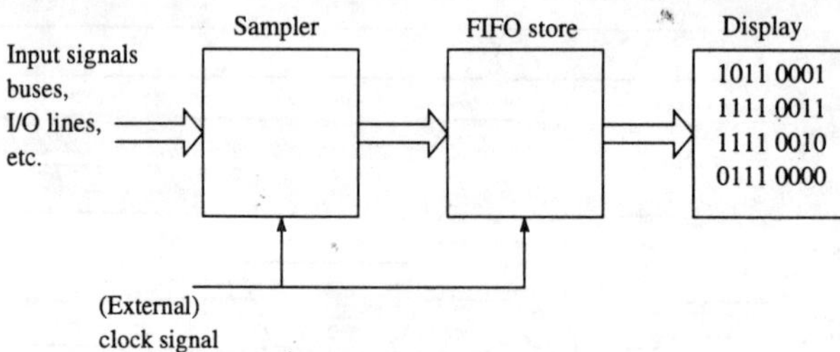

Figure 14.6 Block diagram of logic analyser.

The memory in a logic analyser is in the form of a bank of shift registers organized as a FIFO. Data is shifted from one bank of registers to the next in the chain, following each clock cycle. The capacity of the memory sets the limit on the number of samples that can be taken, 200×32-bit words being typical for an instrument used in measurement on 8-bit systems. The data being captured may be displayed in a variety of different formats, depending on the use to which the information is to be put. In the case of control signals such as R/W, we are interested in bit patterns, so a binary form of display is best. In the case of an address or data bus, however, a hex or octal display is easier to interpret. Some analysers can display disassembled instruction mnemonics, provided the data bus is connected to the appropriate channels of the logic analyser. The display in Fig. 14.7, taken from a screen dump from an HP1650 analyser, is an example of this; it also shows the instruction address and the nature of the bus cycle (see Fig. 14.4 for a photograph of the instrument in use).

680005	– State listing	Invasm
Markers	Off	

Label >	ADDR	68000 Mnemonic	STAT
Base >	Hex	hex	Symbol
–0006	00200C	MOVE.B DO,[A1]	SUPR PGRM READ
–0005	00200E	MOVE.B [A1],DO	SUPR PGRM READ
–0004	003000	FFxx supr data write	SUPR DATA WRITE
–0003	002010	BRA.B 002008	SUPR PGRM READ
–0002	003000	FFxx supr data read	SUPR DATA READ
–0001	002012	0000 unused prefetch	SUPR PGRM READ
+0000	002008	EORI.B #FF,DO	SUPR PGRM READ
+0001	00200A	00FF supr program read	SUPR PGRM READ
+0002	00200C	MOVE.B DO,[A1]	SUPR PGRM READ
+0003	00200E	MOVE.B [A1],DO	SUPR PGRM READ
+0004	003000	00xx supr data write	SUPR DATA WRITE
+0005	002010	BRA.B 002008	SUPR PGRM READ
+0006	003000	00xx supr data read	SUPR DATA READ
+0007	002012	000 unused prefetch	SUPR PGRM READ
+0008	002008	EORI.B #FF,DO	SUPR PGRM READ
+0009	00200A	00FF supr program read	SUPR PGRM READ

Figure 14.7 Screen dump of HP1650 acting as a state analyser.

14.8.4 Triggering logic state analysers

Consider the following problem. A program which contains the following code fragment is to be debugged:

```
Address       instruction
...
...
400418        MOVE.W    (A0),DO
40041A        MOVE.W    (A1)+,D1
40041C        ADD.W     (A2)+,D1
40041E        MOVE.W    D1,(A3)+
400420        SUBI.W    #1,D0
400426        BNE       $40041A
...
...
```

The first instruction accesses a memory chip; if this happens to be faulty, the loop counter value which is loaded into D0 is likely to be wrong, causing the loop to misbehave.

To use a logic analyser to find this bug, it is desirable to start taking samples of the bus signals near the start of the loop, i.e. if the address as indicated by the 24 lines from the analyser connected to the address bus is that of the first instruction. In other words, the data prior to the instructions shown contains no useful information, and should not be stored in the logic analyser's memory. The implication is that we should only start to store the signals being monitored after a certain bit pattern has been observed; in this case the address $400418. The required bit pattern, called a **trigger word**, would have to be stored in the analyser at a prior stage when it is set up. The process of initiating the storage of the data being sampled is termed **triggering**. For each bit in the trigger word there are three possible values: 1, 0 or don't care. For example, if part of a trigger word is 0XX0 (Binary), then the analyser would trigger on any of the states 0, 2, 4 or 6. This might be useful in observing first the general region where problems might arise, then subsequently using logic values in a trigger word that fully defines that region.

In other cases, it may be desirable to sample *up to* the point where a value equal to the trigger word is observed in the data. This would enable events prior to the trigger point to be acquired, as opposed to any *following* events. Most logic analysers make provision for storing data both prior to and after the trigger point. For example, in a system that used interrupts, we might want to trigger when the interrupt vector address is encountered, and examine the events leading up to the interrupt in a calling program (perhaps to examine the data stored on the stack). It might be useful to capture events after the occurrence of the interrupt, so that the behaviour of the interrupt service routine could be analyzed.

A logic analyser is capable of capturing a large volume of data, and commonly only a small proportion of it is helpful to the problem under investigation. For example some nested loops can execute literally thousands of times and generate masses of data, much of which is likely to be irrelevant. Although it is possible to edit the information being acquired by hand, it is better to carry out the 'editing' within the instrument itself. This is known as **selective trace**. One approach is to trace only trigger words. If the trigger word includes some don't care states, an adequate number of states for an analysis can quite often be obtained. For example, if a trigger word is defined as $12X4H, then only states $1204, $1214, ..., $12F4 will be traced. A full trace in this region would have involved all the states from $1200 to $12FF, i.e. 255 states would have to be searched and interpreted in comparison with the 16 states in the selective trace.

14.8.5 Logic timing analysers

Some problems, such as the following, cannot be dealt with by a logic state analyser:

- Cases where it is required to measure time delays between signals (handshaking signals in a serial or parallel interface for example)
- Measurement of propagation delays through gates
- Capturing bad data resulting from glitches

Logic state analysers have limitations here, in the sense that they only sample signals at the incidence of an edge of the clock rate of the system under test, and by implication at the rate of this clock. Any changes in the signals between clock pulses will not be captured. A **logic timing analyser** addresses this situation, having an *internal* clock which is used to sample the data at a relatively rapid rate. The frequency of this clock needs to be high enough to resolve adequately the signals that the instrument is capturing, frequencies of 100 MHz or more being not uncommon. Like a logic state analyser, however, a logic timing analyser suppresses detailed time domain phenomena such as ringing, etc. Also, like a logic state analyser, a timing analyser starts (or stops) capturing data only when preset triggering criteria have been satisfied. It is the logic timing analyser that can be crudely compared to a CRO, since it displays information as waveforms, albeit in a 'cleaned up' form. Most modern logic analysers incorporate both the state and timing functions in one instrument.

Figure 14.8 shows a screen dump from an HP1650 set up as a timing analyser. Note the way the various signals are identified by the labels on the left.

Figure 14.9 shows the effect of sampling on a measured waveform. The edges of the sampled wave coincide with those of the logic analyser clock; and this is a potential source of error if the clock frequency is too low in relation to the clock rate of the data being measured. If the logic analyser clock frequency is very high on the other hand, an excessive memory capacity is required to store samples corresponding to a given time span. With microprocessors like the 68000 with a typical clock frequency of 8 MHz, a sampling rate of around 100 MHz is adequate. With 32-bit microprocessors, which invariably have higher

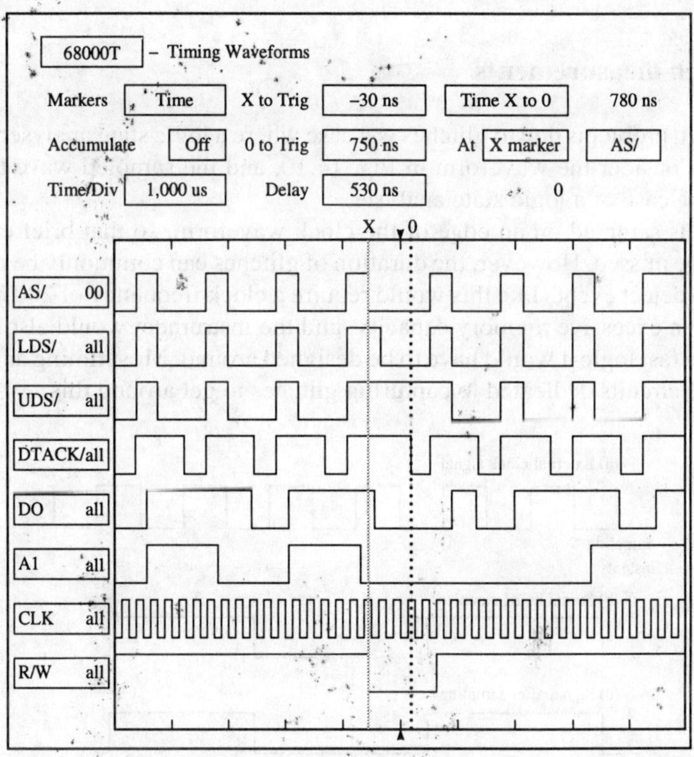

Figure 14.8 Screen dump of HP1650 acting as a timing analyser.

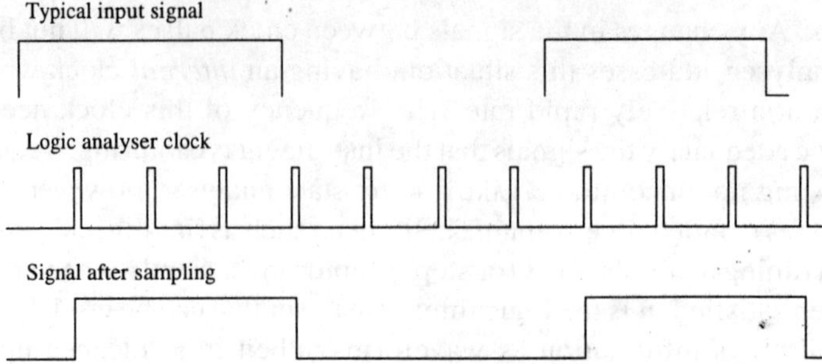

Figure 14.9 Logic timing analyser signals.

clock frequencies, sampling rates greater than this would be needed, so the analyser needs to have very high-speed—and expensive—logic.

Commonly, logic state and timing analysers are used in concert, with the state analyser perhaps being used to identify approximately the location of a problem, after which the timing analyser may be used to home in on the problem by virtue of operating at a higher sampling rate. In some cases, a CRO can be used if the detailed shape of a pulse such as a glitch requires examination of waveforms in their raw (time domain) form. Most logic analysers can in such circumstances generate a signal that can be used to trigger the CRO.

14.8.6 Glitch measurements

Tracking down problems due to glitches is a case where a logic state analyser would not be of any help. Consider the waveform in Fig. 14.10, and the sampled wave that would be obtained in the case of a logic state analyser.

The signal is sampled on an edge of the clock waveform, so that brief events such as glitches will be missed. However, the duration of glitches can commonly be of the order of 5 nsec, and to detect events like this would require a clock frequency of 200 MHz or more. This implies an excessive memory capacity, and the instrument would also be expensive because of the fast logic it would have to be designed around. Most timing analysers incorporate special circuits dedicated to capturing glitches to get around this.

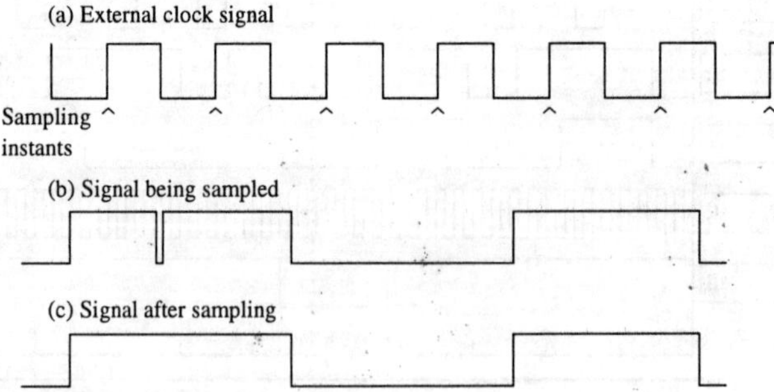

Figure 14.10 Effect of a glitch.

14.8.7 In-circuit emulators

A logic analyser—like a CRO—is a passive instrument in that it captures data but does not in any way *control* the system under test. An in-circuit emulator (ICE) on the other hand extends the debug capability by actually replacing the system's microprocessor as the controlling agent in a system under test. Essentially, an ICE works by generating outputs such as address signals which would normally be generated by the microprocessor, and responding to I/O signals like interrupts, or other input signals. Facilities such as breakpoint insertion or single stepping are available, as with software debug, but the conditions that control facilities like this now extend to hardware events—assertion of an interrupt signal would be one example of an event that could set up a breakpoint. The essential aim with an ICE is to emulate the final environment of the target system in both hardware and software terms, something manifestly not achieved with software debug. We use an ICE when a purely software debug approach to testing is not going to be of any help, as of course it in fact never can be in testing hardware.

Physically, the ICE works by generating (or responding to) signals at a probe that connects to the socket in which the microprocessor chip would normally fit. When the ICE is to be used, the microprocessor chip is unplugged from the circuit board, and replaced with this probe. The ICE, as seen by the target system, behaves in an identical way to the original microprocessor chip, generating the appropriate bus and control signals to set up fetch and execution of instructions just as the microprocessor would have done. The ICE would also in most cases have probes separate from that connected to the microprocessor socket, which can have I/O and other signals connected to it for use in connection with events which are to control breakpoints and other activities. ICEs come in a variety of different types of package, but commonly are a unit that attaches to the host computer used for software development, as well as to the system under test.

An ICE can be used to carry out debug on partially constructed hardware, and it has memory which can replace part or all of the memory in the target system. It is normally possible to map any part of the emulator's memory to any part of the target's memory space. Depending on the state of development of the hardware, some of target's memory may not be yet in place, in which case that in the ICE can be used to emulate it. This provision is also very useful if the memory in the target system is suspect—some lines stuck at 1 in one of the memory chips, say. Such memory can be temporarily replaced by the functional memory inside the emulator. It would be possible to load a test program into the memory of the ICE that could test the functionality of the target's RAM memory. This would involve writing certain bit patterns to the memory under investigation, and then subsequently reading the memory back and comparing the data read with that originally written (Williams, 1987).

14.8.8 Hardware debug—a final word

Several test instruments have been described in this section, although we have tended to concentrate on those aspects of measurements and debug that may have implications for the software in some way. It should be added that many faults can be uncovered with relatively simple instruments such as logic probes, multimeters or CROs. Logic analysers and ICEs should be reserved for the really tough problems where their relative sophistication is required. It might also be said that, as in most aspects relating to measurements and fault finding, a considerable degree of practical experience is required in terms of the behaviour of the software and hardware of the system being developed, the application of appropriate debug strategies and the selection and use of suitable instruments.

14.9 EXERCISES AND REVIEW QUESTIONS

1. Enumerate aspects of the behaviour of a microprocessor-based system where the hardware cannot be fully tested in isolation from the software.
2. Write a RAM test program in 68000 assembler. The program is to use a so-called 'walking one's' bit pattern, that is $1, $2, $4, $8, etc. This means that these bit patterns are to be written to each store location to be tested, and a check made to see if the pattern is actually stored. The program should check that *only* the particular location under test has been changed.
3. You are to test a multitasking program written in C for use on an SBC such as that shown in Fig. 1.2, which uses the services of a real-time executive. The host computer is a VAX minicomputer. Give possible stages in the testing of the software, describing the nature and use of the various hardware and software development tools needed at each stage.
4. Discuss the respective roles of breakpoints and tracing in debug. Quote cases from your own experience in which either technique might be used in preference to the other.
5. You are to extend the capability of a debugger by adding a variant of the trace facility. This is to display a *selection* of register values and the contents and addresses and contents of up to four store locations. Investigate the format of a suitable command line and write the relevant code using 68000 assembler which allows the selection of the registers and store locations required.
6. An interrupt service routine in a 68000 system is to be investigated using a logic state analyser. It is required to capture events just prior to an interrupt as well as events inside the interrupt service routine itself. Discuss in general terms the triggering arrangements that should be used. (*Hint:* You might find it advantageous to scan Appendix 2 before answering this question.)
7. In Sec. 14.6.3 you were invited to detect a second bug in the program. Accept the invitation if you have not already done so.

REFERENCES AND FURTHER READING

Cooling, J. E. (1991) *Software Design for Real-time Systems*, Chapman and Hall, London.
Kane, G., Hawkins, D. and Leventhal, L. (1989) *68000 Assembly Language Programming*, McGraw-Hill, Berkeley, Calif.
Kneen, J. (1980) *Logic Analyzers for Microprocessors*, Hayden, New Jersey.
Korn (1982) *Microprocessor and Small Digital Computer Systems for Engineers and Scientists*, McGraw-Hill.
Lawrence, P. D. and Mauch, K. (1987) *Real-Time Microcomputer System Design—An Introduction*, McGraw-Hill, New York.
Tseng, V. (ed.) (1984) *Microprocessor Development, and Development Systems*, Granada, London.
Williams, G. B. (1987) *Troubleshooting on Microprocessor Based Systems*, Pergamon, Oxford.

The 68000 instruction set

A1.1 68000 DATA ADDRESSING MODES

Notation

ea means effective address
Dn, Dx, Dy means any data register
An, Ax, Ay means any address register
SR means status register
PC means program counter

Register direct addressing

Data register direct ea = Dn
Address register direct ea = An
Status register direct ea = SR

Register indirect addressing

Register indirect   ea = (An)
Postincrement register indirect   ea = (An), An <− An + 1
Predecrement register indirect   An <− An − 1, ea = (An)
Register indirect with offset   ea = (An) + D_{16}
Indexed register indirect with offset ea = (An) + (Rx) + D_8

Program counter relative addressing

Relative with offset   ea = (PC) + D_{16}
Relative with index and offset ea = (PC) + (Rx) + D_8

Absolute data addressing
(A). Absolute short   ea = (Next word)
(B). Absolute long   ea = (Next two words)

Immediate data addressing
Immediate   Data = Next word(s)
Quick immediate   Inherent data

A1.2 CONDITION CODES

These are five bits within the Status Register, their functions are as follows:

N (negative)   Set if the most significant bit of the result is set. Cleared otherwise.
Z (zero)   Set if the result equals zero. Cleared otherwise.
V (overflow)   Set if there was an arithmetic overflow. This implies that the result is not
               representable in the operand size. Cleared otherwise.
C (carry)   Set if a carry is generated out of the most significant bit of the operands for
               an addition. Also set if a borrow is generated in a subtraction. Cleared other-
               wise.
X (extend)   Transparent to data movement. When affected it is set the same as the C bit.
               Used as an operand for multi-precision computations.

The conventions used for the condition code register are as follows:

\#   Set or reset according to the result of the operation
na   Not affected by the operation
0   Cleared to 0
1   Set to 1
u   Undefined, may be 0 or 1

In all cases of conditional Branches & Sets (cc) in instructions may specify the following
conditions:

CC	carry clear	LT	less
CS	carry set	MI	minus
EQ	equal	NE	not equal
GE	greater or equal	PL	plus
GT	greater	VC	no overflow
HI	higher	VS	overflow
LE	less or equal	F	false
LS	low or same	T	true

A1.3 LIST OF INSTRUCTIONS IN ALPHABETICAL ORDER

ABCD   (Add decimal with extend) ABCD Dy,Dx ABCD −(Ay),−(Ax)
Size: (.B)
Description: Add source operand and extend bit to destination operand and store the
                     result in the destination location. Addition performed using BCD

arithmetic. The operands addressed either as data register to data register, or memory to memory using the predecrement address in g mode.

Condition codes: $(X = \#)\ (N = u)\ (Z = \#)\ (V = u)\ (C = \#)$

ADD   (Add binary) ADD <ea>,Dn ADD Dn,<ea>
Size: (.B .W .L)
Description: Add source operand to destination operand and store result in destination location.
Effective address: For source operand all addressing modes allowed except address register direct when operand size is byte. For destination operand only alterable memory addressing modes allowed.
Condition codes: $(X = \#)\ (N = \#)\ (Z = \#)\ (V = \#)\ (C = \#)$

ADDA   (Add address) ADDA <ea>,An
Size: (.W .L)
Description: Add source operand to destination address register and store the result in an address register.
Effective address: All addressing modes allowed for source operand.
Condition codes: Not affected.

ADDI   (Add Immediate) ADDI #<data>,<ea>
Size: (.B .W .L)
Description: Add immediate data to destination operand and store the result in destination operand.
Effective address: Only data alterable modes allowed for destination.
Condition codes: $(X = \#)\ (N = \#)\ (Z = \#)\ (V = \#)\ (C = \#)$

ADDQ   (Add quick) ADDQ #<data>,<ea>
Size: (.B .W .L)
Description: Add immediate data to destination operand and store the result in destination operand. Data range 1–8.
Effective address: Only data alterable modes allowed for destination. If operand size is byte, address register direct not allowed.
Condition code: $(X = \#)\ (N = \#)\ (Z = \#)\ (V = \#)\ (C = \#)$

ADDX   (Add extended) ADDX Dy, Dx ADDX $-$(Ay),$-$(Ax)
Size: (.B .W .L)
Description: Add source operand to destination operand together with extend bit and store result in destination location. Only data register to data register or memory to memory allowed.
Condition code: $(X = \#)\ (N = \#)\ (Z = \#)\ (V = \#)\ (C = \#)$

AND   (Logical AND) AND <ea>, Dn) AND Dn,<ea>
Size: (.B .W .L)
Description: AND source operand to destination operand and store result in destination location.
Effective address: Only data alterable modes allowed for source operand. Only alterable memory modes allowed for destination operand.
Condition codes: $(X = na)\ (N = \#)\ (Z = \#)\ (V = 0)\ (C = 0)$

ANDI   (AND immediate) ANDI #<data>,<ea>
Size: (.B .W .L)
Description: AND immediate data with destination operand and store result in desti-
                         nation location. If operation size is byte or word, status register may be
                         destination but privileged instruction if word.
Effective address: Only data alterable modes (or status register) allowed.
Condition codes: (X = na) (N = #) (Z = #) (V = 0) (C = 0)

ASL   (Arithmetic shift left) ASL Dx,Dy ASL #<data>,Dy ASL <ea>

ASR   (Arithmetic shift right) ASR Dx,Dy ASR #<data>,Dy ASR <ea>
Size: (.B .W .L)
Description: Arithmetically shifts the bits of the operand in the direction specified.
                         Carry bit receives last bit shifted out of operand. When register contents
                         shifted, number of bit positions moved defined as either immediate data
                         (shift range 1–8) or value held in register specified. Memory only
                         shifted one bit position and size restricted to word. For left shift over-
                         flow flag indicates any sign change. For right shift sign bit replicated.
Effective address: Only data alterable address modes allowed.
Condition codes: (X = #) (N = #) (Z = #) (V = #) (C = #)

B(cc)   (Branch conditionally) B(cc) <label>
Size: (.B word)
Description: If the specified condition is met, program execution continues at loca-
                         tion (PC) + displacement. The displacement is a two's complement
                         integer which counts the relative distance in bytes. PC value is current
                         instruction location + two. Either an 8-bit displacement in the instruc-
                         tion word or a 16-bit displacement word used.
Condition codes: Not affected.

BCHG   (Test a bit and change) BCHG Dn,<ea> BCHG #<data>,<ea>
Size: (.B .L)
Description: Bit in destination operand tested and state of specified bit reflected in Z
                         condition code. After test, state of specified bit changed in destination.
                         If register is the destination any of the 32 bits can be specified. For
                         memory location only bits within a byte can be specified. Bit number
                         involved specified either as immediate data or as value in data register.
Effective address: Only data alterable addressing modes allowed.
Condition codes: (X = na) (N = na) (Z = #) (V = na) (C = na)

BCLR   (Test a bit and clear) BCLR Dn,<ea> BCLR #<data>,<ea>
Size: (.B .L)
Description: Bit in destination operand tested and state of specified bit reflected
                         in Z condition code. After test, specified bit cleared in destination. If
                         register the destination any of the 32 bits can be specified. For a
                         memory location only bits within a byte can be specified. Bit number

involved specified either as immediate data or as the value in the data register.

Effective address: Only data alterable addressing modes allowed.

Condition codes: (X = na) (N = na) (Z = #) (V = na) (C = na)

BRA (Branch always) BRA <label>

Size: (.B word)

Description: Program execution continues at location (PC) + displacement. Displacement a two's complement integer which counts relative distance in bytes. Value in PC the current instruction address plus two. Displacement either an 8-bit displacement in instruction word or a 16-bit displacement word.

Condition codes: Not affected.

BSET (Test a bit and set) BSET Dn,<ea> BSET #<data>,<ea>

Operation: (<bit number>) OF Destination) $-> Z; 1 ->$ <bit number> OF Destination

Size: (.B .L)

Description: Bit in destination operand tested and state of the specified bit reflected in Z condition code. After test, specified bit set to 1 in destination. If register is the destination any of the 32 bits can be specified. For a memory location only bits within a byte can be specified. Bit number involved specified either as immediate data or as value in a data register.

Effective address: Only data alterable addressing modes allowed.

Condition codes: (X = na) (N = na) (Z = #) (V = na) (C = na)

BSR (Branch to subroutine) BSR <label>

Size: (.B word)

Description: Address of instruction immediately following the BSR instruction pushed on to system stack. Program execution then continues at location (PC) + displacement. Displacement a two's complement integer which counts relative distances in bytes. Value in PC is current instruction location plus two. Displacement either an 8-bit displacement in instruction word or a 16-bit displacement word.

Condition codes: Not affected.

BTST (Test a bit) BTST Dn,<ea> BTST #<data>,<ea>

Size: (.B .L)

Description: Bit in destination operand tested and state of specified bit reflected in Z condition code. If a register is the destination any of the 32 bits can be specified. For memory location only bits within a byte can be specified. Bit number involved specified either as immediate data or as value in data register.

Effective address: Only data alterable addressing modes allowed.

Condition codes: (X = na) (N = na) (Z = #) (V = na) (C = na)

CHK (Check register against bounds) CHK <ea>,Dn
Size: (Word)
Description: Content of low-order word in data register specified examined, and compared to upper bound. Upper bound is a two's complement integer. If the register value less than zero or greater than the upper bound contained in operand .W then the processor initiates exception processing. The vector number is generated to reference the CHK instruction exception vector.
Effective address: Only data alterable modes allowed.
Condition codes: (X = na) (N = #) (Z = u) (V = u) (C = u)

CLR (Clear an operand) CLR <ea>
Size: (.B .W .L)
Description: All destination bits set to zero.
Effective address: Only data alterable modes allowed. Note instruction performs a dummy read before writing.
Condition codes: (X = na) (N = 0) (Z = 1) (V = 0) (C = 0)

CMP (Compare) CMP <ea>,Dn
Size: (.B .W .L)
Description: Subtract source from destination and set condition codes according to result; destination location not changed.
Effective address: All addressing modes may be used for source operand except when size is a byte, when address register direct is not allowed.
Condition codes: (X = na) (N = #) (Z = #) (V = #) (C = #)

CMPA (Compare address) CMPA <ea>,An
Size: (.W .L)
Description: Subtract source operand from destination address register and set condition codes according to result; address register not changed. If operation size is a word source operand sign extended to .L operand and operation performed on all 32 bits of address register.
Effective address: All addressing modes are allowed.
Condition codes: (X = na) (N = #) (Z = #) (V = #) (C = #)

CMPI (Compare immediate) CMPI #<data>,<ea>
Size: (.B .W .L)
Description: Subtract immediate data from destination operand and set condition codes according to result; destination location not changed.
Effective address: Only the data alterable modes are allowed.
Condition codes: (X = na) (N = #) (Z = #) (V = #) (C = #)

CMPM (Compare memory) CMPM (Ay)+,(Ay)+
Size: (.B .W .L)
Description: Subtract source operand from destination operand and set condition codes according to result; destination location not changed. Only post-increment mode of addressing allowed.
Condition codes: (X = na) (N = #) (Z = #) (V = #) (C = #)

DB(cc)   (Test condition, decrement and branch) DBcc Dn,<label>
               If (Dn <> −1) then PC + d −> PC else PC + 2 −> PC
Size: (Word)
Description: Instruction a looping primitive of three parameters: a condition, a data
               register and a displacement. Instruction first tests condition to deter-
               mine if termination condition for loop has been met and, if so, no oper-
               ation performed. If condition not true, low-order 16 bits of data register
               decremented by one. If result, −1 execution continues with next
               instruction. If count not −1, execution continues at location indicated
               by current value of PC plus sign-extended 16-bit displacement.
Condition codes: Not affected.

DIVS   (Signed divide) DIVS <ea>,Dn
Size: (Word)
Description: Divide destination operand by source operand and store result in desti-
               nation. Destination operand is 32 bits and source operand is 16 bits.
               Operation performed using signed arithmetic. A 32-bit result with quo-
               tient in least-significant 16 bits and remainder in most-significant 16
               bits. Division by zero causes a trap. Overflow may be detected and set
               before the completion of the instruction.
Effective address: Only data addressing modes are allowed for source.
Condition codes: (X = na) (N = #) (Z = #) (V = #) (C = 0)

DIVU   (Unsigned divide) DIVU <ea>,Dn
Size: (Word)
Description: Divide destination operand by source operand and store result in desti-
               nation. Destination operand is 32 bits and source operand is 16 bits.
               Operation performed using unsigned arithmetic. A 32-bit result with
               quotient in least-significant 16 bits and remainder in most-significant
               16 bits. Division by zero causes a trap. Overflow may be detected and
               set before the completion of the instruction. If overflow occurs con-
               dition flagged but operands are unaffected.
Effective address: Only data addressing modes are allowed for source.
Condition codes: (X = na) (N = #) (Z = #) (V = #) (C = 0)

EOR   (Exclusive 'OR') EOR Dn,<ea>
Size: (.B .W .L)
Description: Exclusive OR source operand to destination operand and store result in
               destination location. Source operand always data register.
Effective address: Only data alterable modes allowed for destination.
Condition codes: (X = na) (N = #) (Z = #) (V = 0) (C = 0)

EORI   (Exclusive 'OR' Immediate) EORI #<data>,<ea>
Size: (.B .W .L)
Description: Exclusive OR immediate data with destination operand and store result
               in destination location. If operation size byte or .W status register may
               be destination but privileged instruction if word.

Effective address: Only data alterable modes (or status register) allowed.
Condition codes: (X = na) (N = #) (Z = #) (V = 0) (C = 0)
                    (X = #) (N = #) (Z = #) (V = #) (C = #) if status register.

EXG   (Exchange registers) EXG Rx,Ry
Size: (.L)
Description: Exchange contents of two registers (address or data).
Condition codes: Not affected.

EXT   (Sign extend) EXT Dn
Size: (.W .L)
Description: If operation word sized, bit [7] of specified data register copied to bits [15:8] of that data register. If operation .L sized, bit [15] of designated data register copied to bits [31:16] of that register.
Condition codes: (X = na) (N = #) (Z = #) (V = 0) (C = 0)

JMP   (Jump unconditionally) JMP <ea>
Unsized
Description: Program execution continues at address specified by instruction.
Effective address: Only control addressing modes are allowed.
Condition codes: Not affected.

JSR   (Jump to subroutine) JSR <ea>
Unsized
Description: Address of instruction immediately following the JSR instruction pushed on to system stack. Program execution continues from address specified in instruction.
Effective address: Only control addressing modes are allowed.
Condition codes: Not affected.

Lea   (Load effective address) Lea <ea>,An
Size: (.L)
Description: ea loaded into specified address register.
Effective address: Only control addressing modes are allowed.
Condition codes: Not affected.

LINK   (Link and allocate) LINK An,#<displacement>
Unsized
Description: Current content of specified address register pushed on to stack. After push, address register loaded from updated stack pointer. Finally sign-extended displacement added to stack pointer.
Condition codes: Not affected.

LSL   (Logical Shift left) LSL Dx,Dy LSL #<data>,Dy LSL <ea>
LSR   (Logical Shift right) LSR Dx,Dy LSR #<data>,Dy LSR <ea>
Size: (.B .W .L)
Description: Shifts bits of operand in direction specified. Carry bit receives last bit shifted out of operand. When register contents shifted, number of bit positions moved defined as either immediate data (shift range 1–8) or value held in specified register.
                    Memory only shifted one bit position and size restricted to word. Zeros are shifted into bit positions vacated.

Effective address: Only memory alterable modes allowed.
Condition codes: (X = #) (N = #) (Z = #) (V = 0) (C = #)

MOVE   (Move data from source to destination) MOVE <ea>,<ea>
Size: (.B .W .L)
Description: Move content of source to destination location. Data examined as moved and condition codes set accordingly.
Effective address: *Destination:* Only data alterable modes allowed.
Source: All addressing modes allowed except address register direct when size is a byte.
Condition codes: (X = na) (N = #) (Z = #) (V = 0) (C = 0)

MOVE to CCR   (Move to condition codes) MOVE <ea>,CCR
Size: (Word)
Description: Content of source operand moved to condition codes. Source operand a .W but only low-order byte used to update condition codes. Upper byte ignored.
Effective address: Only data addressing modes allowed.
Condition codes: (X = #) (N = #) (Z = #) (V = #) (C = #)

MOVE to SR   (Move to the status register) MOVE <ea>,SR
Size: (Word)
Description: Contents of source operand moved to status register. Source operand a word and all bits of status register affected.
Effective address: Only data addressing modes are allowed.
Condition codes: (X = #) (N = #) (Z = #) (V = #) (C = #)

MOVE from SR   (Move from the status register) MOVE SR,<ea>
Size: (Word)
Description: Contents of status register moved to destination location. Operand size is word.
Effective address: Only data alterable addressing modes allowed.
Condition codes: Not affected.

MOVE USP   (Move User Stack Pointer) MOVE USP,An MOVE An,USP
Size: (.L)
Description: Contents of user stack pointer transferred to or from specified address register.
Condition codes: Not affected

MOVEA   (Move address) MOVE <ea>,An
Size: (.W .L)
Description: Move contents of source to destination address register. Word size operands sign extended to 32 bits.
Effective address: All addressing modes are allowed.
Condition codes: Not affected.

MOVEM   (Move multiple reg) MOVEM <reg list>,<ea> MOVEM <ea>,<reg list>
Size: (.W .L)
Description: Selected registers transferred to or from consecutive memory locations
                         starting at location specified by effective address.
Effective address: Specifies memory to or from which registers are to be moved. For regis-
                         ter to memory transfer, only control alterable or predecrement address-
                         ing modes allowed. For memory to register transfer, only control or
                         postincrement addressing modes allowed.
Condition codes: Not affected.

MOVEP   (Move peripheral data) MOVEP Dx,d(Ay) MOVEP d(Ay),Dx
Size: (.W .L)
Description: Data transferred between data register and alternate bytes of memory
                         starting at location specified, and incrementing by two. High-order
                         byte of data register transferred first and low-order byte last. Address
                         register indirect plus displacement mode used to address memory. If
                         the address is even, all the transfers are made on the high-order half of
                         the data bus, whereas if the address is odd, the low-order half of the data
                         bus is used.
Condition codes: Not affected.

MOVEQ   (Move quick) MOVEQ #<data>,Dn
Size: (.L)
Description: Move immediate data to data register. Data contained in an 8-bit field
                         within operation word. Data sign-extended to .L operand and all 32 bits
                         transferred to data register.
Condition codes: $(X = na) (N = \#) (Z = \#) (V = 0) (C = 0)$

MULS   (Signed multiply) MULS <ea>,Dn
Size: (Word)
Description: Multiply two signed 16-bit operands yielding a 32-bit signed result.
                         Operation performed using signed arithmetic. Register operand taken
                         from low-order word; upper word unused.
Effective address: Only data addressing modes are allowed.
Condition codes: $(X = na) (N = \#) (Z = \#) (V = 0) (C = 0)$

MULU   (Unsigned multiply) MULU <ea>,Dn
Size: (Word)
Description: Multiply two unsigned 16-bit operands yielding a 32-bit unsigned
                         result. Operation performed using unsigned arithmetic. Register
                         operand taken from low-order word; upper word unused.
Effective address: Only data addressing modes are allowed.
Condition codes: $(X = na) (N = \#) (Z = \#) (V = 0) (C = 0)$

NBCD   (Negate decimal with extend) NBCD <ea>
Size: (.B)
Description: Operand addressed as destination and the extend bit is subtracted from
                         zero. Operation performed using decimal arithmetic. Result saved in
                         destination location.

Effective address: Only data alterable modes are allowed.
Condition codes: (X = #) (N = u) (Z = #) (V = u) (C = #)

NEG (Negate) NEG <ea>
Size: (.B .W .L)
Description: Operand addressed as destination subtracted from zero. (Two's complement.) Result stored in destination location.
Effective address: Only data alterable modes are allowed.
Condition codes: (X = #) (N = #) (Z = #) (V = #) (C = #)

NEGX (Negate with extend) NEGX <ea>
Description: Operand addressed as destination and the extend bit is subtracted from zero. Result stored in destination location.
Effective address: Only data alterable modes allowed.
Condition codes: (X = #) (N = #) (Z = #) (V = #) (C = #)

NOP (No operation) NOP
Unsized
Description: No operation occurs. Processor state, other than program counter, unaffected.
Condition codes: Not affected.

NOT (Logical complement) NOT <ea>
Size: (.B .W .L)
Description: One's complement of destination operand taken and stored in the destination location.
Effective address: Only data alterable modes allowed.
Condition codes: (X = na) (N = #) (Z = #) (V = 0) (C = 0)

OR (Inclusive 'OR') OR <ea>,Dn OR Dn,<ea>
Size: (.B .W .L)
Description: Inclusive OR the source operand to the destination operand and store the result in destination location.
Effective address: For source operand only data alterable modes allowed. For destination operand only alterable memory addressing modes allowed.
Condition codes: (X = na) (N = #) (Z = #) (V = 0) (C = 0)

ORI (Inclusive 'OR' immediate) ORI #<data>,<ea>
Size: (.B .W .L)
Description: Inclusive OR the immediate data with destination operand and store result in destination.
Effective address: Only data alterable modes or status register addressing allowed.
Condition codes: (X = na) (N = #) (Z = #) (V = 0) (C = 0)

Pea (Push effective address) Pea <ea>
Size: (.L)
Description: ea computed and pushed on to stack.
Effective address: Only control addressing modes are allowed.
Condition codes: Not affected.

RESET   (Reset external devices) RESET
Unsized

Description: Reset line asserted causing all external devices to be reset. Processor state, other than program counter, unaffected and execution begins with next instruction.

Condition codes: Not affected.

ROL   (Rotate (no extend) left) ROL Dx,Dy ROL #<data>,Dy ROL <ea>
ROR   (Rotate (no extend) right) ROR Dx,Dy ROR #<data>,Dy ROR <ea>
Size: (.B .W .L)

Description: Rotates operand bits in direction specified. When register rotated, number of places moved specified either as immediate data (shift range 1–8) or as contents of specified data register.
                     Memory only rotated one bit position and size restricted to word.
                     For ROL bits shifted out of high-order bit go to both carry bit and back into low-order bit.
                     For ROR bits shifted out of low-order bit go to both carry bit and back to high-order bit.

Effective address: Only memory alterable addressing modes allowed.

Condition codes: $(X = na) (N = \#) (Z = \#) (V = 0) (C = \#)$

ROXL   (Rotate with extend left) ROXL Dx,Dy ROXL #<data>,Dy ROXL <ea>
ROXR   (Rotate with extend right) ROXR Dx,Dy ROXR #<data>,Dy ROXR <ea>
Size: (.B .W .L)

Description: Rotates operand bits in direction specified. The Extend bit is included in rotation. When register rotated, number of places moved specified either as immediate data (shift range 1–8) or as contents of specified data register.
                     Memory only rotated one bit position and size restricted to word.
                     For ROXL bits shifted out of high-order bit go to both carry and extend bits; previous value of extend bit shifted into low-order bit.
                     For ROR bits shifted out of low-order bit go to both carry and extend bits; previous value of extend bit shifted to high-order bit.

Effective address: Only memory alterable addressing modes allowed.

Condition codes: $(X = \#) (N = \#) (Z = \#) (V = 0) (C = \#)$

RTE   (Return from exception) RTE
Unsized

Description: Status register and program counter popped from system stack. Previous value lost.

Condition codes: Set according to content of word on stack.

RTR   (Return and restore condition codes) RTR
Unsized

Description: Condition codes and program counter popped from stack. Previous values lost.

Condition codes: Set according to content of word on stack.

RTS   (Return from subroutine) RTS
Unsized
Description: Program counter popped from stack. Previous value lost.
Condition codes: Not affected.

SBCD   (Subtract decimal with extend) SBCD Dy,Dx SBCD −(Ay),−(Ax)
Size: (.B)
Description: Subtract source operand and extend bit from destination operand and
                     store result in destination location. Addition performed using BCD
                     arithmetic. Operands addressed either as data register to data register,
                     or memory to memory using predecrement address register indirect
                     mode.
Condition codes: (X = #) (N = u) (Z = #) (V = u) (C = #)

S(cc)   (Set according to condition) S(cc) <ea>
Size: (.B)
Description: Condition tested, if true, byte specified set true (all ones), otherwise
                     byte set false (all zeros).
Condition codes: Not affected.

STOP   (Load status register and stop) STOP #<data>
Unsized
Description: Immediate operand moved into entire status register; program counter
                     advanced to point at next instruction. Processor stops fetching and exe-
                     cuting instructions. Execution resumes when a trace, interrupt or reset
                     exception occurs.
Condition codes: Set according to immediate data.

SUB   (Subtract binary) SUB <ea>,Dn SUb Dn,<ea>
Size: (.B .W .L)
Description: Subtract source operand from destination operand and store result in
                     destination location.
Effective address: For source operand all addressing modes allowed, except if size is byte,
                     when address register direct not allowed. For destination only alterable
                     memory addressing modes allowed.
Condition codes: (X = #) (N = #) (Z = #) (V = #) (C = #)

SUBA   (Subtract address) SUBA <ea>,An
Size: (.W .L)
Description: Subtract source operand from destination address register and store
                     result in address register. Word size source operands extended to 32
                     bits before operation done.
Effective address: All addressing modes allowed.
Condition codes: Not affected.

SUBI   (Subtract immediate) SUBI #<data>,<ea>
Size: (.B .W .L)
Description: Subtract immediate data from destination operand and store result in
                    destination location.
Effective address: Only data alterable addressing modes allowed.
Condition codes: (X = #) (N = #) (Z = #) (V = #) (C = #)

SUBQ   (Subtract quick) SUBQ #<data>,<ea>
Size: (.B .W .L)
Description: Subtract immediate data from destination location. Data range 1–8.
Effective address: Only alterable addressing modes allowed.
Condition codes: (X = #) (N = #) (Z = #) (V = #) (C = #)
                    (Not affected if subtraction from address register.)

SUBX   (Subtract with extend) SUBX Dy,Dx SUBX −(Ay),−(Ax)
Size: (.B .W .L)
Description: Subtract source operand and extend bit from destination operand and
                    store result in destination location. Only data register to data register or
                    memory to memory using predecrement addressing allowed.
Condition codes: (X = #) (N = #) (Z = #) (V = #) (C = #)

SWAP   (Swap register halves) SWAP Dn
Size: (Word)
Description: Exchange the 16-bit halves of data register.
Condition codes: (X = na) (N = #) (Z = #) (V = 0) (C = 0)

TAS   (Test and set an operand) TAS <ea>
Size: (.B)
Description: Current value of address operand tested and N and Z set accordingly.
                    High-order bit of operand set. Operation indivisable (uses read-
                    modify-write memory cycle) to allow synchronization of several
                    processors.
Effective address: Only data alterable modes allowed.
Condition codes: (X = na) (N = #) (Z = #) (V = 0) (C = 0)

TRAP   (Trap) TRAP #<vector no.>
Unsized
Description: Processor initiates exception processing. Vector number generated to
                    reference vector table entry. Sixteen trap vectors available.
Condition codes: Not affected.

TRAPV   (Trap on overflow) TRAPV
Unsized
Description: If overflow condition set, processor initiates exception processing.
                    Vector number generated to reference TRAPV exception vector. If
                    overflow not set no operation performed.
Condition codes: Not affected.

TST   (Test an operand) TST <ea>
Size: (.B .W .L)
Description: Compares operand with zero. No results saved but condition codes set
                    according to result of test.
Effective address: Only data alterable modes allowed.
Condition codes: (X = na) (N = #) (Z = #) (V = 0) (C = 0)

UNLK   (Unlink) UNLK An
Unsized
Description: Stack pointer loaded from specified address register. Address register
                    then loaded with .L word popped from top of stack.
Condition codes: Not affected.

APPENDIX 2

Exception processing in the 68000

A2.1 REVIEW

Interrupt concepts were encountered in the discussions involving operating systems (Sec. 8.5) and elsewhere. We now discuss the mechanisms for dealing with events, of which interrupts are one example, for the specific case of the 68000 microprocessor. Recall that interrupts are a way of getting a computer to respond to input signals from external devices, the mechanism essentially being that the computer monitors the state of the interrupt signal (s), and when one or more of these is asserted, the normal execute cycle is replaced by an interrupt acknowledge cycle. As part of this, the computer outputs signals to interrogate interrupting devices, inviting them to state which caused the interrupt. The devices respond—in the case of the 68000 at least—by placing an identifying code on the data bus. The microprocessor uses this to select the appropriate interrupt service routine for execution.

   The detailed sequence of events followed will be elaborated on later, but, for now, note that interrupts are just one example of what is commonly known as an **exception**, being a number of different types of event by which the normal fetch execute mechanism of a computer can be pre-empted by something comparable to the way that interrupts are processed, as just described.

A2.2 EXCEPTION TYPES

In the case of the 68000 two main categories of exception can be identified: *external* and *internal*.

Interrupts are an example of an external exception. The others are: *bus errors*, initiated when an input signal called BERR is asserted; and *reset*, which causes exception processing when the RESET input signal is asserted.

The types of internal exception are:

- **Address errors**—Memory locations in the 68000 can be addressed as bytes, words (16 bits), or long words (32 bits). If words or long words are being accessed the address used must fall on an even address boundary (Vaughan, 1983). If this is not in fact the case for a particular instruction an address error will occur and the 68000 is designed to initiate exception processing.

- **Privilege violations**—Privileged instructions are STOP, RESET, RTE, MOVE to SR, AND (word) Immediate to SR, EOR (word) Immediate to SR, OR (word) Immediate to SR, MOVE USP (see Appendix 1 for the 68000's instruction set). These instructions should only be executed when the 68000 is in supervisor mode, otherwise there is a privilege violation which will cause exception processing.

- **Illegal and unimplemented opcodes**—It may be that a code is fetched for which there is no valid 68000 instruction. The 68000 would start exception processing in such cases to allow a way of attempting a recovery. Two particular codes starting with the hex digits A or F are termed **unimplemented** rather than illegal, and a somewhat different exception processing would occur in those cases. This is a neat way of allowing programmers to simulate unimplemented instructions in software. Op codes starting with A or F are commonly referred to as **A or F line emulators** respectively.

- **Instruction traps**—These are a form of exception caused by one of the 68000's TRAP instructions and some others listed below. The basic form of the TRAP is used as the standard way for a user task to invoke a service in the operating system. A single parameter is associated with the instruction, and this is used to identify which out of 16 possible exception-handling routines is to be executed. The TRAPV instruction causes an exception if the overflow flag is set. The TRAPV could be used as a way of handling this kind of error condition if it follows instructions that could cause an overflow. The divide instructions DIVS and DIVU will *automatically* cause an exception if any attempt to divide by zero is made. Finally the CHK instruction can cause an instruction trap type of exception.

- **Trace function**—The status register has a bit called T. If this is set, exception processing occurs after each instruction. Normally this would be used for debug purposes when single stepping or some other close control over instruction execution was required.

A2.3 EXCEPTION PRIORITIES

We have noted elsewhere that it is possible for several interrupts to occur at the same time, and therefore a method of allocating priorities must be provided. The same is true of other forms of exception and Table A2.1 lists the priorities for all types of exception in the 68000.

Exceptions are divided into three groups which define three priorities from 0 (high) to 2 (low). Within these different priorities, interrupts have further priorities.

Within this scheme, there is provision for allocating priorities for individual interrupts, and this works as follows. An interrupting device needs to place a three-bit code on the three ILP (Interrupt Request Input) pins indicating to the microprocessor the priority level of the interrupt. Typically this code would be supplied from a priority encoder, the inputs of which would in turn be supplied from the various interrupting devices. Codes range from 0 (highest priority) to 7 (no interrupt pending). The microprocessor compares the interrupt request level to a three-bit code called the interrupt mask, which are three bits forming part of the status register. Only if the request level is less than or equal to the value encoded in the interrupt mask will the interrupt be further processed.

Table A2.1 Exception priorities

Group exception	Effect of exception
0 Reset Bus error Address error	Abort current bus cycle and process exception
1 Trace Interrupt request Illegal/unimplemented Opcode Privilege violation	Complete current instruction and process exception
2 Instruction traps	Exception processing initiated *directly* by the instruction

A2.4 EXCEPTION VECTOR TABLE

The addresses of exception handlers are held as long words in the **exception vector table** (see Table A2.2). The table occupies 1024 bytes of memory starting at location 0, and extending to location $3FF. After the particular exception has been identified, an entry from the exception vector table is loaded into the program counter, and the exception handler starts to execute. Programmers can load the addresses of their exception handlers into the table which might then normally be held in some nonvolatile form of memory, or else the operating system might fill the table when the computer is being booted following power up.

Table A2.2 Exception vector table

Vector address (Hex)	Exception type
0	Reset—Initial supervisor stack pointer
4	Reset—Initial program counter
8	Bus error
C	Address error
10	Illegal instruction
14	Divide by zero
18	CHK instruction
1C	TRAPV instruction
20	Privilege violation
24	Trace
28	Opcode 1010 emulation
2C	Opcode 1111 emulation
30	reserved
34	reserved
…	reserved
…	reserved
5C	reserved
60	Spurious interrupt
64	Level 1 autovector
68	Level 2 autovector
6C	Level 3 autovector
70	Level 4 autovector
74	Level 5 autovector
78	Level 6 autovector
7C	Level 7 autovector
80	TRAP #1 instruction vector
84	TRAP #2 instruction vector
…	
…	
BC	TRAP #7 instruction vector
C0	reserved
C4	reserved
…	reserved
…	reserved
100	User interrupt vector
…	User interrupt vector
…	User interrupt vector
3FC	User interrupt vector

A2.5 DETAILS OF THE SEQUENCE FOLLOWED IN PROCESSING EXCEPTIONS

We only give an outline of the sequence in processing exceptions; for more detail see Kane *et al.* (1988). Only the particular case of interrupts is considered, for which the steps listed below take place. We have noted (see Table A2.1) that this only starts after the completion of the current instruction, and then only if the priority level is below a certain value.

- The contents of the status register are saved.
- The S bit in the status register is saved, thus putting the microprocessor into supervisor mode.
- The T bit in the status register is set, disabling the trace function.
- The interrupt mask bits are changed to the values on the ILP pins.
- The 68000 now outputs signals to devices that may have caused the interrupt.
- The device must now place the interrupt vector on the low eight bits of the data bus.
- With the interrupt vector stored in the microprocessor, the program counter and also the previously saved value of the status register are pushed on to the system stack, ready for the return to the instruction following that at which the interrupt was detected.
- Finally, the interrupt vector is used to point to the appropriate entry in the exception vector table, this being loaded into the program counter. The exception handler then springs into action during which—in many systems at least—further data would be saved, allowing an orderly return to the original program. This is up to the operating system to arrange (see Sec. 9.6).

REFERENCES

Kane, G., Hawkins, D. and Leventhal, L. (1988) *68000 Assembly Language Programming*, Osbourne/McGraw-Hill, Berkeley, Calif.
Vaughan, J. G. (1983) 'Design of an operating system nucleus for real-time applications', *Software and Micro-systems*, vol. 2, no. 3, June, pp. 73–79.

Examples of real-time operating systems

There are a comparatively large number of RTOSs; many have been developed by companies for specific dedicated applications which need to customize the operating system to deal with specific needs, or else perhaps just to avoid the licensing costs incurred with a commercially available RTOS. Here we look at just two of the more popular, commercially available systems, to give illustrative examples of some alternative design approaches to the structure of an RTOS. No attempt is made at a comprehensive description of the two RTOSs discussed here, rather the idea is to highlight some of their more distinctive features.

Note that the two RTOSs described here are dedicated to systems based around the 68xxx microprocessor family. (Intel's iRMX2—another popular RTOS—is dedicated to systems based around the Intel 80xxx family.)

A3.1 OS-9 and OS-9000

OS-9 (Dibble, 1988; Pucket *et al.*, 1986) is a multi-user, multitasking operating system introduced in 1981 originally for computers based around the Motorola 6809 microprocessor—a chip little used today and effectively superseded by Motorola's 68xxx family. However, some users found certain features of the operating system which were suited to industrial real-time control, and by 1983 Microware ported OS-9 to the 68xxx family of microprocessors. OS-9 is now commonly found in systems based around the VME backplane bus. Full support is given for ROMable software, this being particularly important in many embedded applications. Other notable features of OS-9 are:

- High efficiency—achieved by careful memory management methods in preference to relying on disk intensive operations.
- Kernel, I/O modules and device drivers are written in assembler to get good code

efficiency, while modules less critical in a temporal sense are written in C, for example the shell, and the command utilities.

- The design of the operating system is highly modular, and allows individual modules to be deleted or replaced, so that OS-9 can be readily customized to the particular needs of the hardware it is to run. For example, diskless embedded ROM-based systems could dispense with the modules dealing with disks.
- System interfaces very similar to those in UNIX and many UNIX application programs can be readily adapted for use on an OS-9 system, with little or no modification required.

Microware have more recently produced the operating system OS-9000 (Smith, 1990) which can be used in Intel 80386 and 80486 based systems, as well as 68xxx-based systems. OS-9000 is written in C rather than the assembler used to write OS-9, in the interests of portability.

A3.1.1 OS-9 structure

OS-9 has a hierarchical structure with the kernel at the highest level (see Fig. A3.1). The kernel is responsible for I/O management, task management, memory management, and also linking the remaining modules. At the next level there are the three file managers. The **sequential character** file manager handles all non-mass storage devices such as terminals and printers. The **pipe manager** handles intertask communication using buffers to hold data. Finally the **random block file manager** handles disk devices.

The next level comprises the various device drivers which carry out the I/O to physical devices. At this level, users would frequently add their own drivers to customize OS-9 to specific systems. Standard OS-9 systems are usually provided with a disk driver, a serial port driver for terminals, modems, etc., and a driver for parallel printers interfaces.

The lowest level consists of **device descriptors**, which are small tables that associate specific I/O ports with their logical name, device driver and file manager. By using device descriptors, only one copy of each driver is required for each specific type of I/O device even though there may be many devices of any given type.

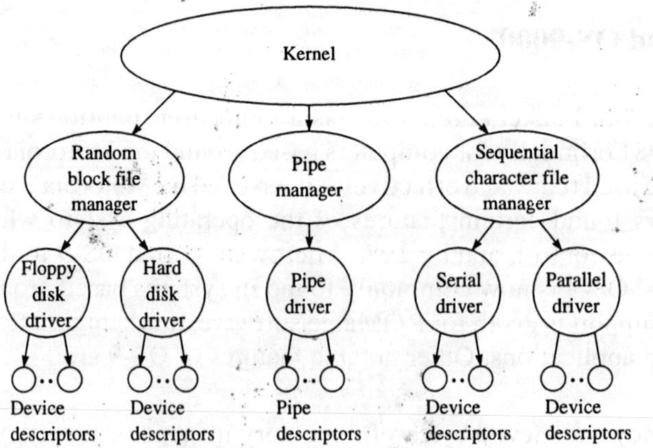

Figure A3.1 Structure of OS-9.

A3.1.2 Tasks in OS-9

Basic concepts relating to task management were covered in Chapter 9, features specific to OS-9 are as follows. When OS-9 adds a task to the queue of runnable tasks, it subsequently uses a mechanism called **ageing** whereby the priority level of a task can be changed dynamically. Low-priority tasks—as they run—are effectively given a higher and higher level of priority and eventually get a chance to use the processor. Task age is recorded as an offset to a counter. This is incremented when a task is added to the queue, ageing all tasks in the queue.

A3.2 PDOS

PDOS is a multitasking RTOS developed by the company Eyring for use on the Motorola 68xxx family of microprocessors. The operating system is sold with an assembler, linker, editor and other utilities needed in developing an application. PDOS includes a ROMable kernel, file manager, monitor and debugger, as well as the tools previously referred to. A comprehensive set of cross-development tools is available for use in target systems, an example being a real-time debugger called Smartbug. Another example is PXRAY, which is a source level debugger.

A3.2.1 The kernel

The kernel is the core module in PDOS responsible for exception handling, task scheduling, I/O management, memory management, management of message queues and a variety of other services. Modules which include the file system, the debugger and the system monitor all call on services that the kernel can carry out. The kernel is comparatively small, of the order of 7 kbytes—an important feature in applications where memory requirements need to be kept to a bare minimum.

A3.2.2 The file system

The file system in PDOS has been structured so as to follow POSIX interface standards. Rapid access to secondary storage devices is another important feature. The file structure is hierarchical; there is provision for multiple device names, each with a root directory and called a **forest**; this gives advantages over singly rooted systems for the following reasons:

- Provision for parallel operations—All file systems impose locks during critical file accesses such as directory updates. By breaking the file system into devices, each tree may have its own lock which allows parallel operation.
- Provision for system maintenance—Multiple partitions may reside on a single device containing a tree. Through the use of a forest, it is easy to back up one tree to another, and maintain parallel file data.
- Device look-up—Separately rooted devices allow for quicker device look-up in comparison with following a single path through subdirectories.
- Foreign file system support—It is possible to use foreign file systems, by having each run as a separate device with its own unique tree. File system 'hooks' have been

provided for the foreign file system support. This allows the PDOS file system to communicate with disks created under MS-DOS, UNIX and other operating systems.

A3.2.3 Intertask communication and synchronization

Tasks in PDOS can communicate by a number of alternative mechanisms. **Event flags**—prioritized semaphores—are single bit flags, globally available to all tasks; they are used to synchronize tasks. Event flags can be set by a task, or from an interrupt service routine. Tasks can be suspended until an event flag is set, following which the task would be rescheduled. This mechanism allows highly efficient and deterministic response to real-world inputs such as interrupts. The time from the execution of an interrupt service routine to the time a task starts to run is better than 30 ms for a 33 MHz 68030-based computer.

Events can be divided into the following categories:

- **Software events** are set from within a task and not affected by task scheduling. A task can suspend itself pending a software event and then be rescheduled when the event is eventually set.
- **Software resetting events** are like software events except that the kernel automatically resets the event when a task is suspended on a rescheduled event.
- **Input events** relate to a task suspending itself if a request is made for a character, and the buffer allocated to store it is in fact empty. Arrival of the character in the buffer would set the corresponding input event.
- **Timing events** are set by a real-time clock at a variety of different intervals (ranging from 100 ms up to 136 years). A task suspended on a timing event is scheduled at the relevant time interval.
- **System resource allocation events** are used to synchronize device drivers and other routines needing ownership of a system resource. This type of event is initially set by the kernel to indicate the availability of the resource, following which no more than one task will be allowed access to the resource. The relevant event needs to be reset by the task using the resource when it is no longer required.

Another mechanism for intertask communication is through shared memory areas in the form of mailboxes. A task places a message into a mailbox, which can then be read by other tasks. Message sizes may be from 4 to 64 bytes.

A3.2.4 Task scheduling

As with any RTOS, PDOS allows for pre-emptive prioritized scheduling of tasks. The range of priority is from 1 (lowest) to 255 (highest). A task of a certain priority level will execute until it suspends itself—waiting for an I/O event for example—or until a higher level priority task becomes runnable. A command and a system call are available for altering task priority. Task pre-emption and scheduling can be initiated by a real-time clock or by interrupt.

A3.2.5 Installable system modules

An installable system module (ISM) is a module that provides some kind of service to tasks, most commonly it is a device driver, i.e. the software interface between an application program and some hardware device. ISMs are structured in such a way as to insulate the hardware from the code of an application; an ISM can be changed with no knock-on effect as far as the application code is concerned. The ISM can be used to control access to devices so as to ensure that two tasks do not access a device simultaneously. One of the other prime characteristics of an ISM is that it can be installed or deinstalled using commands for these two purposes. Installation of an ISM can be achieved by including a command in a file which executes at system boot time.

REFERENCES AND FURTHER READING

Dibble, P. (1988) *OS-9 Insights—An Advanced Programmers Guide to OS-9/68000*, Microware, Des Moines, Iowa.

Intel Corporation (1988) *iRMK Version 1.1 Real-Time Kernel and iRMK286 Release 2.00 Operating System.*

Pucket, D., Falk, L. and Dibble, P. (1986) *The Complete Rainbow Guide to OS-9*, Microware, Des Moines, Iowa.

Smith, B. (1990) 'From a tiny kernel' (article on OS/9000), *Byte*, September, pp. 423–426.

Usenet conference—comp.os.os9.

Vaughan, J. G. (1983) 'Design of an operating system nucleus for real-time applications', *Software and Microsystems*, vol, 2, no. 3, June, pp. 73–79.

INDEX